Lecture Notes in Computer Science 3355

Commenced Publication in 1973
Founding and Former Series Editors:
Gerhard Goos, Juris Hartmanis, and Jan van Leeuwen

Roderick Murray-Smith   Robert Shorten (Eds.)

Switching and Learning in Feedback Systems

European Summer School on Multi-Agent Control
Maynooth, Ireland, September 8-10, 2003
Revised Lectures and Selected Papers

 Springer

Volume Editors

Roderick Murray-Smith
Glasgow University, Dept. of Computing Science
Glasgow G12 8QQ, Scotland, UK
E-mail: rod@dcs.gla.ac.uk

Robert Shorten
Hamilton Institute, NUI Maynooth
Co. Kildare, Ireland
E-mail: robert.shorten@may.ie

Library of Congress Control Number: 2004118125

CR Subject Classification (1998): F.1, C.3, G.3, I.2, I.6

ISSN 0302-9743
ISBN 3-540-24457-3 Springer Berlin Heidelberg New York

Springer is a part of Springer Science+Business Media

springeronline.com

© Springer-Verlag Berlin Heidelberg 2005
Printed in Germany

Typesetting: Camera-ready by author, data conversion by Boller Mediendesign
Printed on acid-free paper   SPIN: 11380887   06/3142   5 4 3 2 1 0

Preface

A central theme in the study of dynamic systems is the modelling and control of uncertain systems. While 'uncertainty' has long been a strong motivating factor behind many techniques developed in the modelling, control, statistics and mathematics communities, the past decade, in particular, has witnessed remarkable progress in this area with the emergence of a number of powerful new methods for both modelling and controlling uncertain dynamic systems. The specific objective of this book is to describe and review some of these exciting new approaches within a single volume. Our approach was to invite some of the leading researchers in this area to contribute to this book by submitting both tutorial papers on their specific area of research, and to submit more focussed research papers to document some of the latest results in the area. We feel that collecting some of the main results together in this manner is particularly important as many of the important ideas that emerged in the past decade were derived in a variety of academic disciplines. By providing both tutorial and research papers we hope to be able to provide the interested reader with sufficient background to appreciate some of the main concepts from a variety of related, but nevertheless distinct fields, and to provide a flavor of how these results are currently being used to cope with 'uncertainty.' It is our sincere hope that the availability of these results within a single volume will lead to further cross-fertilization of ideas and act as a spark for further research in this important area of applied mathematics.

It is a huge challenge to completely characterize methods for dealing with 'uncertainty' in a concise manner, and it is impossible to document all of the methods that have emerged over the past decade. The work that is included in this book is work that to a large extent is due to the widespread availability of cheap computation power. Identification paradigms based upon non-parametric statistics and Monte Carlo simulation that were once considered too impractical to be of interest to engineers, are now the subject of great interest in the community, and form the basis of many practically useful nonlinear system identification techniques. Similarly, complex supervisory (switching) control strategies that were also once considered too complex to manage in practical situations are now providing the basis for the control of uncertain and rapidly time-varying dynamic systems. Switching control strategies can also be considered Multi-Agent or Multiple Model systems, although each research area has tended to use different tools, and to apply their methods to different application areas. It is our hope that this book provides, in particular, a rigorous snapshot of some of the developments that have taken place in these areas over the past decade, and also presents state-of-the art research in selected areas of switching and learning systems. In the context of this overall objective, our aim was to produce a book that would be of use to a graduate researcher wishing to undertake research in this vast field, and that would have both introductory chapters and leading-edge research, together with example applications of the methods. Toward this aim, we divided the book into three parts:

Part I introduces the challenges in switching systems for feedback control systems, with a broad overview by K. Narendra. This is followed by a review of some of the stability issues that arise in supervisory control systems by Shorten, Mason and Wulff. Grancharova and Johansen then survey explicit approaches to constrained optimal control for complex systems, which is of particular interest for practical application.

Part II covers the use of Gaussian Process priors in feedback control contexts. Recently it was observed that when the number of submodels in a multiple-model system increases towards infinity, the system tends towards a Gaussian process. Gaussian Process priors were also found to be well-suited to nonlinear regression tasks, and were very competitive with methods such as artificial neural networks. There has, to date, been very little published work in the combination of Gaussian Process priors in feedback control contexts. This series of papers is intended to provide an overview of the ways in which the approach can be used, and its effectiveness. One of the challenges in the use of Gaussian Process priors which has to be overcome before they are likely to be accepted for control purposes is their high computational cost for medium to large data sets. Quiñonero-Candela and Rasmussen introduce Gaussian Process priors, discuss the link with linear models, and propose a reduced rank GP approach that would reduce the computational load. The topic of efficient GPs for large data sets is continued in the following chapter by Shi et al. which provides an alternative approach, one also of interest in applications where the nonlinear system is convolved with a known system before measurement. An adaptive variant of single-step-ahead model-predictive control is presented by Sbarbaro and Murray-Smith, showing how Gaussian Process priors are well suited for cautious control, while simultaneously learning about a new plant. In order to use Gaussian processes in multiple-step-ahead control, the issue of propagation of uncertainty in time-series predictions needs to be resolved. An approach to this for Gaussian processes is presented in Girard and Murray-Smith. The following chapter by Kocijan provides an illustrative example of the use of Girard's propagation algorithm in a model-predictive control setting, controlling a simulated pH process.

Part III is composed of papers making more specific research contributions or applying switching and learning techniques to a range of application domains. Vilaplana et al. present the results of tests of a new controller for cars equipped with 4-wheel steer-by-wire, and provides a very clear application example of the individual channel design approach. This is followed by two chapters on applications in communications systems: Wong et al. present a geometric approach to designing minimum-variance beamformers that are robust against steering vector uncertainties; Xiao et al. consider allocation of communication resources in wireless communication channels, demonstrating their approach on the design of a networked linear estimator and on the design of a multivariable networked LQG controller. Ragnoli and Leithead present a theoretical contribution, investigating inconsistencies in the theory of linear systems. Tresp and

Yu present an introduction to nonparametric hierarchical Bayesian modelling with Dirichlet distributions, and apply this in a multiagent learning context for a recommendation engine, allowing a principled combination of content-based and collaborative filtering. Roweis and Salakhutdinov investigate simultaneous localization and surveying with multiple agents, based on the use of constrained Hidden Markov models, allowing agents to navigate and learn about an unknown static environment. In the final chapter, Williamson and Murray-Smith present an application of adaptive nonlinear control methods to user interface design. The Hex system is a gestural interface for entering text on a mobile device via a continuous control trajectory. The dynamics of the system depend on the language model and change as new letters are entered such that users are supported without being constrained.

The book documents the work behind presentations at the European Summer School on Multi-Agent Control, held at the Hamilton Institute in Maynooth, Ireland, in September 2003. The meeting was partially supported by the EC-funded research training network *MAC: Multi-Agent Control*, and included many of the outcomes of the project. The participants in the summer school brought insight, techniques and language from very diverse theoretical backgrounds to bear on a range of leading-edge applications. We hope that the publication of this book will bring these exciting cross-disciplinary developments to a broader audience.

October 2004   Roderick Murray-Smith, Robert Shorten
Glasgow and Maynooth

Organization

The Maynooth Summer School was jointly organized by the EC-funded research training network *MAC: Multi-Agent Control*, and the Hamilton Institute, NUI Maynooth.

Conference Chair	Robert Shorten (Hamilton Institute, Ireland)
Network Coordinator	Roderick Murray-Smith (Glasgow University, UK and Hamilton Institute, Ireland)
Local Organizing Committee	Rosemary Hunt
	Oliver Mason
	Kai Wulff (Hamilton Institute, Ireland)

Table of Contents

Switching and Control

Gaussian Processes

Applications of Switching & Learning

From Feedback Control to Complexity Management: A Personal Perspective

Kumpati S. Narendra

Yale Center for Systems Science, Yale University,
New Haven, CT, USA

Abstract. Revolutionary advances in technology have generated numerous complex systems that have become integral parts of our socioeconomic environment. The study of such systems – those which contain many interacting parts – is currently attracting considerable attention. In this paper, the author retraces his personal attempts, over a period of four decades, to develop simple models for adaptation, learning, identification and control using artificial neural networks, and hybrid systems, and goes on to describe how they are providing insights into dealing with complex interconnected systems.

> We dance round in a ring and suppose,
> But the Secret sits in the middle and knows.
> Robert Frost

1 Introduction

The term "system" refers to a collection of components or subsystems that are interconnected in some fashion to achieve an overall objective. By the control of a system we mean qualitatively the ability to alter, direct, or improve its behavior, and a control system is one in which some physical quantities are maintained more or less accurately around prescribed constant or time-varying values.

1.1 Feedback

The distinctive hallmark of control theory, and its single most valuable contribution to science, is the concept of feedback (figure 1), which underlies the whole technology of automatic control.

Every control system, from the simplest (e.g. the thermostat or a simple positioning servo) to the most complex currently in use (e.g. control of unmanned air vehicles) utilizes feedback in one form or another. The essence of the concept involves the triad: measurement, comparison, and correction. That is, measurement of relevant variables, comparison with desired values, and using the errors to correct behavior. As the complexity of systems increased, simple feedback of the output grew into the field of estimation and control, and finally to the control of multivariable, hierarchical, and distributed networks of systems. In

R. Murray-Smith, R. Shorten (Eds.): Switching and Learning, LNCS 3355, pp. 1–30, 2005.

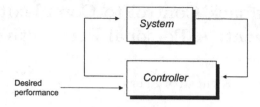

Stage 1. Feed back: Measure, compare, and control

Stage 2. Estimation and Control

Stage 3. Complexity Management

Fig. 1. From Feedback to Complexity Management

the latter, which variables to measure, what information to collect, and what control decisions to make were not often immediately evident. Feedback control had evolved into complexity management.

1.2   What Makes the Problem Difficult?

Systems theory as taught in academia is methodology-driven. Linear and non-linear control theory, optimal control and game theory, stochastic, adaptive, and learning control theories are all members of the effective arsenal of methods developed over the past eight decades by control theorists. Systems practice as carried out in industry is problem-driven, and is tremendously diverse. Such problems are characterized by poor models, distributed sensors and actuators, high-dimensionality decision spaces, multiple subsystems, and complex information patterns. The difficulties that arise in their resolution can be broadly classified under four headings: (i) complexity, (ii) uncertainty, (iii) nonlinearity, and (iv) time-variations.

Computational Complexity It has been known for a long time that the designer's freedom to propose algorithms is limited by the "curse of dimensionality." With the increasing scope of control systems and the resulting rush towards more sophisticated computational architectures, this is assuming greater importance.

Uncertainty The fascinating possibilities implied by the term "uncertainty" were realized by Bellman [1] even as early as 1961. Uncertainty can extend from approximation of complete knowledge (adaptive control) on the one hand to approximation of complete ignorance (learning automata) on the other.

Nonlinearity Even in a purely deterministic context, the presence of nonlinearities in a dynamical system makes the control problem very complex.

Time Variations One of the compelling reasons for considering adaptive and learning methods in the control of practical systems is to compensate for time variations in their dynamic characteristics. Subsystems may fail, parameters may drift with time, and disturbances acting on the system may change with time. Time-varying parameters add substantially to the complexity of the control problem.

Different areas of systems theory are applicable to the above problem areas. Yet, the settings in which each of the approaches can be applied are both limited and disjoint. A combination of the different information processing techniques is needed to achieve a system that performs satisfactorily in a broad domain. Complexity management, which deals with such systems, requires the development of architectural design principles for creating and managing the interaction of the different techniques used. It is the author's belief that dynamical systems theory offers such a framework.

1.3   A Personal Perspective

An efficient survey of the evolution of control theory from simple feedback to complexity management is beyond the scope of a single paper, and is more appropriately the subject for an entire book. The author has tried to circumvent this difficulty by taking a substantially easier course in this article.

Having participated actively for four decades in the development of several subfields in control theory, including stability theory, adaptive control, learning automata, artificial neural networks, and control using multiple models, the author has attempted to provide the reader with some glimpses of his own search for insights into the behavior of simple systems, and how they led in course of time to general principles that were applicable to the control of complex systems.

Only a broad brush, qualitative treatment of the developments in the various fields is provided, and the emphasis throughout the paper is on simple ideas.

2   Adaptive Control

When a suitable mathematical model of a dynamical system is available, powerful analytical techniques exist for computing a control input based on the observed outputs of the system. However, in most practical systems, many parameters are either unknown or vary with time. It is primarily to cope with such uncertainties that the field of adaptive control theory was developed.

In the 1950s and 1960s the research in the field was focused on gradient-based methods for adjusting the control parameters to optimize a performance criterion. In 1966, in a landmark paper, Parks [2] demonstrated that such methods could result in instability, and as a consequence interest shifted to the search for stable methods. Starting around the early 1970s the generally accepted philosophy has been to design the controller to assure the stability of the overall system, and then adjust the parameters of the system within that framework to optimize performance.

Major advances were made in the field in the 1970s and 1980s and it became part of mainstream control theory. Systematic methods for designing adaptive observers and controllers in the 1970s and a detailed investigation of their robustness properties in the 1980s contributed to this. This period also witnessed the study of multivariable adaptive control and stochastic adaptive control. During the following years, interest shifted to nonlinear adaptive control and adaptive control in distributed systems, where many questions remain unresolved and research is now flourishing. In spite of the vast body of literature that exists, the author believes that a relatively small number of ideas have had a significant impact on the evolution of the field. This section mainly focuses on these ideas and describes how they, in turn, led to powerful methods for controlling complex adaptive systems.

2.1   The Adaptive Control Problem

Given a dynamical system whose parameters are known imprecisely, the adaptive control problem can be stated qualitatively as one of designing a controller which will result in the output following a desired output rapidly and with sufficient accuracy. In the ideal case, when no external disturbances are present, the theoretical objective is to make the output error tend to zero asymptotically with time.

Two philosophically different approaches exist for adaptively controlling an unknown plant. In direct adaptive control, no effort is made to identify the unknown plant parameters, and the control parameters are directly adjusted to improve an index of performance. In indirect control, the plant parameters are estimated online and the control parameters are adjusted based on these estimates. The methods outlined in the following sections find application in both direct and indirect control.

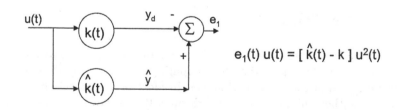

Change parameter $\hat{k}(t)$ in the direction $-\text{sgn}[e_1(t)\, u(t)]$

Fig. 2. Adaptive Control – Static System.

Consider the static system shown in figure 2. The input $u(t)$ is assumed to be bounded, k is an unknown constant, and the output $y_d(t) = ku(t)$. The objective in this simple case is to determine a time-varying function $\hat{k}(t)$ which approximates k such that the output $\hat{y}(t) = \hat{k}(t)u(t)$ asymptotically approaches $y_d(t)$.

Since $y_d(t) = ku(t)$ and $\hat{y}(t) = \hat{k}(t)u(t)$, the output error $e_1(t) = \hat{y}(t) - y_d(t) = (\hat{k}(t) - k)u(t) = \phi(t)u(t)$, where $\phi(t)$ is said to be the parameter error. The key question is whether $\hat{k}(t)$ is to be increased or decreased. Once this is decided, the amount by which it is to be changed (or, how large $\dot{\hat{k}}(t)$ should be) has to be addressed. These are the two principal questions of adaptive control.

In the present case, the product of the error $e_1(t)$ and the input $u(t)$ yields the sign of $\phi(t)$ (since $u^2 \geq 0$), if $u(t) \neq 0$. This implies that the direction in which $\hat{k}(t)$ is adjusted should be opposite to the sign of $e_1(t)u(t)$. The rule for adjusting $\hat{k}(t)$ (also known as the adaptive law) is chosen as

$$\dot{\hat{k}}(t) = \dot{\phi}(t) = -\eta e_1(t)u(t), \tag{1}$$

where η is a positive constant, and is shown to be stable using a Lyapunov function $V(\phi) = \phi^2(t)$ (whose time derivative is $-2\eta e_1^2(t) \leq 0$).

2.2 A Simple Dynamical System

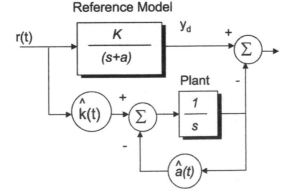

Fig. 3. Adaptive Control – A Simple Dynamic System.

Figure 3 shows the adaptive control of a first-order dynamical system (referred to as the plant). A reference model has a transfer function $W_m(s) = \frac{k}{s+a}$ where

$k, a > 0$ are unknown. The input and output of $W_m(s)$ are called $r(t)$ (the reference input) and $y_d(t)$, respectively. The plant is described by the equation

$$\dot{y}(t) = -\hat{a}(t)y(t) + \hat{k}(t)r(t), \tag{2}$$

where $\hat{a}(t)$ and $\hat{k}(t)$ are adjustable parameters. Once again, the objective is to determine $\hat{a}(t)$ and $\hat{k}(t)$ so that $y(t)$ asymptotically approaches $y_d(t)$. Defining the output error as before, $e(t) = y(t) - y_d(t)$, the error differential equation can be written as

$$\dot{e}(t) = -ae(t) + \phi_a(t)y(t) + \phi_k(t)r(t) \tag{3}$$

where $\phi_a(t)$ and $\phi_k(t)$ are parameter errors $(\hat{a}(t) - a,\ \hat{k}(t) - k)$. As in the static case, it can be shown that the adaptive laws

$$\dot{\hat{k}}(t) = \dot{\phi}_k(t) = -\eta_1 e(t)r(t), \eta_1 > 0$$
$$\dot{\hat{a}}(t) = \dot{\phi}_a(t) = -\eta_2 e(t)y(t), \eta_2 > 0$$

result in stable adaptation and that the output error $e(t)$ tends to zero. The Lyapunov function used to prove this is quadratic in the output error as well as the parameter errors (i.e. $V(e, \phi_a, \phi_k) = \frac{1}{2}(e^2 + \frac{1}{\eta_1}\phi_k^2 + \frac{1}{\eta_2}\phi_a^2)$.

Remark 1. Once again it is seen that each of the adaptive laws involves the product of two signals – the global error $e(t)$ and a local signal which is the input to the parameter being adjusted. This important concept recurs throughout much of adaptive control theory.

Remark 2. So far we have been considering very simple systems. Both the static and dynamic cases can be directly extended to more complex situations as shown below.

2.3   Static Case (A Matrix of Gains)

Let a static system be defined by

$$y_d = Ku \tag{4}$$

where $u(t) \in \mathbb{R}^r$, $K \in \mathbb{R}^{m \times r}$, $y_d(t) \in \mathbb{R}^m$, and the matrix K is unknown. If the objective is to adjust a matrix $\hat{K}(t) \in \mathbb{R}^{m \times r}$ such that $\|y(t) - y_d(t)\| \to 0$ as $t \to \infty$, each parameter $\hat{K}_{ij}$ should be adjusted according to the rule

$$\dot{\hat{K}}_{ij} = -\eta e_i(t)u_j(t), \eta > 0. \tag{5}$$

If $\tilde{K}_{ij} = \hat{K}_{ij}(t) - K_{ij}$ and $\tilde{K} = \hat{K}(t) - K$, the Lyapunov function used to prove global stability and convergence of the output errors to zero is $V(\tilde{k}) = \mathbf{Trace}(\tilde{K}^T \tilde{K})$.

2.4   Dynamic Case (Identification)

Let a linear plant be described by the differential equation

$$\dot{x}_p = A_p x_p + B_p u, \tag{6}$$

$A_p \in \mathbb{R}^{n \times n}$, $B_p \in \mathbb{R}^{n \times m}$, $u \in \mathbb{R}^m$, where the matrix A_p is stable. Let an identification model be represented as

$$\dot{\hat{x}}_p = A_m \hat{x}_p + [\hat{A}_p(t) - A_m] x_p + \hat{B}(t) u$$

$$\dot{\hat{x}}_p = A_m \hat{x}_p + (\hat{A}_p(t) - A_m) x_p + \hat{B}_p(t) u$$

where the matrix A_m is stable and the elements of the time-varying matrices $\hat{A}_p(t)$ and $\hat{B}_p(t)$ are to be adjusted so that $\lim_{t \to \infty} \|e(t)\| = \lim_{t \to \infty} \|\hat{x}_p(t) - x_p(t)\| = 0$.

Using a quadratic Lyapunov function in the output and parameter errors, it can be shown that the adaptive laws

$$\dot{\hat{A}}_p = -e(t) x_p^T(t)$$
$$\dot{\hat{B}}_p = -e(t) u^T(t) \tag{7}$$

will result in $\|e(t)\| \to 0$ as $t \to \infty$.

2.5   Dynamic Case (Control)

In the identification problem discussed above, it was assumed that the plant was stable so that $x_p(t)$ is bounded. We now consider the control problem where the objective is to adaptively stabilize an unstable plant. More precisely, if the reference model and the plant are described by the differential equations

$$\dot{x}_m = A_m x_m + B_m r$$
$$\dot{\hat{x}}_p = A_p x_p + B_m u$$

where $A_m \in \mathbb{R}^{n \times n}$ is stable, $A_p \in \mathbb{R}^{n \times n}$ is unstable, and B_m is a known constant matrix, the objective is to determine the input $u(t)$ such that $\lim_{t \to \infty} \|x_p(t) - x_m(t)\| = 0$. We assume that a feedback matrix K^* exists such that $A_p + B_m K^* = A_m$. To control the system, we choose

$$u(t) = \hat{K}(t) x_p(t) + r(t) \tag{8}$$

and show that adaptive laws

$$\dot{\hat{K}} = -B_m^T e(t) x_p^T(t) \tag{9}$$

result in a stable overall system with the output of the plant tracking the output of the reference model exactly as $t \to \infty$.

2.6   Adaptive Error Models

By the mid 1970s it became clear that all the adaptive laws that had been generated could be obtained in a unified fashion using error models, so that, given an adaptive control problem the adaptive laws could be written by inspection.

In the first step in this method, the given adaptive identification or control problem is recast as an error model which contains only the parameter and output errors. This was due to the realization on the part of control theorists in the early 1970s that only these errors were relevant for a proper formulation of the adaptive control problem. In the context of Lyapunov theory, stability of the overall system had to be demonstrated in the space of the errors (which was also the state space of the error models).

In the second step, the problems that were encountered most often in the field were reduced to three error models. These are shown in figure 4.

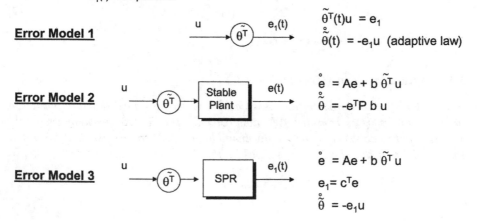

$\tilde{\theta}(t)$ = parameter error

$e(t)$ = state error

$e_1(t)$ = output error

Error Model 1
$$\tilde{\theta}^T(t)u = e_1$$
$$\overset{\circ}{\tilde{\theta}}(t) = -e_1 u \quad \text{(adaptive law)}$$

Error Model 2
$$\overset{\circ}{e} = Ae + b\,\tilde{\theta}^T u$$
$$\overset{\circ}{\tilde{\theta}} = -e^T P\,b\,u$$

Error Model 3
$$\overset{\circ}{e} = Ae + b\,\tilde{\theta}^T u$$
$$e_1 = c^T e$$
$$\overset{\circ}{\tilde{\theta}} = -e_1 u$$

Fig. 4. Error Models.

*In figure 5, (a) corresponds to the static system described earlier. In the second case, (b), the input to a known stable dynamical system is $\phi^T(t)u(t)$ and the output is a vector error signal $e(t)$. In the third case, (c), which is similar to (b), the input is the same as before, but the output $e_1(t)$ is a scalar. However, it is known that the transfer function of the dynamical system is **Strictly Positive Real** (SPR). In all three cases, adaptive control laws for adjusting $\phi(t)$ in a stable fashion have been derived using Lyapunov theory. These are:*

- Model 1: $\dot{\phi}(t) = -e_1(t)u(t)$
- Model 2: $\dot{\phi}(t) = -e^T(t)Pbu(t)$, where P is a suitably chosen positive-definite matrix, and
- Model 3: $\dot{\phi}(t) = -e_1(t)u(t)$.

2.7 Adaptive Observers

In the problem discussed earlier, the state vector of the plant was assumed to be accessible. This made the generation of stable adaptive laws relatively straightforward. However, in most practical cases, this assumption concerning the state vector is not valid, and adaptation has to be carried out using only the accessible outputs. In the 1970s there was a great deal of interest in such problems. Attempts were made to estimate the state vector of the plant in the presence of parametric uncertainty. This led to schemes in which both parameters and state variables were estimated simultaneously, and the devices used were called adaptive observers.

The proof of stability of adaptive observers was substantially more complex than those used earlier. It was soon realized that a proper parameterization of the adaptive observer was needed to generate adaptive laws that were simple and at the same time would assure stability. As indicated below, the basic ideas which eventually led to the design of adaptive observers for both single variable and multivariable systems are both elegant and simple.

Consider the simple example shown in figure 5 (a). The transfer function of a system is $W(s)$ and the input to the system is $\tilde{\theta}^T u(t)$, where $\tilde{\theta}$, $u(t) \in \mathbb{R}^r$ and $\tilde{\theta}$ is an unknown constant vector which has to be estimated. The output of the system is $e_1(t)$. Using this parameterization, simple laws do not exist for estimating $\tilde{\theta}$ in a stable fashion. However, the system shown in (b) consisting of r identical transfer functions $W(s)$ with the input $u(t)$ will result in an output $v(t)$ such that $\tilde{\theta}^T v(t) = e_1(t)$, and hence is dynamically equivalent to it. Mathematically stated, $W(s)\tilde{\theta}^T u(t)$ and $\tilde{\theta}^T W(s)Iu(t)$ yield the same output. The parameterization shown in (b) permits $\tilde{\theta}$ to be estimated by inspection using the first error model, since $\tilde{\theta}^T v = e_1$.

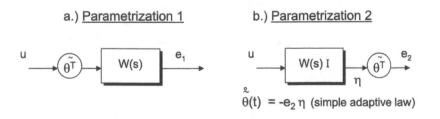

a.) Parametrization 1   **b.) Parametrization 2**

$\dot{\theta}(t) = -e_2 \eta$ (simple adaptive law)

Fig. 5. Parameterization for Estimation.

By reparameterizing the model in (a) so that the vector to be estimated appears at the output simplifies the problem enormously. Without providing all the details, the manner in which the solution to the general adaptive observer problem was obtained is indicated below.

Any n^{th} order linear time-invariant dynamical system can be represented in the form

$$\dot{x} = Ax + gy + bu$$
$$y = x_1 \tag{10}$$

where A is a known stable matrix and $g, b \in \mathbb{R}^n$ are constant vectors. In the adaptive observer problem g and b are unknown. The output y is the first state variable of the system. Representing the unknown $2n$ dimensional vector as

$$\theta^T = [b^T, g^T]$$

the problem is to estimate θ with $\hat{\theta}$. We note that g and b occur at the input of the unknown plant. By reparameterizing it, it can be expressed in the form shown in figure 6, from which θ can be estimated in a straightforward manner.

Plant Parameterization

$$\overset{\circ}{x} = Ax + gy + bu \qquad A = \left(\begin{array}{c|c} k & I \\ \hline & o \end{array} \right)$$

$$y = x_1$$

Unknown parameter vector $\theta = \begin{bmatrix} b \\ g \end{bmatrix}$

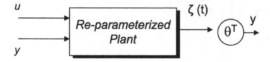

Fig. 6. The Adaptive Observer.

2.8   The Adaptive Control Problem

Another important result obtained in the late 1970s is shown in figure 7 and is concerned with the stable control of an unknown plant. Once again, we note that proper parameterization of the plant and the controller lets simple adaptive control laws to be determined by inspection (using error models).

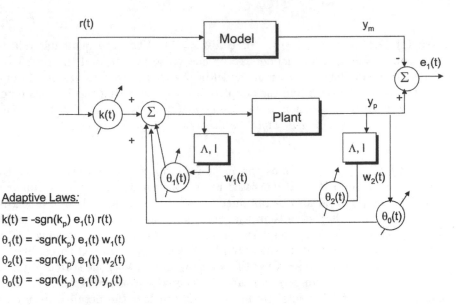

Adaptive Laws:

$k(t) = -\text{sgn}(k_p)\, e_1(t)\, r(t)$

$\theta_1(t) = -\text{sgn}(k_p)\, e_1(t)\, w_1(t)$

$\theta_2(t) = -\text{sgn}(k_p)\, e_1(t)\, w_2(t)$

$\theta_0(t) = -\text{sgn}(k_p)\, e_1(t)\, y_p(t)$

Fig. 7. The Control Problem.

The plant to be controlled is unstable, linear, time-invariant, and of order n, and has zeros in the open left half of the complex plane. A reference model with transfer function $W_m(s)$ has a reference input $r(t)$ and output $y_d(t)$ which is the desired output of the plant. The $2n$ parameters of the plant are unknown and the objective is to determine a control input $u(t)$ such that $y_d(t) - y(t) = e_c(t)$ tends to zero asymptotically with time. In Figure 7, $u(t)$ and $y(t)$ are filtered through identical $(n-1)^{th}$ order stable and controllable systems defined by the pair (Ω, ℓ) to yield outputs $\omega_1(t), \omega_2(t) \in R^{n-1}$. A $2n$-dimensional vector $\omega(t)$ is defined as

$$\omega^T(t) = [r(t), \omega_1^T(t), \omega_2^T(t), y(t)], \tag{11}$$

and a control parameter vector $\theta \in \mathbb{R}^n$ as

$$\theta^T = [k_1, \theta_1^T, \theta_2^T, \theta_0], \tag{12}$$

and the input $u(t)$ to the plant is generated by

$$u(t) = \theta^T \omega(t). \tag{13}$$

As a first step, it is shown that a constant vector θ^* exists such that when $\theta = \theta^*$, the objective of adaptive control is achieved (i.e. error $\to 0$). The second and crucial step is then to determine how θ is to be adjusted so that $\theta(t) \to \theta^*$ as $t \to \infty$.

Case (i) (Relative degree n^* of plant $= 1$) When the plant has relative degree unity, the reference model can be chosen to be SPR and the problem is substantially simplified. The error model of the overall system has the form of error model 3 in figure 4, and the adaptive laws for adjusting the $2n$ parameters can be written by inspection as $\dot{k}_1 = -e_1(t)r(t)$, $\dot{\theta}_1(t) = -e_1(t)\omega_1(t)$, $\dot{\theta}_2(t) = -e_1(t)\omega_2(t)$, and $\dot{\theta}_0(t) = -e_1(t)y(t)$, which is also depicted in figure 7.

Case (ii) ($n^* \geq 2$) Perhaps the most important contribution to adaptive control was made in 1980. It consisted in demonstrating that results similar to those for case (i) could also be obtained for the general case where the relative degree of the plant is greater than one. For a long time (1977-79) it was thought by many that the general adaptive control problem could not be solved. But, once again, a simple strategy suggested by Monopoli [7] enabled three different groups [8], [9], and [10] to arrive at the solution. In line with the stated objective of the paper, we shall merely indicate the changes in the controller structure that yielded this fundamental result, and not consider the detailed arguments that went into the different proofs.

Consider a transfer function $W(s)$ in series with a control parameter error vector $\tilde{\theta}(t)$ as shown in figure 5.

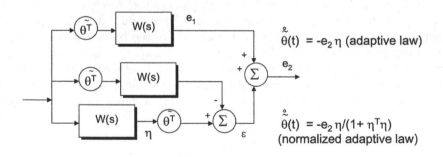

Fig. 8. The Augmented Error.

If $W(s)$ is stable but not SPR, as stated earlier, adaptive rules for adjusting $\tilde{\theta}(t)$ are not available to make $\lim_{t\to\infty} e_1(t) = 0$. The principal difficulty lies in the fact that $\tilde{\theta}(t)$ precedes $W(s)$. Now assume that a network can be constructed which has the structure in Figure 8).

When $\tilde{\theta}(t)$ is a constant $\epsilon_1(t)$ will tend to zero. $\epsilon_1(t)$ is called an auxiliary signal. If the network is connected in parallel to the given plant and $e_1(t)+\epsilon_1(t) = e_2(t)$, $e_2(t)$ is called the augmented error signal and

$$e_2(t) = \tilde{\theta}(t)v(t)$$

where $v(t) = W(s)u$. It is immediately evident that a simple adaptive law can be generated (using the first error model) as

$$\dot{\tilde{\theta}}(t) = -e_2(t)v(t). \tag{14}$$

In [8-10] this adaptive law was shown to result in global stability, and to ensure that

$$\lim_{t\to\infty} e_1(t) = \lim_{t\to\infty} e_2(t) = 0. \tag{15}$$

Remark 3. We note that the controller, as described above, contains three copies of the parameter to be adjusted, i.e. $\tilde{\theta}(t)$. All of them are adjusted in tandem. This idea has also found application in other areas of adaptive control.

2.9   Robust Adaptive Control

Soon after the proof of stability of the idealized adaptive control problem was given, it was shown that such laws are non-robust, i.e. small disturbances could result in instability. It soon became clear that questions related to the robustness of adaptive systems needed a great deal more attention, and the field of robust adaptive control was born.

The three classes of problems that require attention are shown in figure 9.

   i. There may be undesired disturbances affecting the performance of the system.
  ii. Models of the plant to be controlled are rarely perfect and the unmodeled dynamics may have a destabilizing effect.
 iii. The parameters of the plant, which were assumed to be constant in the ideal case, may vary with time.

In all cases, the control laws may have to be modified so that the boundedness of all the signals in the system is preserved.

During the 1980s numerous contributions were made to robust adaptive control. Robust algorithms were developed for both direct and indirect controllers and deterministic and stochastic plants. In spite of the very large number of publications in the area and the ingenious solutions proposed, very few elegant results exist as in the ideal case which provide deep insights into the nature of adaptation. As a result, we shall not discuss robust adaptive control any further but merely assume that the reader can choose an appropriate algorithm for his needs.

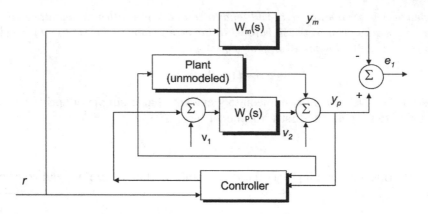

1) Time-varying plant parameters.

2) Bounded external disturbances v_1 and v_2.

3) Unmodeled dynamics - reduced order model and controller.

Fig. 9. Robust Adaptive Control Problems.

2.10   Nonlinear Adaptive Control

In the late 1980s and early 1990s considerable work was done on the adaptive control of special classes of nonlinear systems. The latter were assumed to contain known nonlinearities but with unknown parameters appearing linearly. The adaptive control of such systems is described below.

$$\dot{x}_1 = f_1^0(x_1, x_2) + \theta^T f_1(x_1, x_2)$$
$$\dot{x}_2 = f_2^0(x_1, x_2, x_3) + \theta^T(x_1, x_2, x_3)$$
$$\vdots$$
$$\dot{x}_n = f_n^0(x_1, \ldots, x_n) + \theta^T f_1(x_1, \ldots, x_n) + (g_0(x_1, \ldots, x_n) + \theta^T g(x_1, \ldots, x_n))u \quad (16)$$

A nonlinear dynamical system is represented by the equations 16. It is assumed that all the state variables $x_i(t)$ of the system are accessible and that the nonlinear functions f_i are known. The constant parameter vector θ are unknown, making the problem adaptive. The objective is to determine the control input $u(t)$ such that x_1 asymptotically tracks a desired signal $y_d(t)$.

The determination of explicit quadratic Lyapunov functions to prove the stability of the adaptive system has been shown in [12] and [13] and is theoretically very attractive. However, the adaptive laws are quite complex even for relatively low-order systems, making them not very attractive for practical adaptive control.

2.11 Summary

In summary, the principal ideas to emerge from four decades of research in adaptive control are that algebraic and analytic methods can be suitably combined to generate intuitively appealing and simple adaptive laws for complex systems. The existence of a control parameter vector which can achieve the desired objective must first be established. This is the algebraic part. The parameterizations of the plant and controller play an important role here. In the analytic part, rules for adjusting the control parameter vector are developed. These depend upon a global error signal and a local signal related to the parameter. Proving stability of the overall system is rarely simple or straightforward. However, the results obtained thus far suggest that it may be possible to derive similar adaptive laws even for complex interconnected systems.

3 Learning Automata

In the late 1960s, while the author was working on adaptive control systems, he was introduced to an entirely different paradigm for controlling complex systems in the presence of uncertainty through the contributions of various Russian researchers including Tsetlin and Krylov. For the next twenty years, (1967-1987) he carried out research with his graduate students concurrently in these complementary areas, which culminated in the publication of two books [3,14] in 1987. In this section, the basic ideas of the learning automata approach are introduced.

In the adaptive control systems described in the previous section it was assumed that a mathematical description of the process to be controlled was available, but that the parameters of the model were unknown. While interacting with industrial laboratories, the author frequently encountered systems where the uncertainties were of a higher order and no good models could be developed for them. All that was known was that one out of a finite number of actions could be chosen at any instant to which the plant would probabilistically yield either a good (success) response or a bad (failure) response. The underlying probabilities being unknown, the efficacy of each action could be concluded by actually interacting with the system.

While both adaptive control and learning automata involve feedback, there are fundamental differences between them. Though both involve iterative procedures, updating is done in parameter space in one method and in probability space in the other. Also, the objective in the latter case is to optimize the mathematical expectation of a random functional by the choice of one action out of a finite set. Further, the action space need not be a metric space and hence global rather than local optima can be obtained.

As in the case of adaptive control, our interest in this paper will not be in the detailed discussion of the numerous results in the field, but rather to indicate how simple ideas evolved so that they could be applied to complex systems.

3.1   The Environment and the Automaton

The basic components of the learning automaton are shown in figure 10. The random environment has a finite input set $\alpha = \{\alpha_1, \alpha_2, \ldots, \alpha_r\}$, a finite output set $\{0, 1\}$ where 0 is called failure and 1 success, a set d_i, $i \in \{1, 2, \ldots, r\}$ of reward probabilities corresponding to actions α_i, and a learning algorithm (or automaton) which determines the rule by which an action is to be chosen at instant $(k+1)$ on the basis of all the observations up to time k. The combination of the automaton and the environment is called the learning automaton.

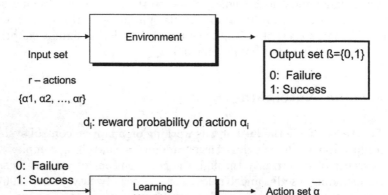

Fig. 10. The Learning Automaton.

In the stochastic learning automaton shown in figure 10, the learning algorithm is an iterative probability-updating scheme. If $P(n)$ is the vector of probabilities of the actions at stage n, and an action $\alpha(n) = \alpha_i$ is chosen resulting in an output $\beta(n) = \beta_j$, the algorithm has the form

$$P(n+1) = T[P(n), \alpha(n), \beta(n)]. \tag{17}$$

The objective then is to determine T such that the overall system has desired properties, e.g. the vector $P(n)$ converges to $P^\star = [0, \ldots, 0, 1, 0, \ldots, 0]^T$, where $\lim_{n \to \infty} P_j(n) = 1$ and corresponds to the optimal action α_j (that for which $d_j = \max_i\{d_i\}$).

3.2   Learning Algorithms

The general philosophy of the learning algorithm is to increase the probability of an action α_j if it results in a success and decrease its probability if it results in a failure, as shown here:

$$P_j(n+1) = P_j(n) + g[P(n)](\text{success})$$
$$P_j(n+1) = P_j(n) - h[P(n)](\text{failure}) \tag{18}$$

How to choose g and h and how the probabilities of the other actions are to be altered is the mathematical problem to be resolved.

3.3 Performance Measures

If the action probabilities are $P_i(k)$, $i \in \{1, 2, \ldots, r\}$ at an instant k, the average probability of success at that instant in $\eta(k) = \sum_{i=1}^{r} P_i(k)d_i$. If the actions are all chosen with the same probability $P_i(k) = \frac{1}{r}$ (for all i), the strategy is called pure chance, and the corresponding probability of success in $\eta_0(k) = \frac{1}{r}\sum_{i=1}^{r} d_i$. The learning algorithm is said to be "expedient" (a term from the Russian literature) if

$$\liminf_{k \to \infty} E[\eta(k)] > \eta_0 \tag{19}$$

(or the outcome is better than chance). Optimality implies that in the limit, the action corresponding to the reward probability $d_{\max}$ is chosen with probability one, or

$$\liminf_{k \to \infty} E[\eta(k)] = d_{\max}. \tag{20}$$

A learning algorithm is said to be absolutely expedient if

$$E[\eta(k+1)|P(k)] \geq \eta(k) \tag{21}$$

with probability one. The sequence $\{P(k)\}_{k \geq 0}$, where $P(k)$ is the vector of action probabilities, is a homogeneous Markov process with stationary transition function. $P(k)$ lies in the r-dimensional simplex S_r where

$$S_r = \left\{ P \mid \sum_{i=1}^{r} P_i = 1, 0 \leq P_i \leq 1 \right\} \tag{22}$$

and $V_r \subset S_r$ is defined as the set of r unit vectors e_i, i.e.

$$V_r = \{e_i | i \in \{1, 2, \ldots, r\} (\text{the } i^{\text{th}} \text{ unit vector})\} \tag{23}$$

3.4 Absorbing and Ergodic Algorithms

If $\{P(k)\}$ converges to V_r with probability one, the algorithm used is called an absorbing algorithm. In this case, one action is chosen with probability one as k tends to infinity. If, on the other hand, $\{P(n)\}$ converges in distribution to a random variable $P^* \in S_r$, independent of the initial stage, it is said to be ergodic.

3.5 A Summary of Important Results

The following are some of the important results which were obtained over a twenty-year period. They are simple enough to be easily modified for use in complex control problems.

L_{RI} **and** L_{RP} **Schemes** If an automaton has two actions with only success and failure as outputs (a *P model*), it was shown independently by Norman and Shapiro [15] and Narendra [16] that a linear reward-inaction scheme (L_{RI}) is absorbing (in such a scheme, the probability of an action is increased if the output is a success and is left unaltered if the output results in a failure). The linear reward-penalty scheme (L_{RP}) was shown to be ergodic and absolutely expedient.

L_{RI} **Scheme ϵ-optimal** In an L_{RI} scheme the probability of convergence to the optimal action can be made as close to one as desired by the choice of the step size.

P, Q, and S Models The results of the two action automata have been extended to the case of r actions, where r is any positive integer. Also, the results have been extended to cases where the automaton has a finite number of values (a **Q model**) or a continuous set of values (an **S model**).

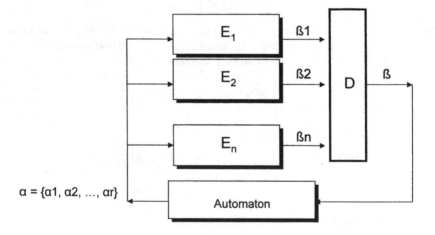

Fig. 11. Multiple Environments.

Multiple Environments If an automaton acts in multiple environments at the same time (figure 11), and the average output of all the environments is used as the output of a composite environment, then the latter can be made ϵ-optimal.

Hierarchy of Automata Learning automata can be connected in an hierarchical fashion (figure 12). In this case, the action set of an automaton at one

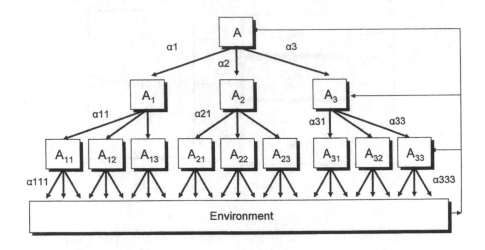

Fig. 12. Hierarchical Automata.

level is a finite set of automata at the level below it. The actions of the last layer of an automata act in an abstract environment whose output is a success or a failure. It has been shown that the parameters of the algorithms at every level can be chosen so that the overall system is ϵ-optimal.

Games of Automata The most significant results with great theoretical consequences and practical utility have been obtained for learning automata operating in a game context. Consider the system shown in figure 13. N automata operate simultaneously in a random environment E. The number of actions available to each automaton may be different. Each automaton is also unaware of the other automata, their number, their action sets, and the learning algorithms they use. It acts exactly as if it were operating in a stationary random environment.

The response of the environment is $\beta(k)$ and depends upon all the actions acting on it at the instant k. However, each automaton updates its action probability vector on the basis of the observed global output $\beta(k)$. Given the strategies of the individual automata, the objective is to determine the asymptotic behavior of the overall system. It has been shown [17] that if all the automata use the L_{RI} scheme, the overall system will be absolutely expedient and can be made ϵ-optimal by the choice of the step sizes of the individual automata.

From the above discussion it is clear that learning automata may prove very attractive for modeling the interactions of many rational players in a dynamic situation. Further, through reformulation of the basic environment, phenomena such as Nash equilibration, Pareto optimality, coalition formation, and implicit bargaining can be interpreted meaningfully in situations modeled by automata games.

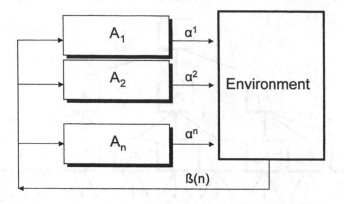

Fig. 13. Games of Automata.

3.6   Time-Varying Environments

So far, we have assumed that the reward probabilities of the environment are constant. However, in practice, if an action is used often, (e.g. a specific route in a routing problem) the probability of failure will increase for that action. Using simple qualitative arguments like these, learning automata were analyzed in environments when the reward probabilities of the environment are functions of the action probabilities. It was shown that the system would evolve to an equilibrium state in which either penalty probabilities or penalty rates due to different options are equalized. [18]

3.7   Estimation Algorithms

In the case of both adaptive systems and learning automata, the updating of the relevant vectors was carried out on the basis of the instantaneous errors. Not surprisingly, questions arose in both fields as to whether one could not do better by using all past information to make a decision at every instant. Instantaneous responses of the system were used for mathematical tractability – to prove stability in adaptive control and to prove stochastic convergence in learning automata. Modifications of the algorithms taking into account past data led to integral algorithms in adaptive control and estimation algorithms in learning automata. While the use of past information improves performance significantly in many cases, it also makes proofs of stability and convergence substantially more difficult.

3.8   Summary

Learning deals with the ability of systems to improve their response based on past experience. In the descriptive learning paradigm, as well as the learning automaton treated in this section, the decision maker updates its strategy for choosing actions on the basis of elicited responses. The learning automaton, which is simple in structure, easily interconnected with other automata, stochastic in nature, and shown to be optimal in hierarchical and distributed structures, appears to have the flexibility and analytical tractability to deal with systems in which large uncertainty exists.

At the same time, since an individual automaton uses very little prior information, its speed of response is, in general, slow, and hence it is ideally suited for situations in which decisions have to be made over longer time scales than the adaptive control systems treated in 2.

4   Artificial Neural Networks for Control

The best-developed part of control theory deals with linear systems and powerful methods for designing controllers for such systems were available even three decades ago. However, as applications became more complex, control theorists had to deal increasingly with nonlinear systems. The resulting problems called for both theoretical principles for designing controllers and practical methods for implementing them. As a consultant to industrial laboratories in the 1980s, the author came across many ingenious schemes for controlling complex nonlinear systems which required appropriate models and methods for the practical realization of controllers. That was when he became interested in artificial neural networks as components in dynamical systems.

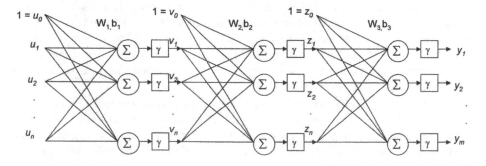

Fig. 14. Multilayer feed-forward neural network.

From a system-theoretic point of view, artificial neural networks are practically implementable, convenient parameterizations of nonlinear maps. During the late 1980s, conclusive proofs were given by numerous authors that multilayer feedforward networks are capable of approximating, in a very precise and satisfactory sense, any continuous function on a compact set. As a result, such networks found wide application in many fields for both function approximation and pattern recognition. In 1990, following extensive simulation studies, the author proposed with his graduate student [19] that neural networks should also be used in dynamic situations, particularly for identification and control. Neural networks, it was argued, are ideally suited to cope with complexity, uncertainty, and nonlinearity – three of the four difficulties mentioned in the introduction – encountered in complex control problems.

For neural networks to be successful in function approximation and pattern recognition, it must be assumed (or theoretically demonstrated) that a nonlinear map exists between the inputs and the outputs. In control problems also, where it is the approximation capabilities of the networks that makes them attractive, it first has to be demonstrated that suitable dynamic nonlinear maps exist. Hence, proving the existence of the appropriate maps and developing methods for approximating them, on-line or off-line, constitute the two parts of neurocontrol.

4.1   Nonlinear Dynamical Systems

Let Σ be a nonlinear dynamical system to be controlled, and let it be represented by the discrete-time state equations

$$\sum : x(k+1) = f[x(k), u(k)], f(0,0) = 0$$
$$y(k) = h[x(k)], h(0) = 0 \tag{24}$$

where $u(k), y(k) \in \mathbb{R}^m, x(k) \in \mathbb{R}^n$ and represent the input, output, and state of Σ at instant k, and f and h are smooth functions. Qualitatively, our objective is to choose $u(k)$ so that all the signals in the system remain bounded and the output $y(k)$ tracks a specified desired signal $y_d(k)$. It is worth noting that the problem as stated here is merely the nonlinear version of the adaptive control problem discussed in 2. As in that problem, our interest is in situations where the equations describing the plant are not known a priori, as well as in those cases where external perturbations are also present.

From a purely mathematical point of view, the precise control of a nonlinear dynamical system is a formidable problem. It becomes substantially more difficult when uncertainty is also present in the system. An approach that proved successful in the mid 1990s was to confine attention to the class of nonlinear systems whose linearizations are well-behaved around the equilibrium state. In such cases the implicit function theorem can be used to assure the existence of appropriate nonlinear maps in some domain containing the equilibrium state. Neural networks can then be used to approximate these maps using the data available concerning the system.

As shown in the following subsections, many of the ideas, developed over a period of three decades for linear adaptive control systems, were successfully extended to the control of nonlinear systems using the above procedure during the 1990s.

4.2   Nonlinear System Representation

Let a single-input, single-output nonlinear system of dimension n be represented by the state equations 24. In the simple case where the state variables $x(k)$ of the system are accessible, the maps f and h can be estimated separately using artificial neural networks and a gradient-based parameter adjustment method such as backpropagation. If, however, only the outputs of the system can be measured, a suitable input-output representation of the plant is needed. An example of such a representation is the NARMA *(nonlinear auto-regressive and moving-average) model, where the output y at time $k+1$ depends upon the values assumed by both the input u and the output y at the previous n instants of time, i.e.*

$$y(k+1) = F[y(k), y(k-1), \ldots y(k-n+1), u(k), \ldots u(k-n+1)] \quad (25)$$

Identification in this context corresponds to approximating the function F. A neural network is particularly suited to carry out this approximation if data in the form of input values and output values are available.

4.3   Control

Assuming that a suitable model of the plant is available in the form

$$\hat{y}(k+1) = \hat{F}[y(k), \ldots y(k-n+1), u(k), \ldots u(k-n+1)] \quad (26)$$

the question that arises is whether a feedback controller can be designed so that $\lim_{k\to\infty} \|\hat{y}(k) - y_d(k)\| = 0$, where $y_d(k)$ is a reference signal. As mentioned earlier, a neural network can be used for this purpose only if it is known that a suitable mapping exists between the available signals and the input $u(k)$. (The control input $u(k)$ is the output of the controller that has to be designed.) In [20] it is shown that if the plant has a well-defined relative degree and zero dynamics that are asymptotically stable and the desired output $y_d(k+1)$ is known at time k, then a mapping ϕ exists such that

$$\Phi[y(k), \ldots y(k-n+1), y_d(k+1), u(k-1), \ldots u(k-n+1)] = u(k). \quad (27)$$

Hence, using $y(k)$ and its past values, the past values of $u(k)$ and the desired signal $y_d(k+1)$, the control input $u(k)$ can be generated to achieve exact asymptotic tracking. (Equations 25-27 can be suitably modified so that the same results also carry over to the more general case where the plant has a relative degree d.)

The simultaneous identification and control of the nonlinear system is shown in Figure 15. The structure of the overall system is strongly motivated by that used in the adaptive control of linear time-invariant systems.

Using both the state representation of the nonlinear plant Σ in equation 24 and the input-output representation in equation 25, several results have been derived for the control of Σ. These include stabilization around the origin, set-point regulation, and asymptotic tracking of a given reference signal. In all cases, the linearization Σ_L of the system plays a central role. Once this was realized, most of the results that are well-known in linear adaptive control were extended to nonlinear systems.

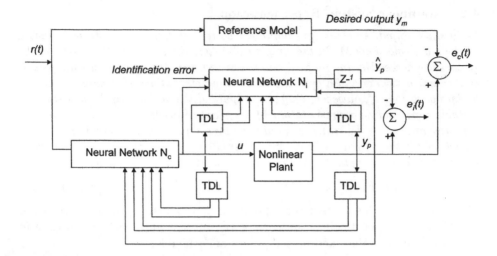

Fig. 15. Control of Nonlinear Dynamical Systems with Neural Networks.

4.4 Summary

The parallel distributed nature of neural networks makes them attractive for coping with complexity. Their ability to approximate nonlinear maps, and the availability of methods for adjusting parameters on the basis of input-output data makes them particularly attractive when unknown nonlinearities are present in a system.

For a neural network to be used in a given context, a mapping must be shown to exist which achieves the desired objective. The neural network is merely used to approximate the nonlinear map. In dynamical systems, nonlinear maps have been derived for the representation of the system as well as the controller, using the implicit function theorem. Hence, all the results derived thus far are valid only in a neighborhood of the equilibrium state.

5   Adaptive Control Using Multiple Models

Among the four reasons that were given in the introduction (i.e. complexity, uncertainty, nonlinearity, and time-variations) which make the control of complex systems difficult, we have seen in the preceding sections that adaptation and learning can deal with uncertainty, while neural networks help to cope with complexity and nonlinearity. The fourth member of the ensemble, i.e. time-variations, however, has not been addressed thus far.

From the very beginning four decades ago, adaptive control theorists have been interested in adaptation in changing environments. However, for the sake of mathematical tractability, they confined their attention to time-invariant systems with unknown parameters. The accepted philosophy was that if an adaptive system was fast and accurate when the unknown plant parameters were constant, it would also prove satisfactory when the parameters varied with time, provided that the variation occurred on a relatively slow time scale. Based on this philosophy, numerous globally stable and robust control algorithms were derived as described in 2, and simulation studies verified that the above assumptions were indeed true if the initial parameter errors were small.

5.1   Reasons for Using Multiple Models

The simulation studies also indicated that when there are large errors in the initial parameter estimates, the tracking error is quite often oscillatory with unacceptably large amplitudes in the transient phase. It was to cope with such situations that the author and his graduate student proposed multiple-model based adaptive control in 1992. [21]

Many other reasons exist besides improving transient response for using multiple models in time-varying environments. A model is merely a representation of the system's behavior in a convenient form, and simplifying assumptions are invariably made to assure mathematical tractability. The best choice of assumptions is different for each system and regime of operation, and thus multiple models arise naturally. Multiple models may also be needed for redundancy and robustness. A third reason is to switch between different models to realize their combined advantages.

5.2   The Structure of the Control System

The control of a linear dynamical system using multiple models is shown in Figure 16. The plant P to be controlled has an input u and an output y. A reference model provides a desired output y_d, and the objective of control is to make the control error $e_c = (y - y_d)$ tend to zero. $M_1, M_2, \ldots, M_N$ are N identification models which are used in parallel with the plant to estimate its parameters. The outputs of the N models are $\hat{y}_i, i \in \{1, 2, \ldots, N\}$ and the estimation error $e_i = \hat{y}_i - y_i$ is used to adjust the parameters of the ith model. Corresponding to each model M_i is a controller C_i, and M_i together with C_i can achieve the desired objective. At every instant, based on a switching criterion, one of the model controller pairs (M_j, C_j) is chosen, and the output u_j of C_j is used as the input at that instant.

Given the prior information about the plant (linear, nonlinear, stochastic, etc.) the design problem is to choose the models M_j and the controllers C_j together with the rules for switching between them, and to demonstrate that the overall system will be stable.

5.3   Tuning, Switching, Switching and Tuning

In classical adaptive control theory (which uses a single model), the parameters of the identification model are changed incrementally, and this results in incremental changes in the controller parameters. When multiple models are used, all

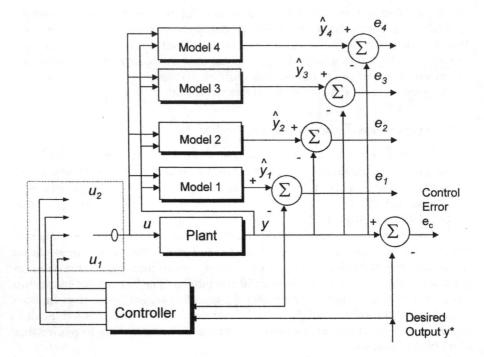

Fig. 16. Control Using Multiple Models.

the identification models are updated incrementally, but the control parameters may change discontinuously due to switching. Thus, the multiple-model based approach frees the control theorist from the traditional view that all adaptation has to be incremental.

Switching is needed for reacting to rapidly changing plant characteristics and avoiding catastrophic failures. Tuning, which is incremental in nature, is relatively slow, and as its name suggests, is desirable for gradually improving the performance of the system. Consequently, both switching and tuning play important roles in the adaptive control of dynamical systems using multiple models.

5.4   Recent Results

The stability of the overall system is of paramount importance in multiple-model adaptive control, as it is in any feedback control system. For linear deterministic and stochastic systems, as well as special classes of nonlinear systems, the stability of the multiple-model based adaptive control has been demonstrated. From a practical standpoint, if the parameters of the identification models are located in regions which are close to those which the time-varying plant parameters can assume, the adaptive controller is found to be very effective. Hence, the location of the models $M_i, i \in \{1, 2, \ldots, N\}$ becomes very important.

5.5 Summary

The success of adaptive control in any specific context depends upon the values of the parameter estimates from which adaptation commences. Multiple-model based adaptive control provides multiple choices for initial conditions and hence is ideally suited for situations where plant parameters vary rapidly and over large domains in parameter space. The success of the approach in such contexts depends upon the location of the fixed models. Work is currently in progress to achieve this in a systematic fashion.

6 Control of Complex Systems

As stated in the introduction, the simple ideas contained in 2-5 have been applied to numerous systems, both simple and complex. Sometimes a single approach such as adaptation or learning may suffice. In more complex cases, combinations of the techniques may be needed. In controlling systems with even greater complexity and uncertainty, the different approaches have to be combined judiciously, and the success of such applications will be limited only by the imagination of the designer. In this section we describe briefly a few problems where the methods described have been applied.

6.1 Use of a Single Paradigm

Adaptive control has been applied to a vast number of problems in engineering where good models of the processes to be controlled exist and the uncertainty is primarily in the values of the parameters. Learning automata (or similar probabilistic learning methods) are attractive when the level of uncertainty is substantially higher and randomness is the dominant attribute of the systems. These methods have been applied successfully to routing in networks, flow control, task scheduling in computer networks, data compression, relaxation labeling, and pattern recognition.

When nonlinearities play a major role, and analytical models are practically nonexistent, but vast amounts of input-output data are available – as in many problems in process control, robotics, manufacturing, and medical instrumentation – neural networks become ideal candidates as identifiers and controllers. In robotics, they have been used in manipulation control, contact control and grasping; in aeronautics, for fault detection and control re-configuration, and in robotics, for autonomous navigation. All these are complex nonlinear multivariable systems which call for nonlinear controllers to meet stringent performance specifications.

Large variations in load, sudden changes in the environment, unanticipated failures in aircraft, and transition control in chemical processes are examples of systems in which parameters vary rapidly with time. Multiple-model based adaptive control has found application in such problems.

6.2   Combining Different Paradigms

Three engineering problems where two of the paradigms suggested earlier were combined to develop efficient controllers are briefly described below.

Operation in Many Regimes When nonlinear dynamical systems operate in many regimes, resulting in rapid variations of parameters, the controllers need to cope with both nonlinearities and time-variations. Control theorists resorted to multiple neural network models for identification and control.

Hierarchical Control in Robots Hierarchical control is ubiquitous in control practice. Two typical examples are robot control and vision-based control. In robot control, the highest level invariably operates on a slower time scale than the rest and decisions are generally probabilistic in nature and based on past performance. Learning appears to be an appropriate technique at this level. Arm control, hand control, and joint control are carried out on a much faster time scale, and robust and adaptive control are the proper choices at these levels.

Learning Automata and Neural Networks Learning schemes operating in random environments can determine, in course of time, the optimal actions at various operating points. Neural networks, on the other hand, are extremely good at interpolation. By combining neural networks with the data provided by learning automata, the optimal actions to be taken in points in continuous domains can be determined.

6.3   Decentralized Control

In all the examples discussed thus far, the control input was generated by a single centralized controller to control a single plant or process in the presence of uncertainty. At the present time, the focus of research in systems theory is on systems in which many subsystems interact, and decision making is distributed among many controllers or agents. The controllers have only partial information about the other subsystems. Decentralized decision making is ubiquitous in both natural (biological) and artificial (man-made) systems and decentralized control is becoming essential in many industrial problems.

Economic systems, communication networks, transportation systems, power systems, and unmanned air vehicles (UAVs) are typical examples of complex interconnected systems with large uncertainties. For example, considering UAVs, the systems have to be robust, adaptive, and self-learning. They have to navigate and land autonomously, be maneuverable, agile, and operate over broad altitude regimes. They must be capable of collecting, collating, and sharing available information for performing cooperative tasks efficiently. Such complex requirements are common to most large decentralized control systems. How to model the overall system, how to choose the appropriate control methodology for each part of the system, and how to assure that the overall system will be stable are some of the challenges faced by the designer.

6.4 Biological Motivation

Biological systems are known to cope easily and effectively with changes in their environments. In the past four decades, as interest shifted to the control of complex systems with increasing uncertainty, efforts were naturally made to incorporate in them characteristics similar to those found in living systems. This resulted in the introduction of words such as adaptation, learning, artificial intelligence and self-organization, (topics treated in this paper) into the control literature. With increased interest in the decentralized control of complex man-made systems in the 21st century, it is generally recognized by the control community that, once again, concepts borrowed from natural systems will be needed to address the problems that are arising.

Examples abound in biological systems where decisions are made in a decentralized fashion and there is seamless integration of the activities of the many agents involved. Through evolution, ants, wasps, bees, fish, and the like have developed extremely efficient methods for finding food, dividing labor, feeding the group, responding to external challenges, and spreading alarm – all of which have applications in engineering systems. This reveals that nature, through evolution, has come up with novel solutions to complex problems, which need to be understood. It is the author's belief that underlying the observed complex behavior of biological communities are simple rules of interaction of individual organisms, just as the stable adaptive control of a complex multivariable system can be traced back to the adaptive properties of a single parameter, and the optimal decision making in automata games can be related to the convergence properties of a simple linear reward-inaction scheme.

Acknowledgement The author would like to thank his student Anthony Di Franco for his valuable and patient assistance throughout the preparation of this paper.

References

1. R. Bellman. Adaptive Processes – A Guided Tour. Princeton, New Jersey: Princeton University Press, 1961.
2. P. C. Parks. "Lyapunov redesign of model reference adaptive control systems," IEEE Trans. Aut. Control. vol. AC-11, pp. 362-367, 1966.
3. Kumpati S. Narendra and Anuradha M. Annaswamy. Stable Adaptive Systems. Prentice Hall, 1989.
4. R. Carroll and D. P. Lindorff. "An adaptive observer for single-input single-output linear systems," IEEE Transactions on Automatic Control. 18:428-435, October 1973.

5. G. Luders and K. S. Narendra. "An adaptive observer and identifier for a linear system," IEEE Trans. Aut. Control. 18:496-499, October 1973.
6. G. Kreisselmeier. "Adaptive observers with exponential rates of convergence," IEEE Trans. Aut. Control. 22:2-8, 1977.
7. R. V. Monopoli. "Model reference adaptive control with an augmented error signal," IEEE Trans. Aut. Control. vol. AC-19, pp. 474-484, 1974.
8. K. S. Narendra, Y-H. Lin, and L. S. Valvani. "Stable adaptive controller design, part ii: proof of stability," IEEE Trans. Aut. Control. vol. AC-25, pp. 440-448, 1980.
9. G. C. Goodwin, P. J. Ramadge, and P. E. Caines. "Discrete-time multivariable adaptive control," IEEE Trans. Aut. Control. vol. AC-25, pp. 449-456, 1980.
10. A. S. Morse. "Global stability of parameter adaptive control systems," IEEE Trans. Aut. Control. vol. AC-25, pp. 433-439, 1980.
11. Petros A. Ioannov and Jing Sun. Robust Adaptive Control. Prentice Hall, 1996.
12. M. Kristic, I. Kannellakopoulos, and P. V. Kokotous. "Nonlinear design of adaptive controllers for linear systems," IEEE Trans. Aut. Control. vol. 39, pp. 738-752, 1994.
13. D. Seto, A. M. Annaswamy, and J. Ballieul. "Adaptive control of nonlinear systems with triangular structure," IEEE Trans. Aut. Control. vol. 39, pp. 1411-1428, 1994.
14. Kumpati Narendra and M. A. L. Thathachar. Learning Automata – An Introduction. Prentice Hall, 1989.
15. M. F. Norman. "On linear models with two absorbing barriers," J. Math. Psych. vol. 5, pp. 225-41, 1968.
16. I. J. Shapiro and K. S. Narendra. "Use of stochastic automata for parameter self-optimization with multi-modal performance critera," IEEE Trans. Sys. Sci. Cyber. SSC-5, pp. 352-360, 1969.
17. K. S. Narendra and R. M. Wheeler. "An n-player sequential stochastic game with identical payoffs," IEEE Trans. Sys. Man Cyber. SMC-13, pp. 1154-58, 1983.
18. K. S. Narendra and M. A. L. Thathachar. "On the behavior of learning automata in a changing environment with application to telephone traffic routing," IEEE Trans. Sys. Man Cyber. SMC-10, pp. 262-69, 1980.
19. K. S. Narendra and K. Parthasarathy. "Identification and control of dynamical systems using neural networks," IEEE Trans. Neur. Networks. vol. 1, pp. 4-27, March 1990.
20. J. B. D. Cabrera and K. S. Narendra. "Issues in the application of neural networks for tracking based on inverse control," IEEE Trans. Aut. Control. 44(11):2007-2027, November 1999.
21. Kumpati S. Narendra and Jeyendran Balakrishnan. "Adaptive control using multiple models," IEEE Trans. Aut. Control. vol. 42 , no. 2, February 1997.

Convex Cones, Lyapunov Functions, and the Stability of Switched Linear Systems

Robert Shorten, Oliver Mason, and Kai Wulff

Hamilton Institute, NUI Maynooth, Ireland
robert.shorten@may.ie

Abstract. Recent research on switched and hybrid systems has resulted in a renewed interest in determining conditions for the existence of a common quadratic Lyapunov function for a finite number of stable LTI systems. While efficient numerical solutions to this problem have existed for some time, compact analytical conditions for determining whether or not such a function exists for a finite number of systems have yet to be obtained. In this paper we present a geometric approach to this problem. By making a simplifying assumption we obtain a compact time-domain condition for the existence of such a function for a pair of LTI systems. We show a number of new and classical Lyapunov results can be obtained using our framework. In particular, we demonstrate that our results can be used to obtain compact time-domain versions of the SISO Kalman-Yacubovich-Popov lemma, the Circle Criterion, and stability multiplier criteria. Finally, we conclude by posing a number of open questions that arise as a result of our approach.

1 Introductory Remarks

The Kalman-Yacubovich-Popov lemma has played an important role in the development of adaptive control algorithms. The classical KYP lemma is closely related to the existence of a common quadratic Lyapunov function for Lur'e type systems and is typically expressed in the form of a constraint on the Nyquist curve of a transfer function. In this paper we present a result on the existence of a quadratic Lyapunov function for a class of time-varying systems [1]. As well as leading to new results in stability theory, we also show that our result reveals an interesting connection between the time and frequency domain [2, 3] for SISO systems. We use this connection to derive a compact time-domain version of the KYP lemma for SISO systems. Further, we show that for SISO systems, time-domain versions of many multiplier criteria can be obtained directly from their frequency domain counterparts [4]. Finally, we present a number of interesting open questions that arise as a result of our work.

2 Mathematical Preliminaries

Throughout, the following notation is adopted: $\mathbb{R}$ and $\mathbb{C}$ denote the fields of real and complex numbers respectively; $\mathbb{R}^n$ denotes the n-dimensional real Euclidean

R. Murray-Smith, R. Shorten (Eds.): Switching and Learning, LNCS 3355, pp. 31–46, 2005.

space; $\mathbb{R}^{n \times n}$ denotes the space of $n \times n$ matrices with real entries; x_i denotes the ith component of the vector x in $\mathbb{R}^n$; a_{ij} denotes the entry in the (i, j) position of the matrix A in $\mathbb{R}^{n \times n}$.

The main results of this paper are based upon Theorem 1. The concepts of weak quadratic Lyapunov functions, strong quadratic Lyapunov functions, and matrix pencils, are central to the statement of this theorem.

(i) **Strong and weak common quadratic Lyapunov functions :**   Consider the set of LTI systems

$$\Sigma_{A_i} : \dot{x} = A_i x, \ i \in \{1, 2, ...M\}. \tag{1}$$

where M is finite and the A_i, $i \in \{1, 2, ...M\}$, are constant Hurwitz matrices in $\mathbb{R}^{n \times n}$ (i.e. the eigenvalues of A_i lie in the open left half of the complex plane and hence the Σ_{A_i} are stable LTI systems). Let the matrix $P = P^T > 0$, $P \in \mathbb{R}^{n \times n}$, be a simultaneous solution to the Lyapunov equations

$$A_i^T P + P A_i = -Q_i, \ i \in \{1, 2, ...M\}. \tag{2}$$

Then, $V(x) = x^T P x$ is a strong quadratic Lyapunov function for the LTI system Σ_{A_i} if $Q_i > 0$, and is said to be a *strong CQLF* for the set of LTI systems Σ_{A_i}, $i \in \{1, ..., M\}$, if $Q_i > 0$ for all i. Similarly, $V(x)$ is a weak quadratic Lyapunov function for the LTI system Σ_{A_i} if $Q_i \geq 0$, and is said to be a *weak CQLF* for the set of LTI systems Σ_{A_i}, $i \in \{1, ..., M\}$, if $Q_i \geq 0$ for all i.

(ii) **The matrix pencil $\sigma_{\gamma[0,\infty)}[A_1, A_2]$:** The matrix pencil $\sigma_{\gamma[0,\infty)}[A_1, A_2]$, for $A_1, A_2 \in \mathbb{R}^{n \times n}$, is the parameterised family of matrices $\sigma_{\gamma[0,\infty)}[A_1, A_2] = A_1 + \gamma A_2$, $\gamma \in [0, \infty)$. We say that the pencil is *non-singular* if $\sigma_{\gamma[0,\infty)}[A_1, A_2]$ is non-singular for all $\gamma \geq 0$. Otherwise the pencil is said to be *singular*. Further, a pencil is said to be _Hurwitz_ if its eigenvalues are in the open left half of the complex plane for all $\gamma \geq 0$. It is important for much of what follows to note that when A_1 is non-singular, the pencil $\sigma_{\gamma[0,\infty)}[A_1, A_2]$ is non-singular if and only if the product $A_1^{-1} A_2$ has no negative eigenvalues.

The relationship between a matrix, its inverse, and a quadratic Lyapunov function will arise in our discussion. In this context we note the following fundamental result.

(iii) **The stability of Σ_A and $\Sigma_{A^{-1}}$ [5, 6]:**
Consider the linear time invariant systems

$$\Sigma_A : \dot{x} = Ax,$$
$$\Sigma_{A^{-1}} : \dot{x} = A^{-1}x,$$

where $A \in \mathbb{R}^{n \times n}$ is Hurwitz. Then, any quadratic Lyapunov function for Σ_A is also a quadratic Lyapunov function for $\Sigma_{A^{-1}}$.

Comment : Suppose that $V(x)$ is a CQLF for the two stable LTI systems Σ_{A_1}, Σ_{A_2}. It is a simple exercise in algebra to verify that the same function $V(x)$ will be a quadratic Lyapunov function for the systems $\Sigma_{\sigma_{\gamma[0,\infty)}[A_1,A_2]}$ and $\Sigma_{\sigma_{\gamma[0,\infty)}[A_1,A_2^{-1}]}$ for all $\gamma \in [0,\infty)$. Hence, $\sigma_{\gamma[0,\infty)}[A_1,A_2]$ and $\sigma_{\gamma[0,\infty)}[A_1,A_2^{-1}]$ are both necessarily Hurwitz for all $\gamma \in [0,\infty)$. Thus the non-singularity of these two pencils is a necessary condition for the existence of a CQLF for the systems Σ_{A_1}, Σ_{A_2}.

Finally, the following observations are useful in deriving the results in Section 5. Lemma 2.1 is a well known result from linear algebra and Lemma 2.2 is a generalisation of a result used by Kalman in [7].

(iv) **Lemma 1.** *[8] Let $A \in \mathbb{R}^{n\times p}$, $B \in \mathbb{R}^{p\times n}$, and let I_n denote the $n \times n$ identity matrix. Then,*

$$det[I_n - AB] = det[I_p - BA]. \tag{3}$$

(v) **Lemma 2.** *[7] Let $A \in \mathbb{R}^{n\times n}$, and $c,b \in \mathbb{R}^{n\times 1}$. Then, the numerator and denominator polynomials of the rational function,*

$$1 + Re\{c^T(j\omega I_n - A)^{-1}b\} = \frac{\Gamma(-\omega^2)}{|M(j\omega)|^2},$$

are given by

$$|M(j\omega)|^2 = det[\omega^2 I_n + A^2],$$
$$\Gamma(-\omega^2) = (1 - c^T A(\omega^2 I_n + A^2)^{-1}b)det[\omega^2 I_n + A^2]. \tag{4}$$

When the matrix A is Hurwitz, $det[\omega^2 I_n + A^2] = det[A]det[\omega^2 A^{-1} + A] \neq 0 \; \forall \; \omega \in \mathbb{R}$.

3   A Result on Common Quadratic Lyapunov Functions

The following theorem considers pairs of stable LTI systems for which no strong CQLF exists, but for which a weak CQLF exists with Q_i, $i \in \{1,2\}$, of rank $n - 1$ in (2), and establishes a set of easily verifiable algebraic conditions, that are satisfied when such a weak CQLF exists[1]. It will be later shown that these conditions are found to play an important role in the question of the existence of strong CQLF's for general LTI systems.

Theorem 1. Let A_1, A_2 be two Hurwitz matrices in $\mathbb{R}^{n\times n}$ such that a solution $P = P^T > 0$ exists to the non-strict Lyapunov Equations

$$A_1^T P + P A_1 = -Q_1 \leq 0, \tag{5}$$

---

[1] This situation corresponds to two stable LTI systems that are on the boundary of having a CQLF as depicted in Figure 1.

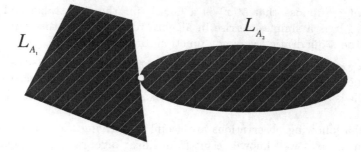

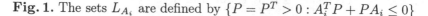

Fig. 1. The sets L_{A_i} are defined by $\{P = P^T > 0 : A_i^T P + PA_i \leq 0\}$

$$A_2^T P + PA_2 = -Q_2 \leq 0, \tag{6}$$

for some positive semi-definite matrices Q_1, Q_2 both of rank $n-1$. Furthermore suppose that no solution exists to the strict Lyapunov equations (2). Under these conditions, at least one of the pencils $\sigma_{\gamma[0,\infty)}[A_1, A_2]$, $\sigma_{\gamma[0,\infty)}[A_1, A_2^{-1}]$ is singular, and at least one of the matrix products $A_1 A_2$ and $A_1 A_2^{-1}$ has a real negative eigenvalue.

Outline of proof: As Q_1 and Q_2 are of rank $n-1$, there are non-zero vectors x_1, x_2 such that

$$x_1^T Q_1 x_1 = 0 \tag{7}$$

$$x_2^T Q_2 x_2 = 0. \tag{8}$$

The proof of Theorem 3.1 is split into two main stages.

Stage 1 : The first stage in the proof is to show that if there exists a positive definite matrix $\overline{P}$ satisfying

$$x_1^T \overline{P} A_1 x_1 < 0 \tag{9}$$

$$x_2^T \overline{P} A_2 x_2 < 0 \tag{10}$$

then a strong positive definite solution exists to (2).

Note that as $x^T PA_1 x$ is a scalar for any x, we can write $x^T Q_1 x = 2x^T PA_1 x$. The same obviously holds for $x^T Q_2 x$.

Now assume that there is some $\overline{P}$ satisfying (9), (10), and, firstly, consider the set

$$\Omega_1 = \{x \in \mathbb{R}^n : \|x\| = 1 \text{ and } x^T \overline{P} A_1 x \geq 0\}.$$

Here $\|x\|$ is the usual Euclidean norm on $\mathbb{R}^n$. The function that takes x to $x^T \overline{P} A_1 x$ is continuous. Thus Ω_1 is closed and bounded, hence compact. Furthermore x_1 (or any non-zero multiple of x_1) is not in Ω_1 and thus $x^T PA_1 x$ is strictly negative on Ω_1.

Let M_1 be the maximum value of $x^T \overline{P} A_1 x$ on Ω_1, and let M_2 be the maximum value of $x^T P A_1 x$ on Ω_1. Then by the final remark in the previous paragraph, $M_2 < 0$. Choose any constant $\delta_1 > 0$ such that

$$\delta_1 < \frac{|M_2|}{M_1 + 1}$$

and consider the positive definite matrix

$$P + \delta_1 \overline{P}.$$

By separately considering the cases $x \in \Omega_1$ and $x \notin \Omega_1$, $\|x\| = 1$, it is easy to see that for all non-zero vectors x of norm 1

$$x^T (A_1^T (P + \delta_1 \overline{P}) + (P + \delta_1 \overline{P}) A_1) x < 0$$

provided $0 < \delta_1 < \frac{|M_2|}{M_1 + 1}$. But we can scale x by any real constant without changing this inequality. Thus $A_1^T (P + \delta_1 \overline{P}) + (P + \delta_1 \overline{P}) A_1$ is negative definite. Let C_1 denote the value $\frac{|M_2|}{M_1 + 1}$.

NOTE: It may appear that we are assuming that the set Ω_1 is non-empty. However, if Ω_1 was empty, then any positive constant δ_1 could be used in the argument above to make $A_1^T (P + \delta_1 \overline{P}) + (P + \delta_1 \overline{P}) A_1$ negative definite.

Now the same argument can be used to guarantee the existence of a positive constant C_2 such that

$$x^T (A_2^T (P + \delta_1 \overline{P}) + (P + \delta_1 \overline{P}) A_2) x < 0.$$

for all non-zero x provided we choose $0 < \delta_1 < C_2$. So, if we choose δ less than the minimum of C_1, C_2, we would have a positive definite matrix

$$P_1 = P + \delta \overline{P}$$

which was a solution of (2).

Stage 2 : So under our assumptions, no positive definite solution $\overline{P}$ exists satisfying Equations (9) and (10). We now show that such a solution $\overline{P}$ would exist unless one of the two pencils $\sigma_{\gamma[0,\infty)}[A_1, A_2]$, $\sigma_{\gamma[0,\infty)}[A_1, A_2^{-1}]$ was singular. Recall ((7), (8)) that there is a positive definite P such that

$$x_1^T P A_1 x_1 = 0 \tag{11}$$

$$x_2^T P A_2 x_2 = 0. \tag{12}$$

Suppose now that there was a Hermitian matrix H such that

$$x_1^T H A_1 x_1 < 0 \tag{13}$$

$$x_2^T H A_2 x_2 < 0. \tag{14}$$

As the set of positive definite matrices is open in the set of Hermitian matrices, we could choose $\epsilon > 0$ such that $P + \epsilon H$ was positive definite. Then $P + \epsilon H$ would satisfy (9), (10). So in fact, there is no Hermitian H satisfying (13), (14). This means that any Hermitian H that makes the expression $x_1^T H A_1 x_1$ negative will make the expression $x_2^T H A_2 x_2$ positive. More formally

$$x_1^T H A_1 x_1 < 0 \iff x_2^T H A_2 x_2 > 0 \tag{15}$$

for Hermitian H. It follows from this that

$$x_1^T H A_1 x_1 = 0 \iff x_2^T H A_2 x_2 = 0.$$

The expressions $x_1^T H A_1 x_1$, $x_2^T H A_2 x_2$, viewed as functions of H, define linear functionals on the space of Hermitian matrices. Moreover, we have seen that the null sets of these functionals are identical. So they must be scalar multiples of each other. Furthermore, (15) implies that they are negative multiples of each other. That is,

$$x_1^T H A_1 x_1 = -k x_2^T H A_2 x_2 \tag{16}$$

with $k > 0$, for all Hermitian matrices H.

It follows from elementary arguments [1] that either $x_1 = \alpha x_2$ and $A_1 x_1 = -(\frac{k}{\alpha}) A_2 x_2$ or $x_1 = \beta A_2 x_2$ and $A_1 x_1 = -(\frac{k}{\beta}) x_2$. Consider the former situation to begin with. Then we have

$$A_1(\alpha x_2) = -(\frac{k}{\alpha}) A_2 x_2$$

$$\Rightarrow (A_1 + (\frac{k}{\alpha^2}) A_2) x_2 = 0$$

and thus the pencil $\sigma_{\gamma[0,\infty)}[A_1, A_2]$ is singular. It follows that the matrix $A_1 A_2^{-1}$ has a negative eigenvalue.

On the other hand, in the latter situation, we have that

$$x_2 = \frac{1}{\beta} A_2^{-1} x_1$$

Thus

$$A_1 x_1 = -(\frac{k}{\beta^2}) A_2^{-1} x_1$$

$$\Rightarrow (A_1 + (\frac{k}{\beta^2}) A_2^{-1}) x_1 = 0$$

Thus, in this case the pencil $\sigma_{\gamma[0,\infty)}[A_1, A_2^{-1}]$ is singular. It follows that the matrix $A_1 A_2$ has a negative eigenvalue. This completes the proof of Theorem 3.1.

$\square$

4   Applications of Main Result

In this section, we describe a number of applications of Theorem 1 and the techniques outlined in the last section. First of all, we shall present two direct applications of the Theorem to the problem of CQLF existence for pairs of exponentially stable LTI systems. The general problem of CQLF existence for families of LTI systems is recognised as an analytical problem of great difficulty. While it can be solved efficiently numerically using linear matrix inequalities [9], closed-form necessary and sufficient conditions for the existence of a CQLF are currently only known for a few special cases of system classes; in particular, for the case of pairs of second order LTI systems [10, 11], and for pairs of n-dimensional systems whose systems matrices differ by a rank 1 matrix [12]. We shall show below that both of these important system classes satisfy the seemingly abstract conditions specified by Theorem 1. Later in the section, we shall see that the same ideas that led to the result of Theorem 1 can be successfully applied to the related question of common diagonal Lyapunov function (CDLF) existence for pairs of exponentially stable positive LTI systems of arbitrary dimension. In fact, we shall present a compact algebraic condition that is necessary and sufficient for a generic pair of such systems to have a CDLF.

(i) Second order systems

We now illustrate the use of Theorem 1 for pairs of stable second order LTI systems.

Let Σ_{A_1} and Σ_{A_2} be stable LTI systems with $A_1, A_2 \in \mathbb{R}^{2 \times 2}$. The following facts follow trivially for second order systems.

(a) If a strong CQLF exists for Σ_{A_1} and Σ_{A_2} then the pencils $\sigma_{\gamma[0,\infty)}[A_1, A_2]$ and $\sigma_{\gamma[0,\infty)}[A_1, A_2^{-1}]$ are necessarily Hurwitz.

(b) If A_1 and A_2 satisfy the non-strict Lyapunov equations (5), (6) then the matrices Q_1 and Q_2 are both rank 1 (rank $n-1$).

(c) If a strong CQLF does not exist for Σ_{A_1} and Σ_{A_2} then a positive constant d exists such that a strong CQLF exists for Σ_{A_1-dI} and Σ_{A_2}. By continuity a non-negative $d_1 < d$ exists such that $A_1 - d_1 I$ and A_2 satisfy Theorem 1 and one of the pencils $\sigma_{\gamma[0,\infty)}[A_1 - d_1 I, A_2]$ and $\sigma_{\gamma[0,\infty)}[A_1 - d_1 I, A_2^{-1}]$ is necessarily singular. Hence, it follows that one of the pencils $\sigma_{\gamma[0,\infty)}[A_1, A_2]$ and $\sigma_{\gamma[0,\infty)}[A_1, A_2^{-1}]$ is not Hurwitz.

Items (a)-(c) establish the following facts. Given two stable second order LTI systems Σ_{A_1} and Σ_{A_2}, a necessary condition for the existence of a strong CQLF is that the pencils $\sigma_{\gamma[0,\infty)}[A_1, A_2]$ and $\sigma_{\gamma[0,\infty)}[A_1, A_2^{-1}]$ are Hurwitz. Conversely, a necessary condition for the non-existence of a strong CQLF is that one of the pencils $\sigma_{\gamma[0,\infty)}[A_1, A_2]$ and $\sigma_{\gamma[0,\infty)}[A_1, A_2^{-1}]$ is not Hurwitz. Together these conditions yield the following known result [10]:

A necessary and sufficient condition for the LTI systems Σ_{A_1} and Σ_{A_2}, $A_1, A_2 \in \mathbb{R}^{2 \times 2}$, to have a strong CQLF is that the pencils $\sigma_{\gamma[0,\infty)}[A_1, A_2]$ and $\sigma_{\gamma[0,\infty)}[A_1, A_2^{-1}]$ are Hurwitz.

(ii) The SISO Circle Criterion

By modifying the argument presented in item (i) above a time domain formulation of the Circle Criterion can be obtained using Theorem 1.

Let Σ_{A_1} and Σ_{A_2} be stable LTI systems with $A_1, A_2 \in \mathbb{R}^{n \times n}$ and $\text{rank}(A_1 - A_2) = 1$. Then, the following facts follow directly [3].

(a) If a strong CQLF exists for Σ_{A_1} and Σ_{A_2} then $\sigma_{\gamma[0,\infty)}[A_1, A_2^{-1}]$ and $\sigma_{\gamma[0,\infty)}[A_1, A_2]$ are necessarily Hurwitz.

(b) If A_1 and A_2 satisfy the non-strict Lyapunov equations (5) (6) then it is shown in [3] that the matrices Q_1 and Q_2 are both generically rank 1 (rank $n - 1$).

(c) If a strong CQLF does not exist for Σ_{A_1} and Σ_{A_2} then a positive constant k exists such that a strong CQLF exists for Σ_{A_1} and $\Sigma_{A_2 + k(A_1 - A_2)}$. By continuity a non-negative $k_1 < k$ exists such that A_1 and $A_2 + k_1(A_1 - A_2)$ satisfy Theorem 1 and the pencil $\sigma_{\gamma[0,\infty)}[A_1^{-1}, A_2 + k_1(A_1 - A_2)]$ is necessarily singular.

(d) Let $A_1, B \in \mathbb{R}^{n \times n}$ with A_1 Hurwitz and $\text{rank}(B) = 1$. Suppose that for some $\lambda_0 > 0$, the matrix product $A_1(A_1 + \lambda_0 B)$ has a negative eigenvalue (the pencil $\sigma_{\gamma[0,\infty)}[A_1^{-1}, A_1 + \lambda_0 B]$ is singular). Then for all $\lambda > \lambda_0$, the product $A_1(A_1 + \lambda B)$ has a negative eigenvalue (the pencil $\sigma_{\gamma[0,\infty)}[A_1^{-1}, A_1 + \lambda B]$ is singular).

Items (a)-(d) establish the following result. Given two stable LTI systems Σ_{A_1} and Σ_{A_2} with $\text{rank}(A_1 - A_2) = 1$, a necessary and sufficient condition for the existence of a strong CQLF is that the pencil $\sigma_{\gamma[0,\infty)}[A_1^{-1}, A_2]$ is non-singular. More formally:

Theorem 2. *Let A, $A + B$ be two Hurwitz matrices in $\mathbb{R}^{n \times n}$ where $\text{rank}(B) = 1$. Then a necessary and sufficient condition for a strong CQLF to exist for the systems Σ_A, Σ_{A+B} is that the matrix product $A(A+B)$ has no negative eigenvalues or equivalently, that the matrix pencil $\sigma_{\gamma[0,\infty)}[A^{-1}, A + B]$ is non-singular.*

(iii) CDLF existence for positive linear systems

Recently, in [13] the same techniques that have been used above to derive Theorem 1 and to obtain the conditions for CQLF existence for pairs of LTI systems with system matrices differing by rank one and second order systems, have been applied to the problem of common diagonal Lyapunov function (CDLF) existence for pairs of so-called *positive* LTI systems. The class of positive systems, whose state variables are constrained to be non-negative for all time, is recognised to be of considerable practical importance, and examples of positive systems commonly occur in areas such as population dynamics, communication systems, pharmaceutics and economics [14]. While, the theory of positive LTI systems is now well-developed [14], recent applications in areas such as congestion control of the Internet and formation flying [15, 16] have indicated the need for a greater understanding of time-varying, and in particular switched positive linear systems. As with general switched linear systems, several fundamental questions relating to the stability of positive switched linear systems are currently unresolved.

Before stating the result on CDLF existence, we need to briefly provide some background on the theory of positive LTI systems. Firstly, it is well-known that an LTI system Σ_A is positive if and only if the system matrix A is a so-called *Metzler* matrix, where a Metlzer matrix A in $\mathbb{R}^{n \times n}$ is one all of whose off-diagonal entries are non-negative, $a_{ij} \geq 0$ for $i \neq j$. A remarkable property of such systems is that a positive LTI system is exponentially stable if and only if it has a *diagonal* Lyapunov function. Thus, for any Metzler, Hurwitz matrix A in $\mathbb{R}^{n \times n}$, there is some positive definite diagonal matrix D such that $A^T D + DA < 0$. This fact gives rise to the problem of determining when a CDLF exists for two or more exponentially stable positive LTI systems. Formally, given the family of exponentially stable positive LTI systems $\Sigma_{A_1}, \ldots, \Sigma_{A_k}$, where $A_i \in \mathbb{R}^{n \times n}$, $1 \leq i \leq k$, if there exists some positive definite diagonal matrix D such that

$$A_i^T D + DA_i < 0 \quad \text{for } 1 \leq i \leq k,$$

then $V(x) = x^T D x$ is a CDLF for the systems $\Sigma_{A_1}, \ldots, \Sigma_{A_k}$. We shall now consider this problem for a pair of exponentially stable positive LTI systems under the mild assumption that their system matrices are irreducible [17, 18]. Now let $\Sigma_{A_1}, \Sigma_{A_2}$ be exponentially stable positive LTI systems, where A_1, A_2 are irreducible, Metzler, Hurwitz matrices in $\mathbb{R}^{n \times n}$. Then the following points were established in [13].

(a) If $\Sigma_{A_1}, \Sigma_{A_2}$ have a CDLF, then so do $\Sigma_{A_1}, \Sigma_{DA_2D}$ for all positive diagonal D in $\mathbb{R}^{n \times n}$. Thus, $A_1 + DA_2D$ must be Hurwitz and hence non-singular for all diagonal $D > 0$.

(b) If there is no CDLF for $\Sigma_{A_1}, \Sigma_{A_2}$ but there exists a non-zero diagonal $D \geq 0$ satisfying

$$A_i^T D + DA_i = Q_i \leq 0 \quad i \in \{1,2\},$$

then Q_1 and Q_2 *must have rank* $n - 1$. This is the crucial stage in the derivation of the condition for CDLF existence given in [13] and is interesting to note in the light of the hypotheses of Theorem 1 above.

(c) Following arguments analogous to those used in the derivation of Theorem 1, it is possible to show that in the situation described in (b) there is some diagonal $D > 0$ such that $A_1 + DA_2D$ is singular.

(d) Finally, if $\Sigma_{A_1}, \Sigma_{A_2}$ have no CDLF, then for $\alpha > 0$ sufficiently large, $\Sigma_{A_1 - \alpha I}$, Σ_{A_2} will have a CDLF. If we then define

$$\alpha_0 = inf\{\alpha > 0 : \Sigma_{A_1 - \alpha I}, \Sigma_{A_2} \text{ have a CDLF }\},$$

then $\Sigma_{A_1 - \alpha_0 I}, \Sigma_{A_2}$ satisfy the conditions of (b). It follows that there is some diagonal $D > 0$ such that $A_1 - \alpha_0 I + DA_2D$ is singular. A suitable rescaling $\overline{D}$ of this D will now make $A_1 + DA_2D$ singular.

Taking points (a)-(d) together, we have the following result giving a necessary and sufficient condition for the existence of a CDLF for a generic pair of exponentially stable positive LTI systems.

Theorem 3. *[13] Let Σ_{A_1}, Σ_{A_2} be exponentially stable positive LTI systems, where A_1, A_2 are irreducible Hurwitz, Metzler matrices in $\mathbb{R}^{n \times n}$. Then a necessary and sufficient condition for $\Sigma_{A_1}, \Sigma_{A_2}$ to have a CDLF is that $A_1 + DA_2D$ is non-singular for all diagonal $D > 0$.*

5 Implications of Main Result for General Multiplier Criteria

Theorem 3.2 provides a time-domain condition for the existence of a CQLF for pairs of LTI systems whose system matrices differ by a rank 1 matrix. Alternative, but equivalent, conditions can be obtained using the SISO circle criterion. This observation raises the question as to whether time domain conditions derived using Theorem 2.1 can be obtained directly from classical frequency domain results. The following theorem provides an affirmative answer to this question.

Theorem 4. *[2, 4] Let $G(j\omega) = \frac{N(j\omega)}{D(j\omega)}$ be a rational transfer function and $K \in \mathbb{R}^+$. Let $\{A, b, c, d\}$ be a controllable realisation of $G(j\omega)$ so that $G(j\omega) = c^T(j\omega I - A)^{-1}b + d$. Let A and $A - \frac{bc^T}{K+d}$ be strictly Hurwitz. Then, a necessary and sufficient condition for*

$$K + Re\left\{G(j\omega)\right\} > 0, \ \forall \, \omega \in \mathbb{R} \cup \{\infty\}, \tag{17}$$

is that the matrix-product $A\left(A - \frac{bc^T}{K+d}\right)$ has no negative real eigenvalues.

Outline of proof: Equation (17) can be written

$$K + d + Re\left\{c^T(j\omega I_n - A)^{-1}b\right\} > 0$$

for all $\omega \in \mathbb{R} \cup \{\infty\}$. In particular $K + d > 0$. Applying Lemma 2 we obtain

$$K + d - 1 + \frac{(1 - c^T A(\omega^2 I_n + A^2)^{-1}b)det\left[\omega^2 I_n + A^2\right]}{det\left[\omega^2 I_n + A^2\right]} > 0$$

$$K + d - c^T A(\omega^2 I_n + A^2)^{-1}b > 0$$

which implies

$$det\left[K + d - c^T A(\omega^2 I_n + A^2)^{-1}b\right] > 0.$$

Applying Lemma 1 yields

$$det\left[(K + d)I_n - (\omega^2 I_n + A^2)^{-1}bc^T A\right] > 0$$

$$det\left[(\omega^2 I_n + A^2)^{-1}\right] det\left[(K + d)(\omega^2 I_n + A^2) - bc^T A\right] > 0$$

$$det\left[(\omega^2 I_n + A^2)^{-1}\right] (K + d)det\left[\omega^2 I_n + A^2 - \frac{1}{K+d}bc^T A\right] > 0$$

$$det\left[\omega^2 I_n + \left(A - \frac{1}{K+d}bc^T\right)A\right] > 0 \tag{18}$$

with the latter following as $K + d > 0$ and A has no imaginary eigenvalues. It follows that a necessary condition for (17) is that the product $A(A - \frac{1}{K+d}bc^T)$ has no negative eigenvalues.

Let $A_1 = A$ and $A_2 = A - \frac{1}{K+d}bc^T$ and suppose that $A_2 A_1$ has no negative real eigenvalue, and that A_1 and A_2 are Hurwitz. It follows using the above argument in reverse that (17) holds. $\square$

Theorem 4 has profound implications for a number of classical frequency domain stability results derived in the context of the Lur'e problem. Lur'e considered the problem of determining the global asymptotic stability of the equilibrium state of the system:

$$\Gamma : \dot{x} = Ax + bu$$
$$u = -f(\sigma, t)$$
$$\sigma = c^T x$$

where $A \in \mathbb{R}^{n \times n}$ is a Hurwitz matrix, $b, c \in \mathbb{R}^{n \times 1}$ and where $k_1 \sigma^2 \leq f(\sigma, t)\sigma \leq k_2 \sigma^2$. In the context of this problem three of the best known results in systems theory were derived; namely; the Kalman-Yacubovich-Popov lemma; the Circle Criterion; and the Popov Criterion. All three of these results establish conditions for stability as a constraint on the rational transfer function $G(j\omega) = c^T(j\omega - A)^{-1}b$.

(i) *The Kalman-Yacubovich-Popov (KYP) Lemma :* The single-input single-output (SISO) version of the Kalman-Yacubovich-Popov lemma [7] is expressed in the form of a strictly positive real (SPR) condition: namely,

$$\gamma + Re\left\{c^T(j\omega I_n - A)^{-1}b\right\} > 0 \ \forall \ \omega \in \mathbb{R},$$

for some $\gamma \in \mathbb{R}^+$. A necessary condition for the above inequality to hold is that both A and $A - \frac{1}{\gamma}bc^T$ are Hurwitz. Hence, given this fact, it follows from Theorem 4 that a time-domain version of the SPR condition for SISO systems is that the matrix $A(A - \frac{1}{\gamma}bc^T)$ does not have any negative real eigenvalues.

(ii) *The Circle Criterion [2]:* The SISO version of the circle criterion is derived directly from the SISO KYP lemma. Here, conditions are derived for the existence of a Lyapunov function $V(x) = x^T Px, P = P^T \in \mathbb{R}^{n \times n}$ for the non-linear Lur'e type system Γ. In the case where $k_1 = 0$ and $k_2 = 1$ a necessary and sufficient condition for the existence of a quadratic Lyapunov function $V(x)$ is that [12]

$$1 + Re\left\{c^T(j\omega I_n - A)^{-1}b\right\} > 0 \ \forall \ \omega \in \mathbb{R}.$$

It follows from Theorem 4 that a time-domain version of the circle criterion with $0 \leq f(\sigma, t) \leq 1$ is that the matrix $A(A - bc^T)$ does not have any negative real eigenvalues (A and $A - bc^T$ are necessarily Hurwitz as before).

(iii) *The Popov Criterion :* The SISO Popov criterion [19] considers the stability of the system Γ where the nonlinearity f is time-invariant. A sufficient condition for the absolute stability of this system is that there exists a strictly positive $\alpha \in \mathbb{R}$ such that

$$\frac{1}{k} + Re\left\{(1 + j\alpha\omega)c^T(j\omega I_n - A)^{-1}b\right\} > 0 \ \forall \ \omega \in \mathbb{R}.$$

It follows from Theorem 4 that a time-domain version of the Popov criterion can be stated as follows: there exists a positive $\alpha \in \mathbb{R}$ such that the matrix $\bar{A}(\bar{A} - \frac{1}{d+\frac{1}{k}}\bar{b}\bar{c})$ does not have any negative real eigenvalues where $\{\bar{A}, \bar{b}, \bar{c}, \bar{d}\}$ is the control canonical form of $(1 + j\alpha\omega)c^T(j\omega I_n - A)^{-1}b$ and that both these matrices are Hurwitz.

Comment : The Popov criterion is an example of a multiplier criterion. Over the past 40 years many authors, including Popov, Zames and Falb, Willems, Narendra and Taylor, and many others [20, 21, 22] have developed stability multiplier criteria that exploit additional assumed properties of the sector nonlinearity f. Roughly speaking, these conditions are expressed in the form of a strictly positive real condition on a function of the form of

$$\gamma + Re\left\{c^T(j\omega I_n - A)^{-1}b\right\} > 0 \ \forall \ \omega \in \mathbb{R}.$$

Consequently, compact time-domain versions of these criteria can be obtained using Theorem 4. A particularly useful consequence of our result is that it leads to very compact conditions for checking strict positive realness of a given LTI system [23].

6   Examples

Example 1. Consider the non-linear system

$$\begin{aligned} \Gamma : \ \dot{x} &= Ax + bu \\ u &= -f(\sigma) \\ \sigma &= c^T x \end{aligned}$$

where

$$A = \begin{bmatrix} 0 & 1 \\ -1 & -2 \end{bmatrix}, \ b = \begin{bmatrix} 0 \\ 1 \end{bmatrix}, \ c = \begin{bmatrix} 1 \\ 0 \end{bmatrix},$$

and where $0 \le f(\sigma)\sigma \le \sigma^2$. This system may be analysed using both the Circle and Popov Criterion.

(i) *The Circle Criterion :* The matrix product $W = A(A - bc^T)$ is given by

$$W = \begin{bmatrix} -2 & -2 \\ 4 & 3 \end{bmatrix}.$$

Since W does not have any negative eigenvalues it follows from the Circle Criterion that the system Γ is absolutely stable.

(ii) *The Popov Criterion :* The control canonical form of $(1 + j\alpha\omega)c^T(j\omega I_n - A)^{-1}b$ is:

$$\bar{A} = \begin{bmatrix} 0 & 1 \\ -1 & -2 \end{bmatrix}, \bar{b} = \begin{bmatrix} 0 \\ 1 \end{bmatrix}, \bar{c} = \begin{bmatrix} 1 \\ \alpha \end{bmatrix}, \bar{d} = 0.$$

The matrix product $W(\alpha) = \bar{A}(\bar{A} - \bar{b}\bar{c}^T)$ is given by

$$W(\alpha) = \begin{bmatrix} -2 & -2-\alpha \\ 4 & 3+2\alpha \end{bmatrix}.$$

The matrix $W(1)$ does not have negative real eigenvalues. Hence, it also follows from the Popov Criterion that the system Γ is absolutely stable.

Example 2. Consider the stable dynamic systems Σ_{A_1} and Σ_{A_2} with:

$$A_1 = \begin{bmatrix} 0 & 1 & 0 \\ 0 & 0 & 1 \\ -1 & -2 & -3 \end{bmatrix}, \quad A_2 = \begin{bmatrix} 0 & 1 & 0 \\ 0 & 0 & 1 \\ -2 & -3 & -1 \end{bmatrix}.$$

The matrix product $A_1 A_2$ is given by:

$$A_1 A_2 = \begin{bmatrix} 0 & 0 & 1 \\ -2 & -3 & -1 \\ 6 & 8 & 1 \end{bmatrix}.$$

A CQLF cannot exist for Σ_{A_1} and Σ_{A_2} as the eigenvalues of $A_1 A_2$ are given by $\lambda_i = \{1, -2, -1\}$.

7   Open Questions

In this section, we briefly discuss two major open questions that arise out of the work described earlier in the paper, and that should form the subject of future research.

Identification of system classes that satisfy Theorem 1:

We have seen above how Theorem 1 unifies, in a certain sense, the results on CQLF existence previously derived for second order systems and systems whose system matrices differ by rank one. In both of these cases, the conditions for CQLF existence are easy to verify and can be interpreted in terms of the dynamics of the associated switched linear systems. Moreover, the form of the conditions given in the conclusions of Theorem 1 suggest that for any class of system to which the Theorem can be applied, it may be possible to derive similarly attractive conditions for CQLF existence using analogous techniques to those employed above. Given the need for verifiable conditions that can be used to determine the stability of switched linear systems, this observation leads to the important problem of identifying further classes of systems to which Theorem 1 can be applied. The discovery of such system classes is likely to lead to other results along the lines of Theorem 2, giving verifiable, dynamically meaningful conditions for CQLF existence.

Extension of convex set techniques to non-quadratic Lyapunov functions:

In general, the existence of a strong CQLF is only a sufficient condition for the exponential stability of switched linear systems under arbitrary switching, and in certain situations, less conservative results may be obtained based on non-quadratic Lyapunov functions. Much recent work in this direction has focussed on Lyapunov functions, defined using vector norms, that are piecewise linear or piecewise quadratic. For the analysis of switching systems, the fact that these functions can be non-smooth is an advantage, as the switching action itself is non-smooth in nature.

The l_1-norm based Lyapunov function

$$V(x) = \|Wx\|_1 \qquad W \in \mathbb{R}^{2 \times 2}$$

was proposed in [24]. Such functions are referred to as *unic* Lyapunov functions. Note that while the LTI system Σ_A has a quadratic Lyapunov function if and only if the eigenvalues of A lie in the open left half plane, a unic Lyapunov function exists for the system if and only if the eigenvalues of A lie within the so-called 45° region

$$\{z \in \mathbb{C} : |Im(z)| < |Re(z)|, Re(z) < 0\}.$$

In [25], conditions for common unic Lyapunov function existence for pairs of second order LTI systems are described. These conditions are related to the matrix-pencil conditions given for CQLF existence for the second-order case in Section 4 above.

The conditions for common unic Lyapunov function existence for second order systems were derived using direct algebraic arguments specific to the second order case. In the light of the work described earlier on the CQLF existence problem, a natural question to ask is whether similar convex-cone based techniques can be applied in this setting to obtain more general results on common unic Lyapunov function existence. In order to do this, we would need to obtain a greater understanding of the set of unic Lyapunov functions corresponding to a given stable LTI system, and then investigate the possibility of obtaining results similar to Theorem 1 for unic Lyapunov functions.

8   Concluding Remarks

In this paper we have presented a result on common quadratic Lyapunov functions. We have shown that this result unifies a number of classical stability results and leads to time-domain versions of a number of known frequency-domain stability criteria. As well as addressing the two open problems described in the paper, future research will proceed in a number of directions; (i) by classifying the classes of systems that satisfy the assumptions of Theorem 1; (ii) by exploring the potential of time-domain stability criteria in for deriving stable adaptive control systems; and (iii) by using the time-domain stability criteria to reinterpret classical frequency domain analysis. Work in all three directions is ongoing and will be reported in future publications.

Acknowledgements This paper first appeared in part at the 12th Yale Workshop on Adaptive and Learning systems [26]. The authors gratefully thank Professor K. S. Narendra for permission to reproduce parts of this paper. This work was partially supported by the European Union funded research training network *Multi-Agent Control*, HPRN-CT-1999-00107[2], by the Enterprise Ireland grant SC/2000/084/Y, and by Science Foundation Ireland under the grant 00/PI.1/C067. Neither the European Union or Enterprise Ireland is responsible for any use of data appearing in this publication.

References

[1] R. Shorten, K. Narendra, and O. Mason, "A result on common Lyapunov functions," *IEEE Transactions on Automatic Control*, vol. 48, no. 1, pp. 110–113, 2003.

[2] R. Shorten and K. Narendra, "On common quadratic Lyapunov functions for pairs of LTI systems whose system matrices are in companion form," *IEEE Transactions on Automatic Control*, vol. 48, no. 4, pp. 618–622, 2003.

[3] R. Shorten, O. Mason, F. O'Cairbre, and P. Curran, "A unifying result for the circle criterion and other stability criteria." Accepted for publication in proceedings of the European Control Conference (extended version submitted to International Journal of Control), 2003.

[4] R. Shorten, P. Curran, and K. Wulff, "On time domain multiplier criteria for SISO systems." Submitted to Automatica, 2003.

[5] R. Loewy, "On ranges of real Lyapunov transformations," *Linear Algebra and its Applications*, vol. 13, no. 1, pp. 79–89, 1976.

[6] G. P. Barker, A. Berman, and R. J. Plemmons, "Positive Diagonal Solutions to the Lyapunov Equations," *Linear and Multilinear Algebra*, vol. 5, no. 3, pp. 249–256, 1978.

[7] R. E. Kalman, "Lyapunov functions for the problem of Lur'e in automatic control," *Proceedings of the national academy of sciences*, vol. 49, no. 2, pp. 201–205, 1963.

[8] T. Kailath, *Linear Systems*. Prentice Hall, New Jersey, 1980.

[9] S. Boyd and Q. Yang, "Structured and simultaneous Lyapunov functions for system stability problems," *International Journal of Control*, vol. 49, no. 6, pp. 2215–2240, 1989.

[10] R. Shorten and K. S. Narendra, "Necessary and sufficient conditions for a CQLF for a finite number of stable second order LTI systems," International Journal of Adaptive Control and Signal Processing, vol. 16, no. 9, pp. 709–728, 2003.

[11] N. Cohen and I. Lewkowicz, "A necessary and sufficient criterion for the stability of a convex set of matrices," *IEEE Transactions on Automatic Control*, vol. 38, no. 4, pp. 611–615, 1993.

---

[2] This work is the sole responsibility of the authors and does not reflect the European Union's opinion

[12] K. S. Narendra and R. M. Goldwyn, "A geometrical criterion for the stability of certain non-linear non-autonomous systems," *IEEE Transactions on Circuit Theory*, vol. 11, no. 3, pp. 406–407, 1964.

[13] O. Mason and R. Shorten, "On the simultaneous diagonal stability of a pair of positive linear systems," *submitted to Linear Algebra and its Applications*, 2004.

[14] L. Farina and S. Rinaldi, *Positive linear systems*. Wiley Interscience Series, 2000.

[15] R. Shorten, D. Leith, J. Foy, and R. Kilduff, "Towards an analysis and design framework for congestion control in communication networks," in *Proceedings of the 12th Yale workshop on adaptive and learning systems*, 2003.

[16] A. Jadbabaie, J. Lin, and A. S. Morse, "Co-ordination of groups of mobile autonomous agents using nearest neighbour rules," *IEEE Transactions on Automatic Control*, vol. 48, no. 6, pp. 988–1001, 2003.

[17] A. Berman and R. Plemmon, *Non-negative matrices in the mathematical sciences*. SIAM classics in applied mathematics, 1994.

[18] R. Horn and C. Johnson, *Matrix Analysis*. Cambridge University Press, 1985.

[19] J. Slotine and W. Li, *Applied Nonlinear Control*. Prentice Hall, 1991.

[20] K. Narendra and J. Taylor, *Frequency Domain Criteria for Absolute Stability*. Academic Press, 1973.

[21] G. Zames and P. L. Falb, "Stability conditions for systems with monotone and slope restricted non-linearities," *SIAM Journal of Control and Optimization*, vol. 6, pp. 89–108, 1968.

[22] J. L. Willems, "The circle criterion and quadratic Lyapunov functions for stability analysis," *IEEE Transactions on Automatic Control*, vol. 18, no. 4, p. 184, 1973.

[23] R. Shorten and C. King, "Spectral conditions for strict positive realness of lti systems." Accepted for publication in IEEE Transactions on Automatic Control, 2004.

[24] K. Wulff, R. N. Shorten, and P. Curran, "On the relationship of matrix-pencil eigenvalue criteria and the choice of Lyapunov function for the analysis of second order switched systems," in *American Control Conference*, 2002.

[25] K. Wulff, R. Shorten, and P. Curran, "On the 45 degree region and the uniform asymptotic stability of classes of second order parameter varying and switched systems," *International Journal of Control*, vol. 75, no. 11, pp. 812–823, 2002.

[26] R. Shorten, K. S. Narendra, O. Mason, and K. Wulff, "On the existence of a common quadratic Lyapunov functions for SISO swithed systems," in *proceedings of Twelftfh Yale Workshop on Adaptive and Learning Systems*, 2003.

Survey of Explicit Approaches to Constrained Optimal Control

Alexandra Grancharova [1,2], Tor Arne Johansen [1]

[1] Department of Engineering Cybernetics
Norwegian University of Science and Technology, N-7491 Trondheim, Norway
{Alexandra.Grancharova, Tor.Arne.Johansen@itk.ntnu.no}
[2] Institute of Control and System Research, Bulgarian Academy of Sciences
Acad. G. Bonchev str., Bl.2, P.O.Box 79, Sofia 1113, Bulgaria
alexandra@icsr.bas.bg

Abstract. This chapter presents a review of the explicit approaches to optimal control. It is organized as follows. Section 1 gives a summary of the main results of the optimal control theory. Section 2 presents briefly the methods for unconstrained optimal state feedback control of linear systems. Sections 3, 4 and 5 consider in details the explicit methods for constrained linear quadratic regulation (LQR) together with several examples. The main motivation behind the explicit solution is that it avoids the need for real-time optimization, and thus allows implementation at high sampling frequencies in real-time systems with high reliability and low software complexity. These sections include formulation of the constrained LQR problem, summary of the implicit approaches, basics of the model predictive control (MPC), description of the exact and the approximate approaches to explicit solution of MPC problems and the experimental evaluation of explicit MPC controller performance for laboratory gas-liquid separation plant.

1 Optimal Control Theory

1.1 General Optimal Control Problem Formulation

Optimal control theory considers the problem of how to control a given system so that it has an optimal in certain sense behaviour. Control can be time-optimal, i.e. reaching the desired state in minimum-time, or reaching this state with minimal energy costs, or achieving maximal productivity in a fixed time.

The general optimal control problem is formulated in the following way [1]: The system is described by a set of non-linear, non-autonomous state equations:

$$\frac{dx}{dt} = f(x,u,t), \tag{1}$$

where x is the n-dimensional vector of state variables, u is the m-dimensional vector of control variables which we wish to choose optimally, f is an n-dimensional vector function and t is time. The initial state of the system is supposed to be known:

R. Murray-Smith, R. Shorten (Eds.): Switching and Learning, LNCS 3355, pp. 47-97, 2005.
© Springer-Verlag Berlin Heidelberg 2005

$$x(0) = x_0.$$ (2)

At any time the system is subject to a set of generally non-linear constraints:

$$g(x,u) \geq 0$$ (3)

$$h(x,u) = 0$$ (4)

and the control variables should be in the admissible range defined by the lower and upper bounds:

$$u_{min,i} \leq u_i \leq u_{max,i}, \quad i = 1,...,m.$$ (5)

The system can be also subject to a terminal constraint which describes the target set:

$$\psi[x(t_f)] \geq 0.$$ (6)

The final time t_f may be fixed (fixed-time control problem) or variable (free end time control problem). A special case is the case of infinite final time.

In order to specify what is meant by optimal, we must select a performance index $I[u(t)]$:

$$I(u) = G[x(t_f)] + \int_0^{t_f} F(x,u)dt$$ (7)

which we wish to minimize.

The optimal control problem is then to find the time varying controls $u(t)$ such that the performance index (7) is minimized while satisfying constraints (3), (4), (5), (6). In the case of free end time problem, the optimal final time t_f must be determined in addition to the control variables in order to minimize the performance index.

It has to be noted that expression (7) is sufficiently general to allow the treatment of a wide class of practical problems. In particular, we can have the following cases:

- minimum-time control problem (minimize the time of transferring the system from the initial state to the desired final state):

$$I(u) = \int_0^{t_f} 1 dt \rightarrow \min.$$ (8)

- maximal productivity problem (maximize the amount of desired product at the final time):

$$I(u) = G[x(t_f)] \rightarrow \max.$$ (9)

This form of the performance index is typical for optimal control of batch processes.

- minimize an integral criterion:

$$I(u) = \int_0^{t_f} F(x,u)dt \rightarrow \min$$ (10)

which can include the deviation from the desired final state and the cost of the control action. This form of the performance index is typical for optimal control of continuous processes.

1.2 Necessary Conditions for Optimality

There exist several methods to solve optimal control problems for general non-linear systems. They are given in Fig.1.

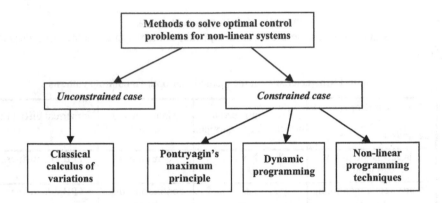

Fig. 1. Methods to solve optimal control problems for non-linear systems.

The main features of the above methods are the following:
1). Classical calculus of variations:
– applied only for unconstrained systems;
– requires continuous-time model;
– there are numerical difficulties associated with solving TPBVP (two point boundary value problem) and with the stiffness of the augmented system of differential equations.
2). Pontryagin's maximum principle:
– takes into account the presence of constraints;
– it can be applied for both continuous and discrete-time description of system dynamics;
– there are numerical difficulties related to solving TPBVP and to the stiffness of the augmented system of equations.
3). Dynamic programming:
There are two versions of dynamic programming:
a) Discrete dynamic programming (based on discretization both in time and state space):
   – takes into account the presence of constraints;
   – requires discrete-time representation of system dynamics;
   – there are numerical difficulties related to curse of dimensionality (even small increase of system dimension, i.e. the number of state variables, the number of

control variables, the number of constraints, leads to an enormous increase of dimensionality of the optimization problem that has to be solved).

b) Differential dynamic programming (Hamilton-Jacobi-Bellman equation):
- takes into account the presence of constraints;
- applicable to continuous-time systems;
- well developed for low-dimensional systems.

4). Non-linear programming techniques
- takes into account the presence of constraints;
- it can be applied for both continuous and discrete-time description of system dynamics;
- leads to high-dimensional optimization problem.

The characteristics of the methods for optimal control synthesis are summarized in Table 1.

Table 1. Characteristics of the methods for optimal control synthesis.

Characteristics → Methods ↓	Takes into account presence of constraints	Mathematical model	Numerical difficulties
Classical calculus of variations	No	continuous	TPBVP, stiffness
Pontryagin's maximum principle	Yes	continuous, discrete	TPBVP, stiffness
Dynamic programming	Yes	continuous, discrete	curse of dimensionality
Non-linear programming techniques	Yes	continuous, discrete	high-dimensional optimization problem

The necessary conditions for optimality (according to Pontryagin's maximum principle [1]) are the following: In order for a control $u(t)$ to be optimal in the sense that it minimizes the performance index (7) while satisfying the system equations (1) and constraints (5), it is necessary that the following condition:

$$\frac{\partial H}{\partial u} = 0 \qquad (11)$$

holds for the unconstrained portion of the path and the Hamiltonian function:

$$H = F(x,u) + \lambda^T f(x,u) \qquad (12)$$

is minimized along constrained portions of the control trajectory. Here λ is an n-dimensional vector of time-dependent Lagrange multipliers which are defined by the equation:

$$\frac{d\lambda}{dt} = -\frac{\partial H}{\partial x} \qquad (13)$$

and the final time condition:

$$\lambda_i(t_f) = \frac{\partial G}{\partial x_i} \qquad (14)$$

for those state variables which are unspecified at the final time t_f. In addition it is necessary that the Hamiltonian function $H(t)$ remains constant along the optimal trajectory and that $H(t)$ takes the constant value of zero when the final time t_f is unspecified.

The necessary conditions of Pontryagin's maximum principle are very similar to those arising from dynamic programming or the classical calculus of variations (when there no constraints).

1.3   Computational Techniques

There are a large variety of computational techniques for determining the optimal control for general non-linear systems. Some of the methods are based on numerically satisfying the necessary conditions for optimality, while others involve more direct search algorithms. The more commonly used algorithms are given in Fig.2. A detailed description of these algorithms is given in [2].

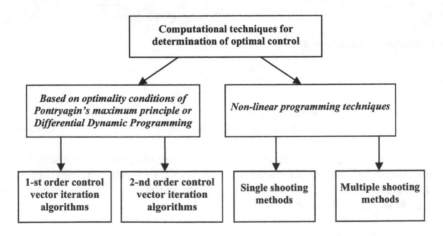

Fig. 2. Computational techniques for optimal control synthesis.

1.3.1 Control Vector Iteration Algorithms [1,2,3]. They are based on the optimality conditions of Pontryagin's maximum principle and represent iterative procedures where at each iteration the system equations (1) are integrated forward in time, the adjoint equations (13) are integrated backward in time and the control vector is updated by applying the optimality condition (11).

Thus, for the *1-st order methods* the following formulae [1,2] is used to improve the control vector (only information about 1-st order derivatives of the Hamiltonian function is used):

$$u^{i+1}(t) = u^i(t) - \varepsilon \frac{\partial H}{\partial U} \ , \quad \varepsilon > 0, \ t \in [0;t_f], \tag{15}$$

while for the *2-nd order methods* the following formulae is applied [3] (it uses information also about 2-nd order derivatives of the Hamiltonian function):

$$u^{i+1}(t) = u^i(t) - \left[\frac{\partial^2 H}{\partial u^2} \right]^{-1} \left[\varepsilon \left(\frac{\partial H}{\partial u} + \left(\frac{\partial f}{\partial u} \right)^T s \right) + \left(\frac{\partial^2 H}{\partial u \partial x} + \left(\frac{\partial f}{\partial u} \right)^T P \right) \delta x \right], \tag{16}$$

$$0 < \varepsilon \le 1, t \in [0;t_f]$$

where:

$$-\frac{ds}{dt} = \left(\frac{\partial f}{\partial x} \right)^T s - \left(\frac{\partial^2 H}{\partial u \partial x} + \left(\frac{\partial f}{\partial u} \right)^T P \right)^T \left[\frac{\partial^2 H}{\partial u^2} \right]^{-1} \left(\frac{\partial H}{\partial u} + \left(\frac{\partial f}{\partial u} \right)^T s \right) \tag{17}$$

$$-\frac{dP}{dt} = \frac{\partial^2 H}{\partial x^2} + \left(\frac{\partial f}{\partial x} \right)^T P + P \frac{\partial f}{\partial x} - \left(\frac{\partial^2 H}{\partial u \partial x} + \left(\frac{\partial f}{\partial u} \right)^T P \right)^T \left[\frac{\partial^2 H}{\partial u^2} \right]^{-1} \left(\frac{\partial^2 H}{\partial u \partial x} + \left(\frac{\partial f}{\partial u} \right)^T P \right) \tag{18}$$

with boundary conditions at t_f:

$$P = \frac{\partial^2 G}{\partial x^2} \ , \ s = 0. \tag{19}$$

In (15) and (16) i is the iteration number.

Similar algorithms are derived by applying *Differential Dynamic Programming* [3].

The disadvantages of the control vector iteration methods is the necessity to solve adjoint system of equations in addition to the original system equations and the big computational efforts related to calculation of the second-order derivatives when applying the second-order methods.

1.3.2 Single Shooting Methods [1,2,4,5,6]. They represent direct search algorithms where control vector is approximated by a set of trial functions of time or by a set of trial functions of state variables.

For open-loop control, each element $u_i(t)$ of control vector is represented by a set of trial functions of time $\phi_{ij}(t)$:

$$u_i(t) = \sum_{j=1}^{l} a_{ij}\phi_{ij}(t) \tag{20}$$

and then standard parameter optimization technique is used to determine the optimal values of coefficients a_{ij}.

For closed-loop control, each element $u_i(t)$ is approximated by a set of trial functions $\phi_{ij}(x)$ of state variables:

$$u_i(x) = \sum_{j=1}^{l} b_{ij}\phi_{ij}(x) \tag{21}$$

The advantages of these approaches are that no adjoint equations need to be solved and standard parameter optimization techniques can be used to determine the coefficients. One disadvantage of these algorithms is that the functional form of the optimal control must be specified in advance which leads to using a more general control function and therefore having a big number of coefficients to be optimized. Another disadvantage is that the boundaries of changing the coefficients are not known a priori.

1.3.3 Multiple Shooting Methods [2,5,6,7]. These methods include approximation of both control and state variables. They consists of dividing the interval of interest $[t_0; t_f]$ (the interval of seeking the optimal solution) into M-number of sub-intervals $[t^{j-1}; t^j]$ where the differential equations describing the system are approximated by algebraic equations and these are added as equality constraints to the optimization problem. Then, standard non-linear parameter optimization techniques are applied to solve the problem.

The parameters to be optimized are:

$$V = [x_1(t_{ij}), x_2(t_{ij}),...,x_n(t_{ij}), u_1(t_j), u_2(t_j),...,u_m(t_j)] \tag{22}$$

$$j = 1,2,...,M, \; i = 1,2,...,N_j, \; t_{ij} \in [t^{j-1}; t^j]$$

and include both control variables and state variables values. A specific feature of these methods is high-dimensionality of the optimization problem since the total number of the optimization parameters is:

$$M(m + N_j n) \tag{23}$$

It has to be noted, however, that even though the multiple shooting methods lead to more parameters to be optimized, the non-linear programming problem is usually better conditioned and easier to solve, in particular if the block diagonal structure of the problem is exploited. For comparison, the application of single shooting methods results in optimization problems with less structure.

1.4   Example

As an example, the time optimal control of a continuous stirred tank reactor (CSTR) is determined in [8] by applying control vector iteration algorithms. In the reactor (shown in Fig.3), a first-order irreversible reaction A→B takes place.

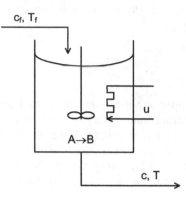

Fig. 3. Continuous stirred tank chemical reactor.

The mathematical model of the CSTR and values of the parameters are taken from the literature [4]. The mass and heat balance of the CSTR expressed through dimensionless concentration y_1 and temperature y_2 are [4]:

$$\frac{dy_1}{dt} = \frac{(1-y_1)}{\theta} - ke^{-\frac{E}{y_2}}y_1 \quad , y_1(0) = 1 \tag{24}$$

$$\frac{dy_2}{dt} = \frac{(y_f - y_2)}{\theta} + ke^{-\frac{E}{y_2}}y_1 - \alpha u(y_2 - y_c) \quad , y_2(0) = y_f, \tag{25}$$

where the dimensionless quantities y_1, y_2, y_c and y_f are defined as follows:

$$y_1 = \frac{c}{c_f} \,, y_2 = \frac{T}{Jc_f} \,, y_c = \frac{T_c}{Jc_f} \,, y_f = \frac{T_f}{Jc_f} \tag{26}$$

Here, c and c_f are the concentrations in the reactor and of the feed stream, T, T_c and T_f are the temperatures respectively in the reactor, of the cooling stream and of the feed stream, J is the heat of the chemical reaction. The reaction rate is given by:

$$r = ke^{-\frac{E}{y_2}}y_1, \tag{27}$$

where k is the pre-exponential multiplier and E is the activation energy of the reaction. The values of the parameters θ, c_f, T_f, T_c, J, α, k, E are taken from [4]. The coolant flowrate is the control variable $u(t)$ and it is constraint to be in the interval:

$$0 \le u(t) \le 1500 \tag{28}$$

The control problem to be solved is to move the reactor from the initial state to the steady state in minimum-time taking into account constraint (28) imposed on the control variable, i.e.:

Initial state :   Steady state :   **(29)**

$$y_1(0) = 1 \quad \xrightarrow{\text{min - time}} \quad y_1^* = 0.4071$$
$$y_2(0) = 3 \qquad\qquad\qquad\quad y_2^* = 3.3025$$
$$u^* = 370$$

The optimal control trajectory determined in [8] by applying control vector iteration algorithm is given in Fig.4 and the corresponding optimal trajectories of the concentration and the temperature in the reactor are shown in Fig.5 and Fig.6.

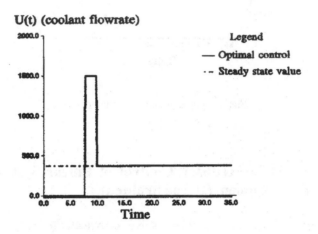

Fig. 4. Optimal control trajectory.

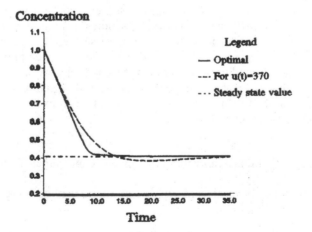

Fig. 5. Optimal trajectory of concentration.

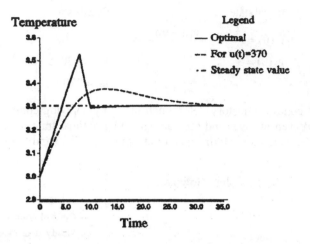

Fig. 6. Optimal trajectory of temperature.

2   Optimal State Feedback Control of Linear Systems – Linear Quadratic Problem (Unconstrained)

Only for very few cases it is possible to solve analytically (exactly) the Hamiltonian system of differential equations (equations (1) and (13)) when the system (1) is non-linear. In the general case, numerical techniques must be applied to obtain an approximate solution of these equations.

However, many systems that we wish to control are already adequately described by linear dynamic models. In this case, it is possible to solve analytically the equations related to the optimality conditions and to synthesize linear feedback controllers that minimize certain performance criteria. It is necessary to distinguish between *terminal controllers* and *regulators*. A *terminal controller* [9] is designed to bring the system close to desired conditions at a terminal time (which may or may not be specified) while exhibiting acceptable behaviour on the way. A *regulator* [9] is designed to keep a stationary system within an acceptable deviation from a reference condition using acceptable amounts of control.

2.1   Synthesis of Terminal Controllers for Linear Continuous-Time Systems

The following problem is considered [9]: Given a time-varying linear system of the form:

$$\dot{x} = A(t)x + B(t)u , \tag{30}$$

we desire to bring it from an initial state $x(t_0)$ to a terminal state:

$$x(t_f) \cong 0 \ , \ t_f = \text{terminal time} , \tag{31}$$

using "acceptable" levels of control $u(t)$ and not exceeding "acceptable" levels of the state $x(t)$ on the way. One way to do this is to minimize a performance index made up of a positive definite quadratic form in the terminal state plus an integral of positive definite quadratic forms in the state and the control:

$$I = \frac{1}{2}\left(x^T S_f x\right)_{t=t_f} + \frac{1}{2}\int_{t_0}^{t_f}\left(x^T Q x + u^T R u\right)dt , \tag{32}$$

where S_f, $Q(t)$ and $R(t)$ are positive definite matrices.

This *linear quadratic problem* represents a classical problem in optimal control theory [9]. The optimal control, i.e. control that minimizes performance index (32) can be obtained by applying the optimality conditions mentioned above. Thus, the system equations (30) have to be solved simultaneously with the Euler-Lagrange equations [9]:

$$\dot{\lambda} = -\frac{\partial H}{\partial x} \ , \qquad \lambda(t_f) = S_f x(t_f) \tag{33}$$

$$\frac{\partial H}{\partial u} = 0 , \tag{34}$$

where the Hamiltonian function is given by:

$$H = \frac{1}{2}x^T Q x + \frac{1}{2}u^T R u + \lambda^T\left(Ax + Bu\right). \tag{35}$$

Performing the differentiations indicated in (33) and (34), we have [9]:

$$\dot{\lambda} = -Qx - A^T \lambda \tag{36}$$

$$u = -R^{-1}B^T \lambda . \tag{37}$$

By making substitutions, we obtain a linear two-point boundary-value problem [9]:

$$\begin{bmatrix} \dot{x} \\ \dot{\lambda} \end{bmatrix} = \begin{bmatrix} A \ , \ -BR^{-1}B^T \\ -Q \ , \ -A^T \end{bmatrix}\begin{bmatrix} x \\ \lambda \end{bmatrix} \ ; \ \begin{matrix} x(t_0) \text{ given} \\ \lambda(t_f) = S_f x(t_f) \end{matrix} \tag{38}$$

There are *three* ways to solve this boundary-value problem [9]: 1). solution by *transition matrix* (by *linear superposition*); 2). solution by the *sweep method* (by *Riccati transformation*); 3). solution by *dynamic programming*.

By applying the sweep method, the solution for $\lambda(t)$ is represented by the form [9]:

$$\lambda(t) = S(t)x(t) , \tag{39}$$

which has been termed the *Riccati transformation* . Here $S(t)$ is a symmetric positive definite $n \times n$ matrix. We may think of (39) as generating a boundary condition for (38) equivalent to the terminal condition $\lambda(t_f) = S_f x(t_f)$, but at earlier times. In effect, the coefficients of the terminal condition are "swept" backward to the initial time. Then, since $x(t_0)$ is known, $\lambda(t_0)$ may be computed from $\lambda(t_0) = S(t_0)x(t_0)$ and (38) can be integrated forward in time as an initial-value problem. Substituting (39) into (38) yields:

$$\dot{S}x + S\dot{x} = -Qx - A^T Sx . \tag{40}$$

By substituting $\dot{x}$ from (38) into (40) and using (39), we have:

$$\left(\dot{S} + SA + A^T S - SBR^{-1}B^T S + Q \right)x = 0 . \tag{41}$$

Since $x(t) \neq 0$, we have:

$$\dot{S} = -SA - A^T S + SBR^{-1}B^T S - Q \tag{42}$$

with the terminal condition:

$$S(t_f) = S_f . \tag{43}$$

Equation (42) is known as *matrix Riccati equation* [9]. The *Riccati* equation must be integrated backward from the terminal time $t = t_f$ to the initial time $t = t_0$. It is then possible to determine $\lambda(t_0)$ as follows:

$$\lambda(t_0) = S(t_0)x(t_0) . \tag{44}$$

Then, the solution $x(t)$ and $\lambda(t)$ can be determined by forward integration of (38).

As result of solving this linear quadratic control problem, we obtain the optimal state feedback control law [9]:

$$u(t) = -K(t)x(t) \tag{45}$$

$$K(t) = \left[R(t) \right]^{-1} B^T (t)S(t) , \tag{46}$$

where $S(t)$ is the solution of *Riccati* equation.

There are some points to note [1]:
- The time-varying gain $K(t)$ can be determined off-line (by solving (42) for $S(t)$) because $K(t)$ does not depend on $x(t)$ or $u(t)$.
- If we let $t_f \to \infty$ and A, B, Q, R are constant matrices, then $S(t)$ becomes a constant matrix and is the solution of the *algebraic Riccati equation* [1]:

$$SBR^{-1}B^T S - SA - A^T S - Q = 0 . \tag{47}$$

Thus $K(t)$ is also a constant matrix and the controller is a constant-gain proportional controller:

$$u(t) = -Kx(t) \tag{48}$$

$$K = R^{-1}B^T S . \tag{49}$$

2.2 Synthesis of Linear Quadratic Regulators for Continuous-Time Systems

A *regulator* is a feedback controller designed to keep a *stationary system* within an acceptable deviation from a reference condition using acceptable amounts of control [9]. The disturbances to the system are often unpredictable, i.e. random. Here, we will consider only deterministic initial disturbances, that is $x(t_0) \neq 0$.

It has been shown above that for a *stationary system* (A, B constant matrices), for constant matrices Q, R in the performance index and for the time period $t_f - t_0$ approaching infinity, the feedback gain matrix is constant:

$$K = R^{-1}B^T S , \tag{50}$$

where S is the solution of the *algebraic Riccati equation*:

$$SBR^{-1}B^T S - SA - A^T S - Q = 0 . \tag{51}$$

In this case, the optimal value of the performance index is [9]:

$$I_{min} = \frac{1}{2}x^T(t_0)Sx(t_0) , \tag{52}$$

and it is independent on time.

Thus, if a *finite* solution S of equation (51) exists and is *positive definite*, then $x(t)$ and $u(t)$ are *bounded* (never become infinite) and:

$$u(t) = -Kx(t) \tag{53}$$

is a *stable* regulator [9].

In general, solution of the quadratic equation (51) will produce more than one value for S. The extra roots can be eliminated by the requirement that $S > 0$. However, another approach is to integrate backward the *differential Riccati equation*:

$$\dot{S} = -SA - A^T S + SBR^{-1}B^T S - Q \tag{54}$$

with the boundary condition:

$$S(t_f) = 0 \tag{55}$$

until:

$$\dot{S} \approx 0 \tag{56}$$

This is a valuable technique for *synthesizing regulators* [9].

2.3   Synthesis of Linear Quadratic Regulators for Discrete-Time Systems

It is more convenient not to consider the behaviour of a continuous-time system at all instants of time t but only at a sequence of instants t_k, k=0, 1, 2, ... This is related to using digital devices (digital controllers, digital computers) when controlling and analyzing continuous-time systems. Discrete-time systems are described by *state difference equation* [10]:

$$x(k+1) = f[x(k), u(k), k], \tag{57}$$

where $x(k)$ is the state and $u(k)$ is the input at time-instant t_k. Similarly, the output at time t_k is given by the *output equation*:

$$y(k) = g[x(k), u(k), k] \tag{58}$$

Linear discrete-time systems are described by state difference equation of the form [10]:

$$x(k+1) = A(k)x(k) + B(k)u(k), \tag{59}$$

where $A(k)$ and $B(k)$ are matrices of appropriate dimensions. The corresponding output equation is:

$$y(k) = C(k)x(k) + D(k)u(k). \tag{60}$$

Then, *the discrete-time regulator problem* is defined as follows [10]: Consider the discrete-time linear system:

$$x(k+1) = A(k)x(k) + B(k)u(k), \tag{61}$$

where:

$$x(k_0) = x_0, \tag{62}$$

with the output variable:

$$y(k) = C(k)x(k). \tag{63}$$

Consider the performance index:

$$I = \sum_{k=k_0}^{k=k_f} \left[y^T(k+1)Qy(k+1) + u^T(k)Ru(k) \right] + x^T(k_f)S_f x(k_f) \tag{64}$$

where $Q(k+1)$ and $R(k+1)$ are *positive definite* matrices for $k = k_0, k_0 + 1, \ldots, k_f - 1$ and S_f is *non-negative definite*. Then the problem of determining the input $u(k)$ for $k = k_0, k_0 + 1, \ldots, k_f - 1$ is called *the discrete-time deterministic linear optimal regulator problem* [10]. If all matrices occurring in the problem formulation are constant, we refer to it as *the time-invariant discrete-time linear optimal regulator problem* [10]. The optimal solution to the formulated problem is given by the following theorem [10]:

Theorem 1:

Consider *the discrete-time deterministic linear optimal regulator problem*. The optimal input is given by:

$$u(k) = -K(k)x(k) \;, \quad k = k_0, k_0 + 1, \; \dots \; , k_f - 1, \tag{65}$$

where:

$$K(k) = \left\{ R(k) + B^T(k)\left[C^T(k+1)Q(k+1)C(k+1) + S(k+1) \right]B(k) \right\}^{-1} \tag{66}$$
$$\cdot B^T(k)\left[C^T(k+1)Q(k+1)C(k+1) + S(k+1) \right]A(k)$$

Here the inverse always exists and the sequence of matrices $S(k), \; k = k_0, k_0 + 1, \; \dots \; , k_f - 1$ satisfies the *matrix difference equation*:

$$S(k) = A^T(k)\left[C^T(k+1)Q(k+1)C(k+1) + S(k+1) \right]\left[A(k) - B(k)K(k) \right] \tag{67}$$
$$k = k_0, k_0 + 1, \; \dots \; , k_f - 1$$

with the terminal condition:

$$S(k_f) = S_f. \tag{68}$$

The value of the criterion (64) achieved with this control law is given by:

$$I_{\min} = x^T(k_0)S(k_0)x(k_0) \tag{69}$$

For *time-invariant* systems the optimal solution to the above problem is given by the following theorem [10]:

Theorem 2:

Consider *the time-invariant discrete-time linear optimal regulator problem*. Then if the system is both stabilizable and detectable the following facts hold:

1. The solution $S(k)$ of the difference equations (66) and (67) with the terminal condition $S(k_f) = S_f$ converges to a constant steady-state solution S as $k_f \to \infty$ for any $S_f \geq 0$.

2. The steady-state optimal control law is time-invariant and asymptotically stable.

3. The steady-state optimal control law minimizes (64) for $k_f \to \infty$ and for all $S_f \geq 0$. The minimal value of the criterion is given by:

$$I_{\min} = x^T(k_0)Sx(k_0) \tag{70}$$

3   Constrained Linear Quadratic Regulation

3.1   Problem Formulation

We consider time-invariant linear discrete-time system described by the state-space equation:

$$x(t+k+1) = Ax(t+k) + Bu(t+k) , \quad k \geq 0, \tag{71}$$

where A and B are the state transition and input distribution matrices. It is assumed that (A, B) is stabilizable.

The control objective is to regulate the state of the system optimally to the origin. Optimality is defined in terms of a *quadratic objective* and *a set of inequality constraints* [11]. The objective is defined over an infinite horizon and is given by:

$$I[x(t),\{u(t),u(t+1), ...\}] = \sum_{k=0}^{\infty} \left[x^T(t+k)Qx(t+k) + u^T(t+k)Ru(t+k) \right] \tag{72}$$

in which $Q \geq 0$ and $R > 0$ are symmetric weighting matrices. The constraints are also defined on an infinite horizon and take the form:

$$Hx(t+k+1) \leq h , \quad k \geq 0 \tag{73}$$

$$Gu(t+k) \leq g , \quad k \geq 0, \tag{74}$$

where vectors h and g (of dimension respectively n_h and n_g) define the constraint levels and H and G are the state and input constraint distribution matrices.

Then, *3 control problems* are formulated [11].

Problem 1 – Unconstrained linear quadratic regulation:

$$\min_{\{u(t),u(t+1), ...\}} \sum_{k=0}^{\infty} \left[x^T(t+k)Qx(t+k) + u^T(t+k)Ru(t+k) \right] \tag{75}$$

subject to:

$$x(t+k+1) = Ax(t+k) + Bu(t+k) , \quad k \geq 0. \tag{76}$$

This problem has been considered in the previous section and it was shown that the solution is the linear feedback control law:

$$u(t+k) = -Kx(t+k) , \quad k \geq 0, \tag{77}$$

where the controller gain matrix K can be calculated from the solution of the discrete algebraic Riccati equation.

Problem 2 - Constrained linear quadratic regulation:

$$\min_{\{u(t),u(t+1), ...\}} \sum_{k=0}^{\infty} \left[x^T(t+k)Qx(t+k) + u^T(t+k)Ru(t+k) \right] \tag{78}$$

subject to:

$$x(t+k+1) = Ax(t+k) + Bu(t+k) , \quad k \geq 0 \tag{79}$$

$$Hx(t+k+1) \leq h , \quad k \geq 0 \tag{80}$$

$$Gu(t+k) \leq g , \quad k \geq 0. \tag{81}$$

Problem 2 is a natural extension of *Problem 1* that includes constraints. The difficulty associated with *Problem 2* is the infinite number of decision variables in the optimization and the infinite number of constraints.

Problem 3 - Model Predictive Control (MPC) Problem:

$$\min_{\{u(t),u(t+1),\ldots u(t+N-1)\}} \sum_{k=0}^{N-1} \left[x^T(t+k)Qx(t+k) + u^T(t+k)Ru(t+k) \right] \tag{82}$$

subject to:

$$x(t+k+1) = Ax(t+k) + Bu(t+k) , \quad k \geq 0 \tag{83}$$

$$Hx(t+k+1) \leq h , \quad k = 0,1,\ldots,N-1 \tag{84}$$

$$Gu(t+k) \leq g , \quad k = 0,1,\ldots,N-1 \tag{85}$$

$$u(t+k) = -Kx(t+k) , \quad k \geq N \tag{86}$$

This form of MPC has a finite number of decision variables, N, and a finite number of constraints, $N(n_h + n_g)$. It can therefore be solved with standard quadratic programming methods. Here, the unconstrained feedback control law (86) is added to the finite set of N decision variables.

3.2 Implicit Approaches

In [11,12,13] *implicit approaches* have been developed to solve the *constrained linear quadratic regulation problem* (*Problem 2*). They are implicit in sense that the optimal control does not have the form of feedback control law, but it is obtained in the form of open-loop time trajectory.

In their pioneering work [12] Sznaier and Damborg showed that *finite-horizon optimization* defined as the *Model Predictive Control problem* also provides the solution to the *infinite-horizon linear quadratic regulation problem with constraints*. This equivalence holds for a certain set of initial conditions, which depends on the length of the finite horizon.

This idea has been developed further by Scokaert and Rawlings [11] and by Chmielewski and Manousiouthakis [13].

In [11] the following definition is made and an algorithm is proposed to solve the *constrained LQR problem*:

Definition 1: Let $X_K \subseteq R^n$ denotes the set of states $x(t)$ for which the unconstrained LQR law, $u(t+k) = -Kx(t+k)$, $k \geq 0$, satisfies (71), (73) and (74).

Algorithm 1 (constrained LQR):

Step 0. Choose a finite horizon N_0, set $N = N_0$.

Step 1. Solve Problem 3 (MPC problem).

Step 2. If $x(t+N) \in X_K$, go to step 4).

Step 3. Increase N, go to step 1).

Step 4. Terminate: $\pi^* = \pi_N$.

Here $\pi_N = \{u(t), u(t+1), \ldots u(t+N-1)\}$ is the optimal control trajectory determined by solving the *MPC problem*, while π^* is the optimal control trajectory that is a solution of the *constrained LQR problem*.

In [11] it has been shown that the algorithm terminates in a finite number of iterations, regardless of the choice of initial horizon in Step 0) and of the heuristics used to increase it in Step 3). In other words, the presented algorithm requires solving a finite number of finite-dimensional positive definite quadratic programs (QP). Also, the *constrained LQR* is shown to be both optimal and stabilizing.

However, the dimension of the QP depends on the initial state and therefore the result is not useful for practical applications where an upper bound on the horizon (respectively of the QP size) which is independent on the initial state is needed for designing the control hardware. In this respect, Chmielewski and Manousiouthakis [13] describe an algorithm which provides a semi-global upper bound. Namely, for any given compact set X_0 of initial conditions, their algorithm provides the horizon N such that the *finite horizon controller* (solution of *Problem 3*) solves the *infinite horizon problem* (*Problem 2*).

3.3   Basics of Model Predictive Control

Model predictive control is an efficient methodology to solve complex constrained multivariable control problems [14,15,16,17]. Here the basics of Model Predictive Control (MPC) are given according to [16].

Consider the problem of regulating to the origin the discrete-time linear time invariant system:

$$x(t+1) = Ax(t) + Bu(t) \tag{87}$$

$$y(t) = Cx(t) \tag{88}$$

while satisfying the following constraints:

$$y_{min} \leq y(t) \leq y_{max}, \quad u_{min} \leq u(t) \leq u_{max} \tag{89}$$

at all time instants $t \geq 0$. In (87) – (89), $x(t) \in \Re^n$, $u(t) \in \Re^m$ and $y(t) \in \Re^p$ are the state, input and output vector respectively, y_{min}, y_{max} and u_{min}, u_{max} are respectively p and m-dimensional vectors and the pair (A, B) is stabilizable.

Model Predictive Control (MPC) solves such a constrained regulation problem in the following way [16]. Assume that a full measurement of the state $x(t)$ is available at the current time t. Then, the optimization problem:

$$\min_{U=\{u_t,u_{t+1},\ldots u_{t+N_u-1}\}} \left\{ I[U,x(t)] = x^T_{t+N_y|t} P x_{t+N_y|t} + \sum_{k=0}^{N_y-1} \left[x^T_{t+k|t} Q x_{t+k|t} + u^T_{t+k} R u_{t+k} \right] \right\} \tag{90}$$

subject to $x_{t|t} = x(t)$ and:

$$y_{\min} \leq y_{t+k|t} \leq y_{\max} \quad, \quad k = 1, \ldots, N_c \tag{91}$$

$$u_{\min} \leq u_{t+k} \leq u_{\max} \quad, \quad k = 0,1, \ldots, N_c \tag{92}$$

$$x_{t+N_y|t} \in \Omega \tag{93}$$

$$x_{t+k+1|t} = A x_{t+k|t} + B u_{t+k} \quad, \quad k \geq 0 \tag{94}$$

$$y_{t+k|t} = C x_{t+k|t} \quad, \quad k \geq 0 \tag{95}$$

$$u_{t+k} = -K x_{t+k|t} \quad, \quad N_u \leq k < N_y \tag{96}$$

is solved at each time t, where $x_{t+k|t}$ denotes the predicted state vector at time $t+k$, obtained by applying the input sequence $u_t, \ldots, u_{t+k-1}$ to model (87), (88) starting from the state $x(t)$.

The name MPC stems from the idea of employing an explicit model of the plant to be controlled which is used to predict the future output behavior. This prediction capability allows solving optimal control problems on line, where tracking error, namely the difference between the predicted output and the desired reference, is minimized over a future horizon [16].

Further, we assume that $Q = Q^T \geq 0$, $R = R^T > 0$, $P \geq 0$, $(Q^{\frac{1}{2}}, A)$ detectable (for instance $Q = C^T C$ with (C, A) detectable), K is some feedback gain, N_y, N_u, N_c are the output, input and constraint horizons, respectively, with $N_u \leq N_y$ and $N_c \leq N_y - 1$, and Ω is a polyhedral terminal set.

One possibility is to choose $K = 0$ and P as the solution of the Lyapunov equation [16]:

$$P = A^T P A + Q. \tag{97}$$

The choice $K = 0$ implies that after N_u time steps the control is turned off and the system is allowed to settle in an open loop manner. This is only meaningful when the system is open-loop stable. With P obtained from (97), $I[U,x(t)]$ measures the settling cost of the system from the present time t to infinity under the assumption that the control is turned off after N_u steps.

Alternatively, one can set $K = K_{LQ}$, where K_{LQ} and P are the solution of the unconstrained infinite horizon LQR problem with weights Q and R [16]:

$$K_{LQ} = -\left(R + B^T P B\right)^{-1} B^T P A \tag{98}$$

$$P = \left(A + B K_{LQ}\right)^T P\left(A + B K_{LQ}\right) + K_{LQ}^{\ T} R K_{LQ} + Q . \tag{99}$$

This choice of K implies that after N_u time steps the control is switched to the unconstrained LQR. With P obtained from (99), $I[U, x(t)]$ measures the settling cost of the system from the present time t to infinity under this control assumption.

The *MPC control law* is based on the following idea [16]: At time t compute the optimal solution to problem (90) – (96) (the optimal input sequence):

$$U^*(t) = \left\{u_t^*, \ldots, u_{t+N_u-1}^*\right\} \tag{100}$$

and apply to the system only the first input from the sequence:

$$u(t) = u_t^* . \tag{101}$$

The remaining optimal inputs are discarded and a new optimal control problem is solved at time $t+1$, based on the new state $x(t+1)$. Such a control strategy is also referred to as *moving* or *receding horizon*. This idea is illustrated in Fig.7.

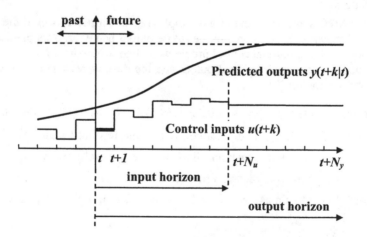

Fig. 7. Receding horizon strategy: only the first input of the computed optimal input sequence is implemented.

As new measurements are collected from the plant at each time t, the receding horizon mechanism provides the controller with the desired feedback characteristics.

The stability of MPC feedback loops was investigated by numerous researchers [14,15,16]. Stability is in general a complex function of the various tuning parameters N_u, N_y, N_c P, Q and R [16]. For applications it is most useful to impose some conditions on N_y, N_c and P so that stability is guaranteed for all $Q \geq 0$, $R > 0$. Then Q and R can be freely chosen as tuning parameters to affect performance. The constraint (93) is sometimes called "stability constraint" and it explicitly forces the state vector to reach an invariant set at the end of the prediction horizon. The stability result is formulated in the following theorem [16]:

Theorem 3:

Let $N_y = \infty$, $K = 0$ or $K = K_{LQ}$, and $N_c < \infty$ be sufficiently large for guaranteeing existence of feasible input sequences at each time step. Then the MPC law (90) – (101) asymptotically stabilizes the system (87) – (88) while enforcing the fulfillment of the constraints (89) from all initial states $x(0)$ such that (90) – (96) is feasible at $t = 0$.

$\square$

4   Exact Approaches to Explicit Solution of MPC Problems

4.1   MPC Computation

By substituting [16,17]:

$$x_{t+k|t} = A^k x(t) + \sum_{j=0}^{k-1} A^j B u_{t+k-1-j} \tag{102}$$

in the optimization problem (90) – (96), this can be rewritten in the form:

$$V^*[x(t)] = \frac{1}{2} x^T(t) Y x(t) + \min_U \left\{ \frac{1}{2} U^T H U + x^T(t) F U \right\} \tag{103}$$

subject to:

$$GU \leq W + Ex(t), \tag{104}$$

where the column vector $U \equiv \left[u_t^T, \ldots, u_{t+N_u-1}^T \right]^T \in \Re^s$, $s = mN_u$, is the optimization vector, $H = H^T > 0$ and H, F, Y, G, W, E are easily obtained from Q, R and (90) – (102).

The optimization problem (103) – (104) is a quadratic program (QP). Because the problem depends on the current state $x(t)$, the implementation of MPC requires the on-line solution of a QP at each time step. Although efficient QP solvers based on active-set methods and interior point methods are available, computing the input $u(t)$ demands significant on-line computation effort. For this reason, the application of MPC has been limited to "slow " and/or "small "processes.

Bemporad *et al.* [16,17] have proposed a new approach to implement MPC, where the computation effort is moved off-line. The MPC formulation described in the previous section provides the control action $u(t)$ as a function of $x(t)$ *implicitly* defined by (103) – (104). By treating $x(t)$ as a vector of parameters, the goal is to solve (103) – (104) off-line with respect to all the values of $x(t)$ of interest and make this dependence *explicit* [16,17].

In terms of operations research, mathematical programs which depend only on one scalar parameter are referred to as *parametric programs*, while problems depending on a vector of parameters as *multi-parametric programs*. According to this terminology, (103) – (104) is a *multi-parametric Quadratic Program* (mp-QP).

Bemporad *et al.* [16,17] have suggested an algorithm to solve mp-QP problems. Once the multi-parametric problem (103) – (104) has been solved off-line, i.e. the solution:

$$U_t^* = U^*[x(t)] \tag{105}$$

of (103) – (104) has been found, the model predictive controller (90) – (96) is available *explicitly*, as the optimal input $u(t)$ consists simply of the first m components of $U^*[x(t)]$:

$$u(t) = [I\ 0\ ...\ 0]U^*[x(t)]. \tag{106}$$

It has been shown by Bemporad *et al.* [16,17] that the solution $U^*(x)$ of the mp-QP problem is a *continuous* and *piecewise affine* function of x. Therefore, the same properties are inherited by the controller.

4.2   Exact Approach to Explicit Solution of MPC Problems

Bemporad *et al.* [16,17] have developed an algorithm to express the solution $U^*(x)$ and the minimum value $V^*(x) = I[U^*(x)]$ as an *explicit* function of the parameters x and to characterize the analytical properties of these functions. In particular they have proved that the solution $U^*(x)$ is a *continuous piecewise affine* function of x in the following sense [16,17]:

Definition 2:

A function $z(x) : X \mapsto \mathfrak{R}^s$, where $X \subseteq \mathfrak{R}^n$ is a polyhedral set, is *piecewise affine* if it is possible to partition X into convex polyhedral regions, CR_i, and $z(x) = H^i x + k^i$, $\forall x \in CR_i$.

Piecewise quadraticity is defined analogously by letting $z(x)$ be a quadratic function $x^T W^i x + H^i x + k^i$.

In [16,17] it is defined:

$$z \equiv U + H^{-1} F^T x(t), \tag{107}$$

where $z \in \mathfrak{R}^s$ and the problem (103) – (104) is transformed to the equivalent problem:

$$V_z^*(x) = \min_z \frac{1}{2} z^T H z \tag{108}$$

subject to:

$$Gz \leq W + Sx(t), \tag{109}$$

where $S \equiv E + GH^{-1}F^T$ and $V_z^*(x) = V^*(x) - \frac{1}{2} x^T \left(Y - FH^{-1}F^T \right) x$.

The solution of mp-QP problems can be approached by employing the principles of parametric nonlinear programming and in particular the first-order Karush-Kuhn-Tucker (KKT) optimality conditions [18], which lead to the Basic Sensitivity Theorem.

Instead, Bemporad *et al.* [16,17] have adopted a more direct approach which exploits the *linearity* of the constraints and the fact that the function to be minimized is *quadratic*. The approach [16,17] is described as follows. In order to start solving the mp-QP problem, an initial vector x_0 inside the polyhedral set X of parameters is needed, such that the QP problem (108) – (109) is feasible for $x = x_0$. Such a vector can be found for instance by solving the *linear program* (LP) [16,17]:

$$\max_{x,z,\varepsilon} \varepsilon \tag{110}$$

subject to:

$$Gz - Sx + \varepsilon \leq W \tag{111}$$

$$\varepsilon \geq 0 \tag{112}$$

$$x \in X. \tag{113}$$

If the LP is infeasible, then the QP problem (108) – (109) is infeasible for all $x \in X$. Otherwise, it is fixed $x = x_0$ and the QP problem (108) – (109) is solved in order to obtain the corresponding optimal solution z_0. Such a solution is unique because $H > 0$ and therefore uniquely determines a set of active constraints $\widetilde{G} z_0 = \widetilde{S} x_0 + \widetilde{W}$ out of the constraints in (108) – (109). Let $\widetilde{G}$, $\widetilde{S}$ and $\widetilde{W}$ denote the rows of G, S and W corresponding to the active constraints. Then, the following theorem is proved [16,17]:

Theorem 4:

Let $H > 0$. Consider a combination of active constraints $\widetilde{G}$, $\widetilde{S}$, $\widetilde{W}$ and assume that the rows of $\widetilde{G}$ are linearly independent. Let CR_0 be the set of all vectors x for which such a combination is active at the optimum (CR_0 is referred to as critical region).

Then, the optimal z and the associated vector of Lagrange multipliers λ are uniquely defined affine functions of x over CR_0.

Proof [16]:
The first-order KKT conditions for the mp-QP are given by:

$$Hz + G^T\lambda = 0 , \quad \lambda \in \mathfrak{R}^q \tag{114}$$

$$\lambda_i\left(G^iz - W^i - S^ix\right) = 0 , \quad i = 1, \dots, q \tag{115}$$

$$\lambda \geq 0 , \tag{116}$$

where the superscript i denotes the i-th row. Equality (114) is solved for z:

$$z = -H^{-1}G^T\lambda \tag{117}$$

and the result is substituted into (115) to obtain the complementary slackness condition:

$$\lambda\left(-GH^{-1}G^T\lambda - W - Sx\right) = 0 . \tag{118}$$

Let $\check{\lambda}$ and $\tilde{\lambda}$ denote the Lagrange multipliers corresponding to *inactive* and *active* constraints, respectively. For *inactive* constraints $\check{\lambda} = 0$. For *active* constraints $-\tilde{G}H^{-1}\tilde{G}^T\tilde{\lambda} - \tilde{W} - \tilde{S}x = 0$ and therefore:

$$\tilde{\lambda} = -\left(\tilde{G}H^{-1}\tilde{G}^T\right)^{-1}\left(\tilde{W} + \tilde{S}x\right), \tag{119}$$

where $\tilde{G}$, $\tilde{W}$, $\tilde{S}$ correspond to the set of active constraints and $\left(\tilde{G}H^{-1}\tilde{G}^T\right)^{-1}$ exists because the rows of $\tilde{G}$ are linearly independent. Thus λ is an affine function of x. By substituting $\tilde{\lambda}$ from (119) into (117), it is obtained:

$$z = H^{-1}\tilde{G}^T\left(\tilde{G}H^{-1}\tilde{G}^T\right)^{-1}\left(\tilde{W} + \tilde{S}x\right) \tag{120}$$

and it is noted that z is also an affine function of x.

Theorem 4 characterizes the solution only locally in the neighborhood of a specific x_0, as it does not provide the construction of the set CR_0 where this characterization remains valid. On the other hand, this region can be characterized immediately [16,17]. The variable z from (117) must satisfy the constraints (109):

$$GH^{-1}\tilde{G}^T\left(\tilde{G}H^{-1}\tilde{G}^T\right)^{-1}\left(\tilde{W} + \tilde{S}x\right) \leq W + Sx \tag{121}$$

and by (116) the Lagrange multipliers in (119) must remain nonnegative:

$$-\left(\tilde{G}H^{-1}\tilde{G}^T\right)^{-1}\left(\tilde{W} + \tilde{S}x\right) \geq 0 \tag{122}$$

as x varies. After removing the redundant inequalities from (121) and (122), a compact representation of CR_0 is obtained. Obviously, CR_0 is a polyhedron in the x-space and represents the largest set of $x \in X$ such that the combination of *active* constraints at the minimizer remains unchanged (Fig.8(a)). Then, the algorithm in [16,17] continues with the division of the parameter space as in Fig.8(b) and (c) by reversing one by one the hyperplanes defining the critical region CR_0. Iteratively each mew region R_i is subdivided in a similar way as was done with X. As noted in [19], the main drawback of this algorithm is that the regions R_i are not related to optimality, as they can split some of the critical regions like CR_1 in Fig.8(d). A consequence is that CR_1 will be detected at least twice.

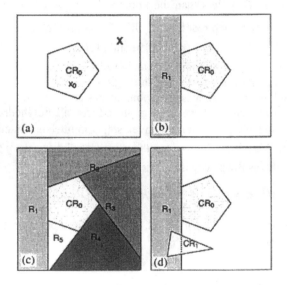

Fig. 8. State space exploration strategy of Bemporad et al [16,17].

The properties of the set of feasible parameters $X_f \subseteq X$ (i.e. the set of parameters $x \in X$ such that a feasible solution $z^*(x)$ exists to the optimization problem (108) – (109)), the value function $V_z^*(x)$ and the solution $z^*(x)$ are formulated in the following theorem [16]:

Theorem 5:

Consider the multi-parametric quadratic program (108) – (109) and let $H > 0$, X convex. Then the set of feasible parameters $X_f \subseteq X$ is convex, the optimizer $z^*(x) : X_f \mapsto \Re^s$ is continuous and piecewise affine and the value function $V_z^*(x) : X_f \mapsto \Re$ is continuous, convex and piecewise quadratic.

Based on the above results, the main steps of the off-line mp-QP solver are outlined in the following algorithm [16]:

Algorithm 2 (exact mp-QP):

Step 1. Let the current region be the whole set $X \subseteq \mathfrak{R}^n$.

Step 2. Choose a vector x_0 in the current region by solving the linear program $(110) - (113)$.

Step 3. For $x = x_0$, compute the corresponding optimal solution (z_0, λ_0) by solving a QP.

Step 4. Determine the set of active constraints when $z = z_0$, $x = x_0$, and build $\tilde{G}$, $\tilde{W}$, $\tilde{S}$.

Step 5. If $r = \text{rank}\,\tilde{G}$ is less than the number l of rows of $\tilde{G}$, take a subset of r linearly independent rows and redefine $\tilde{G}$, $\tilde{W}$, $\tilde{S}$ accordingly.

Step 6. Determine $\tilde{\lambda}(x)$, $z(x)$ from (119) and (120).

Step 7. Characterize the *CR* from (121) and (122).

Step 8. Define and partition the rest of the region.

Step 9. For each nonempty new sub-region, go to step 2.

Step 10. When all regions have been explored, for all polyhedral regions where $z(x)$ is the same and whose union is a convex set, compute such a union.

In conclusion, Algorithm 2 provides the *explicit* solution $u = f(x)$ to the MPC problem $(90) - (96)$, as the piecewise affine function:

$$u = K_1^i x + k_2^i \quad \text{if} \quad H^i x \leq h^i, \, i = 1, \dots, N_{MPC}, \tag{123}$$

where the polyhedral sets $\{H^i x \leq h^i\}$, $i = 1, \dots, N_{MPC}$ are a partition of the given set of states X.

4.3 Efficient Implementation of the Exact Approach to Explicit Solution of MPC Problems

4.3.1 Main Theoretical Result. The approach of Tøndel *et al.* [19] *modifies* the *explicit* approach of Bemporad *et al.* [16,17] by analyzing several properties of the geometry of the polyhedral partition and its relation to the combination of active constraints at the optimum of the quadratic program. Based on that, they derive a new exploration strategy for sub-dividing the parameter space, which:
1. avoids unnecessary partitioning;
2. avoids the solution to LP problems for determining an interior point in each new region of the parameter space;
3. avoids the solution to the QP problem for such an interior point.

As a consequence, there is a significant improvement of efficiency with respect to the algorithm of Bemporad *et al.* [16,17].

Before describing the main idea of the approach [19], some definitions are made [19]:

Definition 3:

Let $z^*(x)$ be the optimal solution to (108) – (109) for a given x. We define *active constraints* the constraints with $G^i z^*(x) - W^i - S^i x = 0$ and *inactive constraints* the constraints with $G^i z^*(x) - W^i - S^i x < 0$. The *optimal active set* $A^*(x)$ is the set of indices of active constraints at the optimum $A^*(x) = \{i \mid G^i z^*(x) = W^i + S^i x\}$. We also define as *weakly active constraint* an active constraint with an associated zero Lagrange multiplier λ^i and as *strongly active constraint* an active constraint with a positive Lagrange multiplier λ^i.

Definition 4:

For an active set, we say that the *linear independence constraint qualification (LICQ)* holds if the set of active constraint gradients are linearly independent, i.e. $\tilde{G}$ has full row rank.

Below, the linear expression of the PWL function $z^*(x)$ over the critical region CR_k is denoted by $z_k^*(x)$. In general, a superscript index is used to denote a row of a matrix or element of a vector.

Definition 5:

Let a polyhedron $X \subset \Re^n$ be represented by the linear inequalities $A_0 x \le b$. Let the i-th hyperplane, $A_0^i x = b^i$ be denoted by Ψ. If $X \cap \Psi$ is (n-1)-dimensional then $X \cap \Psi$ is called a *facet* of the polyhedron.

Definition 6:

Two polyhedra are called *neighboring* polyhedra if they have a common facet.

Definition 7:

Let a polyhedron X be represented by $A_0 x \le b$. We say that $A_0^i x \le b^i$ is *redundant* if $A_0^j x \le b^j \; \forall j \ne i \Rightarrow A_0^i x \le b^i$ (i.e. it can be removed from the description of the polyhedron). The inequality i is *redundant with degree h* if it is redundant but there exists a h-dimensional subset Y of X such that $A_0^i x = b^i$ for all $x \in Y$.

Let us consider a hyperplane defining the common facet between two polyhedra CR_0, CR_i in the optimal partition of the state space. There are two different kinds of hyperplanes [19]. The first (Type I) are those described by (121), which represent a non-active constraint that becomes active at the optimum as x moves from CR_0 to CR_i. This means that if a polyhedron is bounded by a hyperplane which originates from (121), the corresponding constraint will be activated on the other side of the facet defined by this hyperplane. In addition, the corresponding Lagrange multiplier may become positive. The other kind (Type II) of hyperplanes which bounds the polyhedra are those described by (122). In this case, the corresponding constraint will be non-active on the other side of the facet defined by this hyperplane. This is formulated in the following theorem [19]:

Theorem 6:

Consider an optimal active set $\{i_1, i_2, \dots, i_k\}$ and its corresponding n-minimal representation of the critical region CR_0 obtained by (121) – (122) after removing redundant inequalities. Let CR_i be a full-dimensional neighboring critical region to CR_0 and assume LICQ holds on their common facet $\Phi = CR_0 \cap \Psi$ where Ψ is the separating hyperplane between CR_0 and CR_i. Moreover, assume that there are no constraints which are weakly active at the optimizer $z^*(x)$ for all $x \in CR_0$. Then:

Type I. If Ψ is given by $G^{i_{k+1}} z_0^*(x) = W^{i_{k+1}} + S^{i_{k+1}} x$, then the optimal active set in CR_i is $\{i_1, \dots, i_k, i_{k+1}\}$.

Type II. If Ψ is given by $\lambda_0^{i_k}(x) = 0$, then the optimal active set in CR_i is $\{i_1, \dots, i_{k-1}\}$.

4.3.2   Example. The example is taken from [19]. Consider the double integrator [20]:

$$A = \begin{bmatrix} 1 & T_s \\ 0 & 1 \end{bmatrix}, \quad B = \begin{bmatrix} T_s^2 \\ T_s \end{bmatrix}, \tag{124}$$

where the sampling interval is $T_s = 0.05$ and consider the MPC problem over the prediction horizon $N=2$ with cost matrices:

$$Q = \begin{bmatrix} 1 & 0 \\ 0 & 0 \end{bmatrix}, \quad R = 1. \tag{125}$$

The constraints in the system are:

$$-1 \leq u \leq 1 \tag{126}$$

$$-0.5 \leq x_2 \leq 0.5. \tag{127}$$

The mp-QP associated with this problem has the form (108) – (109) with H, F, G, W, S given in [19]:

$$H = \begin{bmatrix} 1.079 & 0.076 \\ 0.076 & 1.073 \end{bmatrix}, \quad F = \begin{bmatrix} 1.109 & 1.036 \\ 1.573 & 1.517 \end{bmatrix} \tag{128}$$

$$G^T = \begin{bmatrix} 1 & 0 & -1 & 0 & 0.05 & 0.05 & -0.05 & -0.05 \\ 0 & 1 & 0 & -1 & 0 & 0.05 & 0 & -0.05 \end{bmatrix} \tag{129}$$

$$W^T = \begin{bmatrix} 1 & 1 & 1 & 1 & 0.5 & 0.5 & 0.5 & 0.5 \end{bmatrix} \tag{130}$$

$$S^T = \begin{bmatrix} 1.0 & 0.9 & -1.0 & -0.9 & 0.1 & 0.1 & -0.1 & -0.1 \\ 1.4 & 1.3 & -1.4 & -1.3 & -0.9 & -0.9 & 0.9 & 0.9 \end{bmatrix} \tag{131}$$

The partitioning starts with finding the region where no constraints are active. As the mp-QP is created from a feasible MPC problem, the empty active set will be optimal in some full-dimensional region ($A_0 = \varnothing$ and $\widetilde{G}$, $\widetilde{W}$ and $\widetilde{S}$ are empty matrices, $z^*(x) = 0$ and the first component of $U^*(x)$ is the unconstrained LQR gain). This critical region is then described by $0 \le W + Sx$ which contains 8 inequalities. Two of these inequalities are redundant with degree 0 (#2 and #4), the remaining 6 hyperplanes are facet inequalities of the polyhedron (see Fig.9(a)).

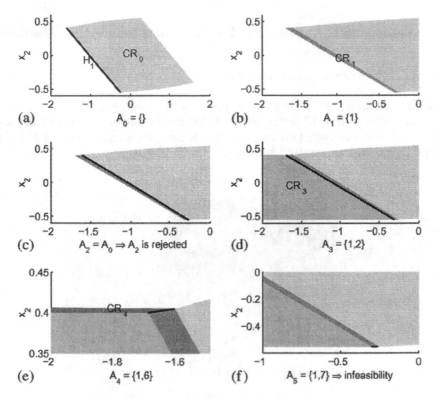

Fig. 9. Critical regions for double integrator.

By crossing the facet given by Ψ_1, defined by inequality 1 and of Type I, as predicted by Theorem 6 the optimal active set across this facet is $A_1 = \{1\}$, which leads to the critical region CR_1 (see Fig.9(b)). After removing redundant inequalities we are left with an n-minimal representation of CR_1 containing 4 facets. The first of these is of Type II, $\lambda^1(x) = 0$. The other three are of Type I. These are inequalities #2, #6 and #7. Consider first the other side of the facet which comes from $\lambda^1(x) = 0$, see Fig.9(c). The region should not have constraint 1 active, so the optimal active set is $A_2 = \varnothing$. This is the same combination of active constraints as A_0, as expected, so A_2 is not pursued. Next, consider crossing the respective facets of inequalities 2, 6 and 7, see Fig.9(d)–9(f). This results in three different active sets: $A_3 = \{1,2\}$, $A_4 = \{1,6\}$ and $A_5 = \{1,7\}$. The sets A_3 and A_4 lead to new polyhedra as shown in the figures. The combination A_5 leads to an interesting case of "degeneracy". The associated matrix $\tilde{G}$ has linearly dependent rows, which violates the LICQ assumption. In this case, A_5 leads to an infeasible part of the state space.

5   Approximate Approach to Explicit Solution of MPC Problems

5.1   Complexity of the Exact Approaches

Consider the same double integrator example as in section 4.3.2. Fig.10 shows the partition for horizon N=10 corresponding to the exact solution provided by the algorithm [19]. We observe that the exact solution is fairly complex, containing 191 polyhedral critical regions, many of them of very small volume.

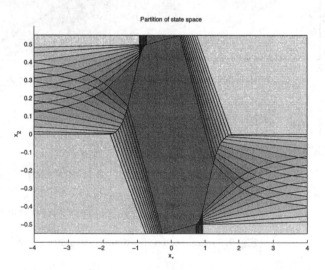

Fig. 10. Polyhedral partition of state space for the double integrator with N=10.

5.2   Main Idea of the Approximate Approach

Here we suggest an entirely different approach to compute sub-optimal explicit MPC solutions [21,22,23]. The idea is to require that the state space partition is represented as a search tree, i.e. to consist of orthogonal hypercubes organized in a hierarchical data-structure that allows *extremely fast real-time search*. The computational complexity with the suggested approach is logarithmic with respect to the number of regions, while a general polyhedral partitioning leads to a computational complexity that is linear with respect to the number of regions, if no additional data structures are built. The optimal solution is computed explicitly using quadratic programming (QP) only at the vertices of these hypercubes, and an approximate solution valid in the whole hypercube is computed based on this data. A hypercube is partitioned into two or more smaller hypercubes only if this is necessary to achieve the desired local accuracy of the solution. This makes the idea similar to storing the pre-computed QP solutions at the various states in a multi-resolution lookup table.

Unlike any other method mentioned above, that all relies on the linearity of the problem to build polyhedral regions and a PWL (piece-wise linear) solution, the suggested method is straightforward to be extended to nonlinear constrained MPC problems by replacing the QP with a nonlinear program.

5.3   Approximate mp-QP Algorithm

We restrict our attention to a hypercube $X \subset \Re^n$ where we seek to approximate the optimal PWL solution $z^*(x)$ to the mp-QP problem (108) – (109). In order to minimize the real-time computational complexity we require that the state space partition is orthogonal and can be represented as a search tree (generalized quad-tree [24], Fig.11 (left)), such that the search complexity is logarithmic with respect to the number of regions.

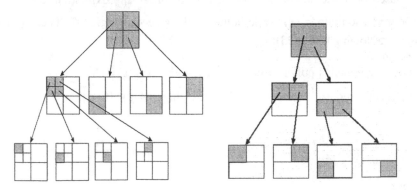

Fig. 11. Partition of a rectangular region in a 2-dimensional state space. Left: quad-tree partition. Right: k - d tree partition.

The orthogonal search tree is a hierarchical data structure where a hypercube can be hierarchically subdivided into smaller hybercubes allowing the local resolution to be adapted, as shown in Fig.11. When searching the tree, only n scalar comparisons are required at each level.

The improved version of the approximate mp-QP algorithm is based on a k - d tree partition of the state space (Fig.11 (right)) as a more flexible and powerful alternative to the generalized quad-tree (Fig.11 (left)). With the k - d tree [24], a hyper-rectangle is split into two equal parts and thus only *one* scalar comparison is required at each level when searching the tree. Also, the k - d tree allows the incorporation of heuristic rules that split the hyper-rectangle at the axis along which the change of error is maximal (before splitting).

There are two versions of the approximate mp-QP algorithm, based respectively on the cost function approximation error $\varepsilon_{cost}(x)$ [21,22], and on the control input approximation error $\varepsilon_{input}(x)$ [23]. The approximation error in the cost function is:

$$\varepsilon_{cost}(x) = \hat{V}_z(x) - V_z^*(x) \tag{132}$$

where $\hat{V}_z(x) = \frac{1}{2}\hat{z}_0^T(x)H\hat{z}_0(x)$ and $V_z^*(x) = \frac{1}{2}z^{*T}(x)Hz^*(x)$ are respectively the sub-optimal and the optimal costs, and $\hat{z}_0(x)$ and $z^*(x)$ are the sub-optimal and the optimal PWL solutions. The approximation error in the control input is:

$$\varepsilon_{input}(x) = \left(z^*(x) - \hat{z}_0(x)\right)^T \Sigma\left(z^*(x) - \hat{z}_0(x)\right) \tag{133}$$

where $\Sigma \geq 0$ is a weighting matrix which typically has non-zero weight only on the components of the solution corresponding to the first sample of the trajectory.

Initially the algorithm will consider the whole region $X_0 = X$. The main idea of the approximate mp-QP algorithm is to compute the solution of the problem (108) – (109) at the 2^n vertices of a considered hyper-rectangle X_0 by solving up to 2^n QPs. Based on these solutions, a feasible local linear approximation $\hat{z}_0(x)$ to the PWL optimal solution $z^*(x)$, valid in the whole hyper-rectangle X_0, is computed by using the following result [25]:

Lemma 1:
Consider the bounded polyhedron $X_0 \subseteq X_f$ with vertices $\{v_1, v_2, \dots, v_M\}$ (here X_f is the feasible set: $X_f = \{x(t) \in \mathfrak{R}^n \mid \exists U \text{ satisfying}(104)\}$). If K_0 and g_0 solve the QP:

$$\min_{K_0, g_0} \sum_{i=1}^{M} \left(z^*(v_i) - K_0 v_i - g_0\right)^T H\left(z^*(v_i) - K_0 v_i - g_0\right) \tag{134}$$

subject to:

$$G(K_0 v_i + g_0) \leq Sv_i + W , \; i \in \{1, 2, \dots, M\}, \tag{135}$$

then the least squares approximation $\hat{z}_0(x) = K_0 x + g_0$ is feasible for the mp-QP (108) – (109) for all $x \in X_0$.

If the maximal approximation error ε_0 in the hyper-rectangle X_0 is smaller than some prescribed tolerance $\bar{\varepsilon} > 0$, no further refinement of X_0 is needed. Otherwise, X_0 is partitioned into two hyper-rectangles and the procedure described above is repeated for each of these. If the approximation error in the cost function is considered, the upper bound ε_0 is determined by using the method proposed in [21,22]. If the approximation error in the control input is being used, then the maximal value ε_0 is determined in the way given in [23].

In order to reduce the complexity of the partition, the heuristic rule described in [23] is applied when splitting the hyper-rectangle X_0. The rule attempts to split the hyper-rectangle at the axis along which the change of the approximation error is maximal (before splitting), because it is reasonable to hope this is how the largest reduction of the error can be made. The heuristic rule uses information about the error ($\varepsilon_{cost}(x)$ or $\varepsilon_{input}(x)$) in the hyper-rectangle $X_0^d \subset X_0$ that contains a finite number of representative points in X_0, typically the vertices of one or more hyper-rectangles contained in the interior of X_0.

Heuristic splitting rule:
Split the hyper-rectangle X_0 by a hyperplane through its center and orthogonal to the axis x_j where the total absolute change of the approximation error measured both at the facet centers of X_0 and the vertices of X_0^d is maximal.

It has been shown in [23] that the use of such heuristics reduces the complexity of the partition significantly.

The complexity is further reduced by implementing control input trajectory parameterization as it is described in [26]. The idea is to use an input trajectory parameterization with less degrees of freedom in order to reduce the dimension of the optimization problem. The most common approach is to pre-determine the time-instants at which the control input u_i is allowed to change (input blocking):

$$N_{change}^{u_i} = [1 \ N_1^{u_i} \ N_2^{u_i} \ ... \ N_l^{u_i}]. \tag{136}$$

The following approximate mp-QP algorithm is taken from [22]:
Algorithm 3 (approximate mp-QP):
Step 1. Initialize the partition to the whole hyper-rectangle, i.e. $P = \{X\}$. Mark the hyper-rectangle X as unexplored.

Step 2. Select any unexplored hyper-rectangle $X_0 \in P$. If no such hyper-rectangle exists, go to step 8.

Step 3. Compute the solution to the QP (108) – (109) for x fixed to each of the 2^n vertices of the hyper-rectangle X_0. If all QPs have a feasible solution, go to step 5. Otherwise, go to step 4.

Step 4. Compute the size of X_0 using some metric. If it is smaller than some given tolerance, mark X_0 infeasible and explored. Go to step 2. Otherwise, go to step 7.

Step 5. Compute an affine state feedback $\hat{z}_0$ using Lemma 1, as an approximation to be used in X_0. If no feasible solution was found, go to step 7.

Step 6. Compute the error bound ε_0 in X_0. If $\varepsilon_0 \leq \bar{\varepsilon}$, mark X_0 as explored and feasible and go to step 2.

Step 7. Split the hyper-rectangle X_0 into two hyper-rectangles X_1 and X_2 by applying the heuristic splitting rule. Mark them unexplored, remove X_0 from P, add X_1 and X_2 to P, and go to step 2.

Step 8. If necessary, split the hyper-rectangles containing the origin such that $z^*(x) = 0$ is optimal everywhere in these hyper-rectangles. Terminate.

This algorithm will terminate with a PWL function that is an approximation to the PWL exact solution and is defined on an inner approximation $\underline{X}_f$ of the set $X \cap X_f$. The set $\underline{X}_f$ is represented as a union of hyper-rectangles.

5.4   Stability of the PWL Approximate Solution

It is shown in [22] that under some assumptions on the terminal set Ω and the tolerance $\bar{\varepsilon}$ the approximate explicit MPC will make the origin asymptotically stable.

Let Γ be the largest hyper-rectangle containing the origin in its interior where the solution computed by the approximate explicit MPC is $u^*(x) = Kx$, i.e. exactly the unconstrained LQR feedback. It is straightforward to show that Algorithm 3 leads to a non-empty Γ due to step 8. Let the terminal set Ω be the maximal output admissible set [27] for the linear system $x(t+1) = (A + BK)x(t)$ contained in the polyhedral set:

$$\Psi = \{x \in \Gamma \mid u_{min} \le Kx \le u_{max},\, y_{min} \le Cx \le y_{max}\} \qquad (137)$$

The set Ω is a polyhedron with a finite number of facets and can be easily computed, since $A + BK$ is Hurwitz and Γ is bounded because X is bounded [27]. The stability result is formulated in the following theorem [22]:

Theorem 7:

Consider the mp-QP problem (108) – (109) with $H > 0$ defined on a hypercube X such that $X_f \subseteq X$. Define $\Sigma = Q + K^T RK$, assume $\Sigma > 0$, and let γ be the largest positive number for which the ellipsoid $E = \{x \in X_f \mid x^T \Sigma x \le \gamma\}$ is contained in Ω. Moreover, assume the tolerance $\bar{\varepsilon}$ satisfies:

$$0 < \bar{\varepsilon} \le \frac{\gamma + x_0^T \Sigma x_0}{2} \qquad (138)$$

where $x_0 = \arg\min_{x \in X_0} x^T \Sigma x$. Then the approximate explicit MPC computed by Algorithm 3 in closed loop with (87) makes the origin asymptotically stable for all $x(0) \in X_f$, and the state and input trajectories are feasible.

5.5   Example

Consider the double integrator from section 4.3.2. With horizon N=10 and $\gamma = 0.13$, Algorithm 3 gives the quad-tree partition in Fig.12 with 214 regions. The method based on the cost function approximation error [22] is used. The sets E and Ω, together with the control and state trajectories obtained with the exact and the approximate approaches are shown in Fig.13. The solid and dashed curves show an exact and approximate trajectory, respectively. We observe that the discrepancy between them is negligible.

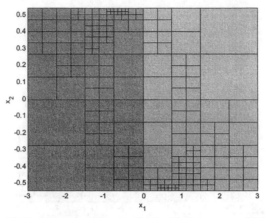

Fig. 12. Quad-tree partition for the double integrator with $N=10$.

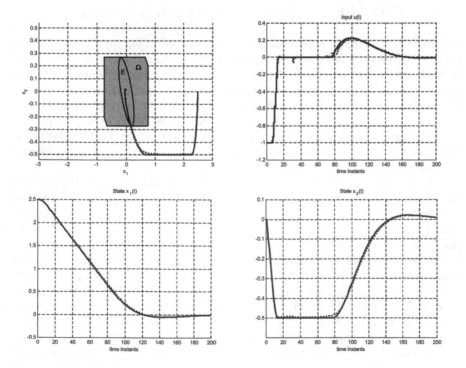

Fig. 13. Top, left: Sets E and Ω. Top, right: Control input. Bottom, left: State x_1. Bottom, right: State x_2.

The $k - d$ tree partition for the double integrator, obtained by applying the method based on the control input approximation error [23] and using the heuristic splitting rule, is given in Fig.14. It can be seen that a significant reduction in complexity is achieved (the state space partition has 97 regions). The relative tolerance in control input error is $\sqrt{\bar{\varepsilon}_r} = 0.5$.

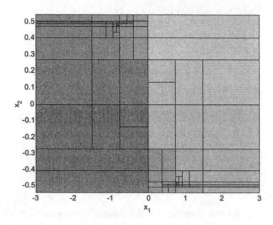

Fig. 14. $k - d$ tree partition for the double integrator with $N=10$.

It is interesting to compare the structure of the partitions of the approximate PWL ex-plicit MPC feedback laws with the partitions of the exact PWL explicit MPC feed-back law, as shown in Fig.10 for the case of horizon $N=10$. In parts of the state space where the exact partition contains several smaller regions while the approximate parti-tion contains only a few large regions, the explanation is that the approximate ap-proach only considers the first sample of the control input and is able to reduce com-plexity. In parts of the state space where the opposite is true, i.e. the approximate partition is more complex, this is due to a structural mismatch because the orthogonal-ity of the hyperplanes of the approximate partition is enforced.

The exact partition in Fig.10 contains 191 polyhedral regions and is thus of compa-rable complexity to the approximate partitions. Still, it is clear that there will be sig-nificantly higher demand for real-time processing capacity and computer memory, since all hyperplanes in the partition are different and they are not orthogonal. This also holds if a search tree is constructed from the exact partition as proposed in [26]. In this case there will be 9 levels in the tree and 60 arithmetic operations are required to compute the exact solution, while about 3150 numbers must be stored in real-time computer memory. With the suggested approach, 18 arithmetic operations are suffi-cient, and only about 700 numbers must be stored for the partition with 97 regions. Of course, the price to be paid for this complexity reduction is an approximation error.

As in [21] we remark that there is a significant difference between the exact and approximate approaches when the complexity of the partition is viewed as a function of the horizon. While the number of regions with the exact approach seems to give a very rapid growth with N, [19], the approximate approach gives a partition complex-ity that is almost independent of the horizon N. One reason for this is that in the approximate approach it is taken into account that we only need the first sample of the input trajectory in order to implement the MPC.

5.6   Robust Approximate Explicit Model Predictive Control in the Presence of Bounded Disturbances

Some of the exact mp-QP approaches have been further extended to ensure robustness of the explicit MPC controllers against disturbances [28,29,30,31]. In [29] it is assumed that the disturbance input belongs to a compact polyhedral set, and the approach in [32] is applied to ensure feasible operation of the MPC controller that minimizes the nominal value of the performance index. This work has been further extended to proportional integral controllers [31]. In [28], an approach to explicit solution of robust MPC problems based on a min-max formulation with a performance index expressed in ∞-norm has been proposed. It has to be mentioned however, that solution obtained by optimizing the worst value of the performance criterion can be quite conservative. In [30], it is supposed that the uncertainty set is a polytope and it is described how a class of uncertain quadratic and linear optimization problems can be converted to a multi-parametric quadratic programming (mp-QP) or multi-parametric linear programming (mp-LP) problems by solving as many linear programs (LPs) as there are constraints in the optimization problem without uncertainty. It is also shown in [30] that if the uncertainty set is given by upper and lower bounds only, then this transformation can be done by simply computing the 1-norms of the rows of the matrix by which the uncertainty enters the constraint.

In this section, an approximate mp-QP approach to explicit solution of constrained linear MPC problems in the presence of bounded disturbances is described [33]. It is based on an orthogonal search tree structure of the state space partition and thus represents an extension of the approximate mp-QP approach [22]. Like in [29], the explicit MPC controller avoids conservativeness by minimizing the nominal value of the performance index and it is robust in the sense that all constraints are satisfied for all possible disturbance realizations within the specified range. Here we consider a special case where the set of the disturbance inputs represents a hyper-rectangle that includes the origin in its interior. Based on this assumption, the conditions which guarantee feasible operation of the MPC controller are derived in a way similar to that in [30] and the original mp-QP problem with disturbance input is converted into an mp-QP problem without disturbances.

Problem formulation:
Consider the linear discrete-time system:

$$x(t+1) = Ax(t) + Bu(t) + T\theta(t) \tag{139}$$
$$y(t) = Cx(t)$$

where $x(t) \in \Re^n$, $u(t) \in \Re^m$, and $y(t) \in \Re^p$ are the state, input and output variable, $\theta(t)$ is the disturbance input that is assumed to belong to a bounded polyhedral set $\theta(t) \in \Theta^A \subset \Re^s$. Also, $A \in \Re^{n \times n}$, $B \in \Re^{n \times m}$, $C \in \Re^{p \times n}$ and $T \in \Re^{n \times s}$. Let $\Theta \equiv \left[\theta_t^T, ..., \theta_{t+N-1}^T \right]^T \in \Theta^B$ is a disturbance realization, with

$\Theta \in \Theta^B = \left\{\Theta^A \times \Theta^A ... \times \Theta^A\right\} \subset \Re^{sN}$. It is assumed that a full measurement of the state $x(t)$ is available at the current time t. Then, for the current $x(t)$, MPC solves the optimization problem:

$$V^*(x(t),\Theta) = \min_{U=\{u_t, ..., u_{t+N-1}\}} J(U,x(t),\Theta) \tag{140}$$

subject to $x_{t|t} = x(t)$ and:

$$y_{\min} \le y_{t+k|t} \le y_{\max}, k = 1, ..., N \tag{141}$$

$$u_{\min} \le u_{t+k} \le u_{\max}, k = 0,1, ..., N-1 \tag{142}$$

$$x_{t+N|t} \in \Omega \tag{143}$$

$$x_{t+k+1|t} = Ax_{t+k|t} + Bu_{t+k} + T\theta_{t+k}, \quad \theta_{t+k} \in \Theta^A, \quad k \ge 0 \tag{144}$$

$$y_{t+k|t} = Cx_{t+k|t}, k \ge 0 \tag{145}$$

with the cost function given by:

$$J(U,x(t),\Theta) = \sum_{k=0}^{N-1}\left[x^T{}_{t+k|t}Qx_{t+k|t} + u^T{}_{t+k}Ru_{t+k}\right] + x^T{}_{t+N|t}Px_{t+N|t} \tag{146}$$

and symmetric $R > 0$, $Q \ge 0$. We assume (A,B) is stabilizable, $(A,\sqrt{Q})$ is observable, Ω is a polyhedral terminal set, and the final cost matrix $P > 0$ is the solution of the associated algebraic Riccati equation. It is also assumed $u_{\max} > 0 > u_{\min}$, $y_{\max} > 0 > y_{\min}$, such that the origin is an interior point in the feasible set $X_f = \{x(t) \in \Re^n \mid \exists U \text{ satisfying} (141) - (145)\}$. Here, we consider the nominal optimization criterion:

$$V^*_{nom}(x(t)) = \min_{U=\{u_t, ..., u_{t+N-1}\}} J(U,x(t),\theta^N) \tag{147}$$

corresponding to $\theta(t) = \theta^N = 0$, where θ^N is the nominal value of the disturbance input. In this problem formulation, the *robustness* is defined in terms of satisfaction of the output and input constraints (141) and (142) under all possible disturbance realizations $\Theta \in \Theta^B$ that influence the state of the system (equation (144)).

By substituting:

$$x_{t+k|t} = A^k x(t) + \sum_{j=0}^{k-1} A^j Bu_{t+k-1-j} + \sum_{j=0}^{k-1} A^j T\theta_{t+k-1-j} \tag{148}$$

in the constraints $(141) - (145)$, they can be represented in the form:

$$GU \leq W + E_1 x(t) + E_2 \Theta \quad , \forall \Theta \in \Theta^B , \tag{149}$$

where $U \equiv \left[u_t^T , \dots , u_{t+N-1}^T \right]^T \in \Re^{mN}$ is the optimization vector and $\Theta \in \Theta^B$ is the disturbance realization. Then the nominal optimization criterion (147) is rewritten as:

$$V_{nom}^* (x(t)) = \frac{1}{2} x^T (t) Y x(t) + \min_U \left\{ \frac{1}{2} U^T H U + x^T (t) F U \right\} \tag{150}$$

We apply the same idea as in [30] of pre-stabilizing (139) with a linear state feedback gain and optimizing over a sequence of perturbations to this control law. Thus, we define:

$$U \equiv -H^{-1} F^T x(t) + z \tag{151}$$

where $z \in \Re^{mN}$ is the control input perturbation. Then, the optimization problem (150) subject to constraint (149) is transformed into the following mp-QP problem:

$$V_{z,nom}^* (x) = \min_z \frac{1}{2} z^T H z \tag{152}$$

subject to:

$$Gz \leq W + S_1 x(t) + S_2 \Theta \quad , \forall \Theta \in \Theta^B . \tag{153}$$

Assumption 1:
The disturbance input set:

$$\Theta^A = \left\{ \theta \in \Re^s | \theta^L \leq \theta \leq \theta^U \right\} \tag{154}$$

represents a hyper-rectangle that includes the origin in its interior.

Definition 8:
Consider the i-th constraint defined by G^i, W^i, S_1^i, S_2^i rows of the matrices G, W, S_1, S_2. The worst disturbance realization for the i-th constraint, denoted by $\tilde{\Theta}^i \in \Theta^B$ is one which solves the linear program:

$$S_2^i \tilde{\Theta}^i = \min_{\Theta \in \Theta^B} \{ S_2^i \Theta \} \tag{155}$$

Remark 1:
The linear program (155) can be easily solved by exploiting the fact that disturbance input set is a hyper-rectangle. Thus:

$$\min_{\Theta \in \Theta^B} \{ S_2^i \Theta \} = \min_{\Theta \in \Theta^B} \left\{ \sum_{j=1}^{sN} S_{2,j}^i \Theta_j \right\} = \sum_{j=1}^{sN} \min_{\Theta_j^L \leq \Theta_j \leq \Theta_j^U} \{ S_{2,j}^i \Theta_j \} = \sum_{j=1}^{sN} \min \left\{ S_{2,j}^i \Theta_j^L , S_{2,j}^i \Theta_j^U \right\} \tag{156}$$

where $S_{2,j}^i$ is the j-th element of the row vector S_2^i, Θ_j is the j-th element of the column vector of disturbance realization $\Theta \in \Theta^B \subset \Re^{sN}$, and Θ_j^L, Θ_j^U are respectively the lower and upper bounds of Θ_j.

Lemma 2:
If there exists an affine function $z(x)$ that satisfies the following constraint:

$$Gz \le \widetilde{W} + S_1 x, \tag{157}$$

where the i-th row of the matrix $\widetilde{W}$ is determined by:

$$\widetilde{W}^i = W^i + S_2^i \widetilde{\Theta}^i \tag{158}$$

and where $\widetilde{\Theta}^i \in \Theta^B$ is the worst disturbance realization for the i-th constraint, then this implies that $z(x)$ will satisfy constraint (153) for all possible disturbance realizations $\Theta \in \Theta^B$. Such $z(x)$ is referred to as *robustly feasible*.

In this way, the constraint (157) which ensures robust feasibility can be easily constructed. Then, the original mp-QP problem (152) – (153) becomes:

$$V_{z,nom}^*(x) = \min_z \frac{1}{2} z^T H z \tag{159}$$

subject to:

$$Gz \le \widetilde{W} + S_1 x(t), \tag{160}$$

where $\widetilde{W}$ is determined by (158). Thus the original mp-QP problem with disturbance input (problem (152) – (153)) is reformulated as an mp-QP problem without disturbance (problem (159) – (160)) and therefore the approximate approach [22] for explicit solution of mp-QP problems can easily be applied to this problem. It has to be stressed that the approximate approach [22] guarantees that the optimal solution is feasible in sense that it will satisfy constraint (160). This directly implies by Lemma 2 above that constraint (153) of the original mp-QP problem will be satisfied for all possible disturbance realizations. This is summarized in the following Lemma:

Lemma 3 (feasible control in the presence of disturbance):
Consider the bounded polyhedron X_0 with vertices $\{v_1, v_2, \ldots, v_M\}$. If K_0 and g_0 solve the QP:

$$\min_{K_0, g_0} \sum_{i=1}^{M} \left(z^*(v_i) - K_0 v_i - g_0 \right)^T H \left(z^*(v_i) - K_0 v_i - g_0 \right) \tag{161}$$

subject to:

$$G(K_0 v_i + g_0) \le \widetilde{W} + S_1 v_i, \ i \in \{1, 2, \ldots, M\}, \tag{162}$$

then the least squares approximation $\hat{z}_0(x) = K_0 x + g_0$ is *robustly feasible* for the mp-QP (152) – (153) for all $x \in X_0$ and all disturbance realizations $\Theta \in \Theta^B$.

Example:
Consider the double integrator:

$$x(t+1) = Ax(t) + Bu(t) + T\theta(t) \tag{163}$$

with:

$$A = \begin{bmatrix} 1 & T_s \\ 0 & 1 \end{bmatrix}, \quad B = \begin{bmatrix} T_s^2 \\ T_s \end{bmatrix}, \quad T = \begin{bmatrix} 1 & 0 \\ 0 & 1 \end{bmatrix} \tag{164}$$

where the sampling interval is $T_s = 0.3$. Consider the MPC problem with horizon $N = 30$. The cost matrices are $Q = \mathrm{diag}(1,0)$, $R = 1$, and the matrix $P > 0$ is given as the solution of the algebraic Riccati equation. The constraints are:

$$-1 \leq u \leq 1 \tag{165}$$

$$-0.5 \leq x_2 \leq 0.5. \tag{166}$$

The disturbance vector $\theta = [\theta_1 \ \theta_2]^T$ has the following bounds:

$$-0.01 \leq \theta_1(t) \leq 0.01 \tag{167}$$

$$-0.015 \leq \theta_2(t) \leq 0.015 \tag{168}$$

The approximation tolerance $\bar{\varepsilon} > 0$ is chosen according to Theorem 7, with $\gamma = 0.1$. The state space partition of the robust approximate MPC controller is shown in Fig.15.

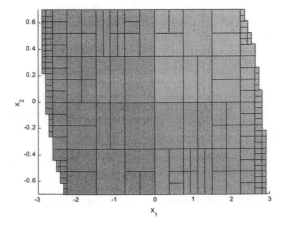

Fig. 15. $k - d$ tree partition of the robust MPC.

It has 172 regions and 11 levels of search. With one scalar comparison required at each level of the k-d tree, 11 arithmetic operations are required in the worst case to determine which region the state belongs to. Totally, 15 arithmetic operations are needed in real-time to compute the control input with this MPC controller (11 comparisons, 2 multiplications and 2 additions).

In Fig.16, the sets Ω, E and S (S is the terminal region to which the state converges), and disturbance realizations with constant magnitude are given. In Fig.17, the control and state trajectories obtained with the robust MPC under these disturbances are shown (the trajectories with the exact mp-QP approach are given for comparison). The approximate and the exact state trajectories are also depicted in Fig.16, where it can be seen that with the increase of time the state enters and remains in the terminal region S. It can be seen from the above figures that the robust MPC keeps all constraints imposed on the system.

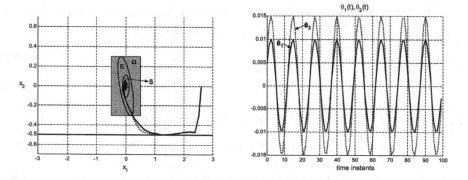

Fig. 16. Left: The sets Ω, E, S, the approximate (the solid curve) and the exact (the dashed curve) state trajectories. Right: Disturbance inputs with constant magnitude.

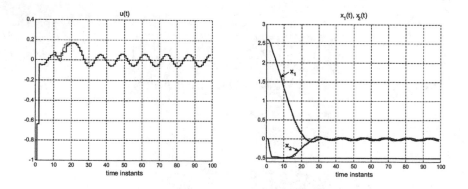

Fig. 17. Control input and state trajectories for the robust MPC (the solid curves are with the approximate controller and the dashed curves are with the exact controller).

In Fig.18, the sets Ω, E and S, and disturbance realizations with decreasing magnitude are given.

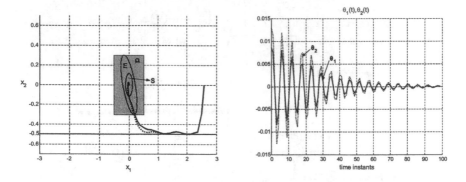

Fig. 18. Left: The sets Ω, E, S, the approximate (the solid curve) and the exact (the dashed curve) state trajectories. Right: Disturbance inputs with decreasing magnitude.

In Fig.19, the closed loop response of the robust MPC under these disturbances is shown. This response is also depicted in Fig.18, where it can be noticed that in the case of decreasing magnitude disturbance, the state approaches the origin when time increases. Again, it can be seen that the robust MPC keeps all constraints imposed on the system.

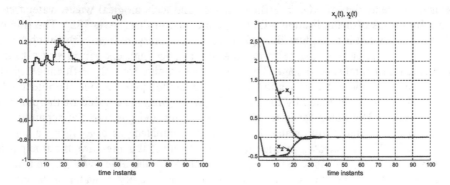

Fig. 19. Control input and state trajectories for the robust MPC (the solid curves are with the approximate controller and the dashed curves are with the exact controller).

5.7   Explicit Model Predictive Control of Gas-Liquid Separation Plant

This section considers the design of approximate explicit MPC controller for a gas-liquid separation plant (Fig.20) and the experimental evaluation of controller performance [34].

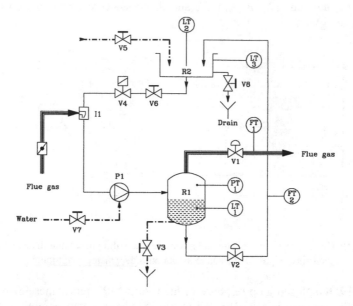

Fig. 20. Process scheme of the separation unit.

The gas-liquid separation is a sub-process within a semi-industrial installation which is used for reduction of NO_x in effluent gasses and technological waste water treatment by means of neutralisation with CO_2 contained in flue gasses [35]. The role of the separation unit is to capture flue gasses under low pressure from effluent channels by means of water flow and to carry them over under high enough pressure to the downstream (neutralisation) stage. The flue gasses coming from the effluent channels are "pooled" by the water flow into the water circulation pipe through the injector I_1. The water flow is generated by the pump P_1 (water ring). The speed of the pump is kept constant. The pump feeds the mixture of water and gas into the separator R_1 where gas is separated from water. Hence, the accumulated gas in R_1 forms a sort of "gas cushion" with increased internal pressure. Owing to this pressure, the flue gas is blown out from R_1 into the next neutralization unit. On the other side the "cushion" forces water to circulate back to the reservoir R_2. The quantity of water in the circuit is constant. If for some reason additional water is needed, the water supply path through the valve V_5 is utilized.

The complete non-linear model of the gas-liquid separator is given in [35]. A linearized model can be obtained from the existing non-linear model:

$$\begin{bmatrix} \Delta \dot{p}_1 \\ \Delta \dot{h}_1 \end{bmatrix} = A_c \begin{bmatrix} \Delta p_1 \\ \Delta h_1 \end{bmatrix} + B_c \begin{bmatrix} \Delta v_1 \\ \Delta v_2 \end{bmatrix}, \tag{169}$$

where Δp_1 and Δh_1 denote the change of separator gas pressure p_1 and liquid level h_1 from the steady-state values ($\Delta p_1 = p_1 - p_{1s}$, $\Delta h_1 = h_1 - h_{1s}$), and Δv_1 and Δv_2 are respectively the changes in the positions v_1 and v_2 of the two valves

($\Delta v_1 = v_1 - v_{1s}$, $\Delta v_2 = v_2 - v_{2s}$). The linear model corresponds to the following steady state:

$$p_{1s} = 0.5\,bar\,,\ h_{1s} = 1.4\,m\,,\ v_{1s} = 0.4152\,,\ v_{2s} = 0.7462 \qquad (170)$$

and the way to compute the elements of the matrices A_c and B_c is given in details in [35]. From the continuous-time model, a linear discrete-time model corresponding to sampling interval $T_s = 1s$ is obtained, with the following state and control matrices:

$$A = \begin{bmatrix} 0.9719 & -0.0001 \\ -0.0006 & 0.9999 \end{bmatrix},\ B = \begin{bmatrix} -0.0832 & -0.0041 \\ 0 & -0.0023 \end{bmatrix}. \qquad (171)$$

The state variables are $x_1 = \Delta p_1\,[bar]$ and $x_2 = \Delta h_1\,[m]$, and the control variables are $u_1 = \Delta v_1$ and $u_2 = \Delta v_2$. The following input and rate constraints are imposed on the valve positions v_1 and v_2:

$$0 \le v_1 \le 1\,,\ 0 \le v_2 \le 0.8625 \qquad (172)$$

$$-0.33 \le \dot{v}_1 \le 0.66\,,\ -0.33 \le \dot{v}_2 \le 0.66\,, \qquad (173)$$

which by taking into account the steady state values (170) are represented as the following constraints on the control inputs u_1 and u_2:

$$-0.4152 \le u_1(t+k) \le 0.5848 \qquad (174)$$
$$-0.7462 \le u_2(t+k) \le 0.1163$$
$$k = 0,1,...,N-1$$

$$-0.33T_s \le u_1(t+k) - u_1(t+k-1) \le 0.66T_s \qquad (175)$$
$$-0.33T_s \le u_2(t+k) - u_2(t+k-1) \le 0.66T_s\,.$$
$$k = 0,1,...,N-1$$

In order to avoid the steady state offset of the model predictive controller, two more states are added to the model (171), which take into account the integral error:

$$x_3(t+1) = x_3(t) + T_s x_1(t)\,,\ x_4(t+1) = x_4(t) + T_s x_2(t)\,. \qquad (176)$$

Thus, the linear discrete-time model of the gas-liquid separation unit becomes:

$$A = \begin{bmatrix} 0.9719 & -0.0001 & 0 & 0 \\ -0.0006 & 0.9999 & 0 & 0 \\ T_s & 0 & 1 & 0 \\ 0 & T_s & 0 & 1 \end{bmatrix},\ B = \begin{bmatrix} -0.0832 & -0.0041 \\ 0 & -0.0023 \\ 0 & 0 \\ 0 & 0 \end{bmatrix}. \qquad (177)$$

The approximate mp-QP approach described in section 5.3 (based on cost function approximation error) is applied to design an explicit MPC controller for the gas-liquid separation plant [34]. The MPC controller solves the optimization problem (90) subject to the system equation (177) and the input constraints (174). The rate constraints (175) are not taken into account during the design of the MPC controller. Instead, a rate limiter is placed at the output of the controller in its real-time implementation, that guarantees the satisfaction of the rate constraints. In (90), P is chosen as the solution of the discrete algebraic Riccati equation and the cost matrices are:

$$Q = \text{diag}\{0.05, 100, 0.005, 0.0001\} \,, \quad R = \text{diag}\{1, 1\} \,. \tag{178}$$

The horizon is $N = 500$ and the time instants at which the input variables can change are:

$$N_{u_1} = [1 \ 5 \ 10 \ 15 \ 20 \ 25 \ 30 \ 35 \ 40 \ 45 \ 50 \ 100 \ 102 \tag{179}$$
$$104 \ 106 \ 108 \ 110 \ 300 \ 302 \ 304 \ 306 \ 308 \ 310]$$

$$N_{u_2} = [1 \ 5 \ 10 \ 15 \ 20 \ 25 \ 30 \ 35 \ 40 \ 45 \ 50 \ 100 \ 300] \tag{180}$$

which makes totally 36 optimization variables. The state space to be partitioned is 4-dimensional and is defined by $X = [-0.5, 0.5] \times [-0.2, 0.2] \times [-3, 3] \times [-10, 60]$. The size of the regions on each of the state variables is restricted to be larger than $\Delta_{x1} = 0.01$, $\Delta_{x2} = 0.004$, $\Delta_{x3} = 0.06$ and $\Delta_{x4} = 0.7$. The approximation tolerance $\bar{\varepsilon} > 0$ is chosen according to Theorem 7 with $\gamma = 0.5$.

The resulting MPC controller has 2693 regions in its state space partition and 24 levels of search. With one scalar comparison required at each level of the k-d tree, 24 arithmetic operations are required in the worst case to determine which region the state belongs to. Totally, 40 arithmetic operations are needed in real-time to compute the two control inputs with this MPC controller (24 comparisons, 8 multiplications and 8 additions).

The real-time experiments were pursued in the environment schematically shown in Fig.21 [34]. This environment encompasses supervisory control on two levels: upper level with Factory Link SCADA system and lower procedural and basic control levels implemented in two PLCs. This is one of the possible configurations of control, which can be found in industry. User-friendly experimentation with the process plant is enabled through interface with Matlab/Simulink environment. This interface enables PLC access with Matlab/Simulink using DDE protocol via Serial Communication Link RS232 or TCP/IPv4 over Ethernet IEEE802.3. Control algorithms for experimentation can be prepared in Matlab code or as Simulink blocks and extended with functions/blocks, which access PLC. This interface also enables user-friendly data acquisition for Matlab users. In our case all control schemes were put together as Simulink blocks and tested at the plant operating points.

In Fig.22 and Fig.23, the real-time performance of the approximate explicit MPC controller, in closed-loop with the plant is shown. The set point is $p_1^* = p_{1s} = 0.5 \, bar$ and $h_1^* = h_{1s} = 1.4 \, m$. The set point changes are handled by using the new set-point values p_1^* and h_1^* when determining the values of the state variables $x_1 = p_1 - p_1^*$ and $x_2 = h_1 - h_1^*$, where p_1 and h_1 are the measured variables.

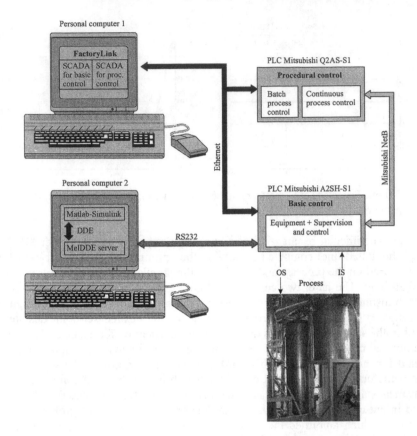

Fig. 21. Scheme showing environment for control and experimentation.

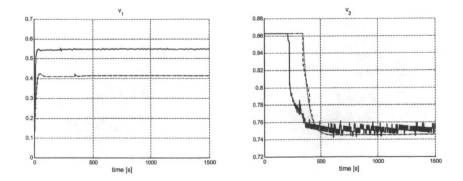

Fig. 22. Left: Trajectory of v_1 (position of valve 1). Right: Trajectory of v_2 (position of valve 2).

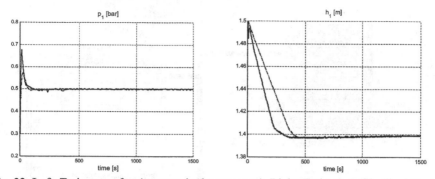

Fig. 23. Left: Trajectory of p_1 (pressure in the separator). Right: Trajectory of h_1 (liquid level in the separator).

The experimental results (the solid line) are compared with the exact MPC trajectory (the dotted line) computed by solving the optimization problem at each time instant, based on the process model and with the simulated approximate trajectory (the dashed line). The latter two curves (the dotted curve and the dashed curve) are difficult to distinguish since they are almost matching. It can be seen from the figures that the explicit MPC controller brings the plant to the desired set-point despite of the error in the steady state process model and the transient performance is close to that of the optimal trajectory. It can be seen from Fig.22 that there is a slight chattering of the signal for the second valve. This can be explained as follows. The signal we depict has come out of analog digital converter which has 10 bit A/D converter resolution that means approximately 0.1% of quantization noise on the range 0 to 1 which was used in our case. Afterwards, this signal has gone through the controller with gain of about 10 which amplified the quantization noise to about 1% (as it can be seen in Fig.22). This is not a problem for the valve and actually even helps beating the small dead zone contained in valve and makes it react faster when the change of control signal occurs.

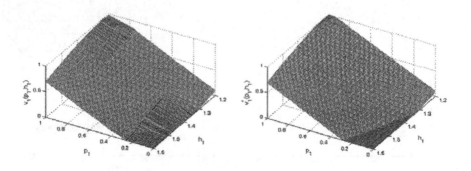

Fig. 24. Left: PWL *approximate* feedback control law for valve position v_1. Right: PWL *exact* feedback control law for valve position v_1.

In Fig.24, the approximate and the exact PWL feedback control laws for valve position v_1 are shown. The approximate and the exact control laws for valve position v_2

are shown in Fig.25. These control laws correspond to value zero of the two integral errors in the system equation (177), i.e. $x_3 = 0$ and $x_4 = 0$.

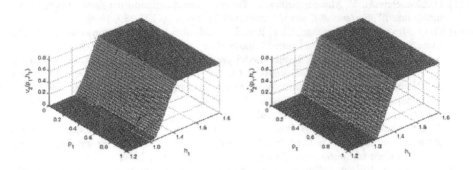

Fig. 25. Left: PWL *approximate* feedback control law for valve position v_2. Right: PWL *exact* feedback control law for valve position v_2.

References

[1]   W. Ray, *Advanced process control*, McGraw-Hill Book Company, 1981.

[2]   S. Strand, "On algorithms for constrained optimal control problems", *Report 89-33-W*, Norwegian University of Science and Technology, 1989.

[3]   D. Jacobson, "Second-order and second-variation methods for determining optimal control: A comparative study using differential dynamic programming", *International Journal of Control*, vol.7, No.2, pp.175-196, 1968.

[4]   G. Hicks, W. Ray, "Approximation methods for optimal control synthesis", *The Canadian Journal of Chemical Engineering*, vol.49, pp.522-528, 1971.

[5]   D. Kraft, "On converting optimal control problems into nonlinear programming problems", *In Computational Mathematical Programming*, Ed. K. Schittkowski, NATO ASI Series, Springer-Verlag, vol.F15, pp.261-280, 1985.

[6]   A. Barclay, P. Gill, J. Ben Rosen, "SQP methods and their application to numerical optimal control", *Report NA 97-3*, Department of Mathematics, University of California, San Diego, 1997.

[7]   J. Renfro, A. Morshedi, O. Asbjornsen, "Simultaneous optimization and solution of systems described by differential-algebraic equations", *Computers and Chemical Engineering*, vol.11, No.5, pp.503-517, 1987.

[8]   A. Grancharova, "Optimal dynamic control of technological systems in the absence and in the presence of uncertainty", *Ph.D. Thesis*, University of Chemical Technology and Metallurgy, Sofia, 1997.

[9]   A. Bryson, Y. Ho, "Applied optimal control: Optimization, estimation and control", Blaisdell Publ. Co., Waltham, Massachusetts, 1969.

[10] H. Kwakernaak, R. Sivan, *Linear optimal control systems*, Wiley-Interscience, New York, 1972.

[11] P. O. M. Scokaert, J. B. Rawlings, "Constrained linear quadratic regulation", *IEEE Trans. Automatic Control*, vol.43, pp.1163-1169, 1998.

[12] M. Sznaier, M. J. Damborg, "Suboptimal control of linear systems with state and control inequality constraints", *Proceedings of 26-th IEEE Conference on Decision and Control*, vol.1, pp.761-762, 1987.

[13] D. Chmielewski, V. Manousiouthakis, "On constrained infinite-time linear quadratic optimal control", *Systems & Control Letters*, vol.29, No.3, pp.121-130, 1996.

[14] D. Q. Mayne, J. B. Rawlings, C. V. Rao, P. O. M. Scokaert, "Constrained model predictive control: Stability and optimality", *Automatica*, vol.36, pp.789-814, 2000.

[15] A. Bemporad, M. Morari, "Robust model predictive control: A survey", In A. Garulli, A.Tesi, A.Vicino, editors, *Robustness in Identification and Control*, number 245 in Lecture Notes in Control and Information Sciences, Springer-Verlag, pp.207-226, 1999.

[16] A. Bemporad, M. Morari, V. Dua, E. N. Pistikopoulos, "The explicit linear quadratic regulator for constrained systems", *Automatica*, vol.38, pp.3-20, 2002.

[17] A. Bemporad, M. Morari, V. Dua, E. N. Pistikopoulos, "The explicit solution of model predictive control via multiparametric quadratic programming", *Proceedings of the American Control Conference*, Chicago, Illinois, pp.872-876, 2000.

[18] J. Nocedal, S. Wright, *"Numerical optimization"*, Springer-Verlag New York, Inc., 1999.

[19] P. Tøndel, T. A. Johansen, A. Bemporad, "An algorithm for multi-parametric quadratic programming and explicit MPC solutions", *Automatica*, vol.39, pp.489-497, 2003.

[20] T. A. Johansen, I. Petersen, O. Slupphaug, "Explicit sub-optimal linear quadratic regulation with state and input constraints", *Automatica*, vol.38, pp.1099-1111, 2002.

[21] T. A. Johansen, A. Grancharova, "Approximate explicit model predictive control implemented via orthogonal search tree partitioning", *Proc. of 15-th IFAC World Congress, Barcelona, Spain*, session T-We-M17, 2002.

[22] T. A. Johansen, A. Grancharova, "Approximate explicit constrained linear model predictive control via orthogonal search tree", *IEEE Trans. Automatic Control*, vol.48, pp.810-815, 2003.

[23] A. Grancharova, T. A. Johansen, "Approximate explicit model predictive control incorporating heuristics", *Proc. of IEEE International Symposium on Computer Aided Control System Design*, Glasgow, Scotland, U.K., pp.92-97, 2002.

[24] J. L. Bentley, "Multidimensional binary search trees used for associative searching", *Communications of the ACM*, vol.18, pp.509–517, 1975.

[25] A. Bemporad, C. Filippi, "Suboptimal explicit RHC via approximate quadratic programming", *Optim. Theory Applicat.*, vol.117, pp.5-38, 2003.

[26] P. Tøndel, T. A. Johansen, "Complexity reduction in explicit linear model predictive control", *Proc. of 15-th IFAC World Congress, Barcelona, Spain*, session T-We-M17, 2002.

[27] E. G. Gilbert, K. T. Tan, "Linear systems with state and control constraints: The theory and application of maximal output admissible sets", *IEEE Trans. Automatic Control*, vol.36, pp.1008–1020, 1991.

[28] A. Bemporad, F. Borrelli, M. Morari, "Robust model predictive control: Piecewise linear explicit solution", *Proc. of European Control Conference*, Porto, Portugal, pp 939-944, 2001.

[29] N. M. P. Kakalis, V. Dua, V. Sakizlis, J. D. Perkins, E. N. Pistikopoulos, "A parametric optimisation approach for robust MPC", *Proc. of 15-th IFAC World Congress*, Barcelona, Spain, 2002.

[30] E. C. Kerrigan, J. M. Maciejowski, "On robust optimization and the optimal control of constrained linear systems with bounded state disturbances", *Proc. of European Control Conference*, Cambridge, U.K. 2003.

[31] V. Sakizlis, N. M. P. Kakalis, V. Dua, J. D. Perkins, E. N. Pistikopoulos, "Design of robust model-based controllers via parametric programming", *Automatica*, vol.40, pp.189-201, 2004.

[32] I. E. Grossmann, K. P. Halemane, R. E. Swaney, "Optimization strategies for flexible chemical processes", *Computers and Chemical Engineering*, vol.7, pp.439-462, 1983.

[33] A. Grancharova, T. A. Johansen, "Design of robust explicit model predictive controller via orthogonal search tree partitioning", *Proc. of European Control Conference*, Cambridge, U.K. 2003.

[34] A. Grancharova, T. A. Johansen, J. Kocijan, "Explicit model predictive control of gas-liquid separation plant", *Proc. of European Control Conference*, Cambridge, U.K., 2003.

[35] D. Vrančić, D. Juričić, J. Petrovčič. "Measurements and mathematical modelling of a semi-industrial liquid-gas separator for the purpose of fault diagnosis", *Report DP-7260*, University of Ljubljana, J. Stefan Institute, Ljubljana, 1995.

Analysis of Some Methods for Reduced Rank Gaussian Process Regression

Joaquin Quiñonero-Candela[1,2] and Carl Edward Rasmussen[2]

[1] Informatics and Mathematical Modelling, Technical University of Denmark,
Richard Petersens Plads, B321, 2800 Kongens Lyngby, Denmark
jqc@imm.dtu.dk
[2] Max Planck Institute for Biological Cybernetics,
Spemann straße 38, 72076 Tübingen, Germany
carl@tuebingen.mpg.de

Abstract. While there is strong motivation for using Gaussian Processes (GPs) due to their excellent performance in regression and classification problems, their computational complexity makes them impractical when the size of the training set exceeds a few thousand cases. This has motivated the recent proliferation of a number of cost-effective approximations to GPs, both for classification and for regression. In this paper we analyze one popular approximation to GPs for regression: the reduced rank approximation. While generally GPs are equivalent to infinite linear models, we show that Reduced Rank Gaussian Processes (RRGPs) are equivalent to finite sparse linear models. We also introduce the concept of degenerate GPs and show that they correspond to inappropriate priors. We show how to modify the RRGP to prevent it from being degenerate at test time. Training RRGPs consists both in learning the covariance function hyperparameters and the support set. We propose a method for learning hyperparameters for a given support set. We also review the Sparse Greedy GP (SGGP) approximation (Smola and Bartlett, 2001), which is a way of learning the support set for given hyperparameters based on approximating the posterior. We propose an alternative method to the SGGP that has better generalization capabilities. Finally we make experiments to compare the different ways of training a RRGP. We provide some Matlab code for learning RRGPs.

1 Motivation and Organization of the Paper

Gaussian Processes (GPs) have state of the art performance in regression and classification problems, but they suffer from high computational cost for learning and predictions. For a training set containing n cases, the complexity of training is $\mathcal{O}(n^3)$ and that of making a prediction is $\mathcal{O}(n)$ for computing the predictive mean, and $\mathcal{O}(n^2)$ for computing the predictive variance.

A few computationally effective approximations to GPs have recently been proposed. These include the sparse iterative schemes of Csató (2002), Csató and Opper (2002), Seeger (2003), and Lawrence et al. (2003), all based

R. Murray-Smith, R. Shorten (Eds.): Switching and Learning, LNCS 3355, pp. 98–127, 2005.
© Springer-Verlag Berlin Heidelberg 2005

on minimizing KL divergences between approximating and true posterior; Smola and Schölkopf (2000) and Smola and Bartlett (2001) based on low rank approximate posterior, Gibbs and MacKay (1997) and Williams and Seeger (2001) on matrix approximations and Tresp (2000) on neglecting correlations. Subsets of regressors (Wahba et al., 1999) and the Relevance Vector Machine (Tipping, 2001) can also be cast as sparse linear approximations to GPs. Schwaighofer and Tresp (2003) provide a very interesting yet brief comparison of some of these approximations to GPs. They only address the quality of the approximations in terms of the predictive mean, ignoring the predictive uncertainties, and leaving some theoretical questions unanswered, like the goodness of approximating the maximum of the posterior.

In this paper we analyze sparse linear or equivalently reduced rank approximations to GPs that we will call Reduced Rank Gaussian Processes (RRGPs). We introduce the concept of degenerate Gaussian Processes and explain that they correspond to inappropriate priors over functions (for example, the predictive variance shrinks as the test points move far from the training set). We show that if not used with care at prediction time, RRGP approximations result in degenerate GPs. We give a solution to this problem, consisting in augmenting the finite linear model at test time. This guarantees that the RRGP approach corresponds to an appropriate prior. Our analysis of RRGPs should be of interest in general for better understanding the infinite nature of Gaussian Processes and the limitations of diverse approximations (in particular of those based solely on the posterior distribution).

Learning RRGPs implies both selecting a support set, and learning the hyperparameters of the covariance function. Doing both simultaneously proves to be difficult in practice and questionable theoretically. Smola and Bartlett (2001) proposed the Sparse Greedy Gaussian Process (SGGP), a method for learning the support set for given hyperparameters of the covariance function based on approximating the posterior. We show that approximating the posterior is unsatisfactory, since it fails to guarantee generalization, and propose a theoretically more sound greedy algorithm for support set selection based on maximizing the marginal likelihood. We show that the SGGP relates to our method in that approximating the posterior reduces to partially maximizing the marginal likelihood. We illustrate our analysis with an example. We propose an approach for learning the hyperparameters of the covariance function of RRGPs for a given support set, originally introduced by Rasmussen (2002). We also provide Matlab code in Appendix B for this method.

We make experiments where we compare learning based on selecting the support set to learning based on inferring the hyperparameters. We give special importance to evaluating the quality of the different approximations to computing predictive variances.

The paper is organized as follows. We give a brief introduction to GPs in Sect. 2. In Sect. 3 we establish the equivalence between GPs and linear models, showing that in the general case GPs are equivalent to *infinite* linear models. We also present degenerate GPs. In Sect. 4 introduce RRGPs and address the issue of training them. In Sect. 5 we present the experiments we conducted. We give some discussion in Sect. 6.

2   Introduction to Gaussian Processes

In inference with parametric models prior distributions are often imposed over the model parameters, which can be seen as a means of imposing regularity and improving generalization. The form of the parametric model, together with the form of the prior distribution on the parameters result in a (often implicit) prior assumption on the joint distribution of the function values. At prediction time the quality of the predictive uncertainty will depend on the prior over functions. Unfortunately, for probabilistic parametric models this prior is defined in an indirect way, and this in many cases results in priors with undesired properties. An example of a model with a peculiar prior over functions is the Relevance Vector Machine introduced by Tipping (2001) for which the predictive variance shrinks for a query point far away from the training inputs. If this property of the predictive variance is undesired, then one concludes that the prior over functions was undesirable in the first place, and one would have been happy to be able to directly define a prior over functions.

Gaussian Processes (GPs) are non-parametric models where a Gaussian process[3] prior is directly defined over function values. The direct use of Gaussian Processes as priors over functions was motivated by Neal (1996) as he was studying priors over weights for artificial neural networks. A model equivalent to GPs, *kriging*, has since long been used for analysis of spatial data in Geostatistics (Cressie, 1993). In a more formal way, in a GP the function outputs $f(x_i)$ are a collection random variables indexed by the inputs x_i. Any finite subset of outputs has a joint multivariate Gaussian distribution (for an introduction on GPs, and thorough comparison with Neural Networks see (Rasmussen, 1996)). Given a set of training inputs $\{x_i | i = 1, \ldots, n\} \subset \mathbb{R}^D$ (organized as rows in matrix X), the joint prior distribution of the corresponding function outputs $\mathbf{f} = [f(x_1), \ldots, f(x_n)]^\top$ is Gaussian $p(\mathbf{f}|X, \theta) \sim \mathcal{N}(0, K)$, with zero mean (this is a common and arbitrary choice) and *covariance matrix* $K_{ij} = K(x_i, x_j)$. The GP is entirely determined by the *covariance function* $K(x_i, x_j)$ with parameters θ.

An example of covariance function that is very commonly used is the squared exponential:

$$K(x_i, x_j) = \theta_{D+1}^2 \exp\left(-\frac{1}{2} \sum_{d=1}^{D} \frac{1}{\theta_d^2} \left(X_{id} - X_{jd} \right)^2 \right) . \tag{1}$$

θ_{D+1} relates to the amplitude of the functions generated by the GP, and θ_d is a lengthscale in the d-th dimension that allows for Automatic Relevance Determination (ARD) (MacKay, 1994; Neal, 1996): if some input dimensions are

---

[3] We will use the expression "Gaussian Process" (both with capital first letter) or "GP" to designate the non-parametric model where a Gaussian process prior is defined over function values

un-informative about the covariance between observed training targets, their associated θ_d will be made large (or effectively infinite) and the corresponding input dimension will be effectively pruned from the model. We will call the parameters of the covariance function *hyperparameters*, since they are the parameters of the prior.

In general, inference requires choosing a parametric form of the covariance function, and either estimating the corresponding parameters θ (which is named by some Maximum Likelihood II, or second level of inference) or integrating them out (often through MCMC). We will make the common assumption of Gaussian independent identically distributed output noise, of variance σ^2. The training outputs $\mathbf{y} = [y_1, \ldots, y_n]^\top$ (or targets) are thus related to the function evaluated at the training inputs by a likelihood distribution[4] $p(\mathbf{y}|\mathbf{f}, \sigma^2) \sim \mathcal{N}(\mathbf{f}, \sigma^2 I)$, where I is the identity matrix. The posterior distribution over function values is useful for making predictions. It is obtained by applying Bayes' rule:[5]

$$p(\mathbf{f}|\mathbf{y}, X, \theta, \sigma^2) = \frac{p(\mathbf{y}|\mathbf{f}, \sigma^2)\, p(\mathbf{f}|X, \theta)}{p(\mathbf{y}|X, \theta, \sigma^2)}$$
$$\sim \mathcal{N}\left(K^\top \left(K + \sigma^2 I\right)^{-1} \mathbf{y}, K - K^\top \left(K + \sigma^2 I\right)^{-1} K\right) . \tag{2}$$

The mean of the posterior does not need to coincide with the training targets. This would be the case however, if the estimated noise variance happened to be zero, in which case the posterior at the training cases would be a delta function centered on the targets.

Consider now that we observe a new input $\mathbf{x}_*$ and would like to know the distribution of $f(\mathbf{x}_*)$ (that we will write as f_* for convenience) conditioned on the observed data, and on a particular value of the hyperparameters and of the output noise variance. The first thing to do is to write the augmented prior over the function values at the training inputs and the new function value at the new test input:

$$p\left(\begin{bmatrix}\mathbf{f}\\f_*\end{bmatrix}\middle|\, \mathbf{x}_*, X, \theta\right) \sim \mathcal{N}\left(0, \begin{bmatrix}K & \mathbf{k}_*\\\mathbf{k}_*^\top & k_{**}\end{bmatrix}\right) , \tag{3}$$

where $\mathbf{k}_* = [K(\mathbf{x}_*, \mathbf{x}_1), \ldots, K(\mathbf{x}_*, \mathbf{x}_n)]^\top$ and $k_{**} = K(\mathbf{x}_*, \mathbf{x}_*)$. Then we can write the distribution of f_* conditioned on the training function outputs:

$$p(f_*|\mathbf{f}, \mathbf{x}_*, X, \theta) \sim \mathcal{N}\left(\mathbf{k}*^\top K^{-1}\mathbf{f}, k_{**} - \mathbf{k}*^\top K^{-1}\mathbf{k}_*\right) . \tag{4}$$

The predictive distribution of f_* is obtained by integrating out the training function values $\mathbf{f}$ from (4) over the posterior distribution (2). The predictive distribution is Gaussian:

$$p(f_*|\mathbf{y}, \mathbf{x}_*, X, \theta, \sigma^2) = \int p(f_*|\mathbf{f}, \mathbf{x}_*, X, \theta)\, p(\mathbf{f}|\mathbf{y}, X, \theta, \sigma^2)\, \mathrm{d}\mathbf{f}$$
$$\sim \mathcal{N}\left(m(\mathbf{x}_*), v(\mathbf{x}_*)\right) , \tag{5}$$

---

[4] Notice that learning cannot be achieved from the likelihood alone: defining a prior over function values is essential to learning.

[5] In Sect. A.2 some algebra useful for deriving (2) is given: notice that the likelihood $p(\mathbf{y}|\mathbf{f}, \sigma^2)$ is also Gaussian in $\mathbf{f}$ with mean $\mathbf{y}$.

with mean and variance given by:

$$m(\boldsymbol{x}_*) = \mathbf{k}_*^\top \left(K + \sigma^2 I\right)^{-1} \mathbf{y} \ , \quad v(\boldsymbol{x}_*) = k_{**} - \mathbf{k}_*^\top \left(K + \sigma^2 I\right)^{-1} \mathbf{k}_* \ . \qquad (6)$$

Another way of obtaining the predictive distribution of f_* is to augment the evidence with a new element y_* corresponding to the noisy version of f_* and to then write the conditional distribution of y_* given the training targets $\mathbf{y}$. The variance of the predictive distribution of y_* is equal to that of the predictive distribution of f_* (6) plus the noise variance σ^2, while the means are identical (the noise has zero mean).

Both if one chooses to learn the hyperparameters or to be Bayesian and do integration, the marginal likelihood of the hyperparameters (or evidence of the observed targets)[6] must be computed. In the first case this quantity will be maximized with respect to the hyperparameters, and in the second case it will be part of the posterior distribution from which the hyperparameters will be sampled. The evidence is obtained by averaging the likelihood over the prior distribution on the function values:

$$p(\mathbf{y}|X, \theta, \sigma^2) = \int p(\mathbf{y}|\mathbf{f})\, p(\mathbf{f}|X, \theta)\, \mathrm{d}\mathbf{f} \sim \mathcal{N}\left(0, K + \sigma^2 I\right) \ . \qquad (7)$$

Notice that the evidence only differs from the prior over function values in a "ridge" term added to the covariance, that corresponds to the additive Gaussian i.i.d. output noise. Maximum likelihood II learning involves estimating the hyperparameters θ and the noise variance σ^2 by minimizing (usually for convenience) the negative log evidence. Let $Q \equiv \left(K + \sigma^2 I\right)$. The cost function and its derivatives are given by:

$$\mathcal{L} = \frac{1}{2}\log|Q| + \frac{1}{2}\mathbf{y}^\top Q^{-1}\mathbf{y} \ ,$$

$$\frac{\partial \mathcal{L}}{\partial \theta_i} = \frac{1}{2}\mathrm{Tr}\left(Q^{-1}\frac{\partial Q}{\partial \theta_i}\right) - \mathbf{y}^\top Q^{-1}\frac{\partial Q}{\partial \theta_i}Q^{-1}\mathbf{y} \ , \qquad (8)$$

$$\frac{\partial \mathcal{L}}{\partial \sigma^2} = \frac{1}{2}\mathrm{Tr}\left(Q^{-1}\right) - \mathbf{y}^\top Q^{-1}Q^{-1}\mathbf{y} \ ,$$

and one can use some gradient descent algorithm to minimize $\mathcal{L}$ (conjugate gradient gives good results, Rasmussen, 1996).

For Gaussian processes, the computational cost of learning is marked by the need to invert matrix Q and therefore scales with the cube of the number of training cases ($\mathcal{O}(n^3)$). If Q^{-1} is known (obtained from the learning process), the computational cost of making predictions is $\mathcal{O}(n)$ for computing the predictive mean, and $\mathcal{O}(n^2)$ for the predictive variance for each test case. There is a need for approximations that simplify the computational cost if Gaussian Processes are to be used with large training datasets.

---

[6] We will from now on use indistinctly "marginal likelihood" or "evidence" to refer to this distribution.

3 Gaussian Processes as Linear Models

Gaussian Processes correspond to parametric models with an infinite number of parameters. Williams (1997a) showed that infinite neural networks with certain transfer functions and the appropriate priors on the weights are equivalent to Gaussian Processes with a particular "neural network" covariance function. Conversely, any Gaussian Process is equivalent to a parametric model, that can be infinite.

In Sect(s). 3.1 and 3.2 we establish the equivalence between GPs and linear models. For the common case of GPs with covariance functions that cannot be expressed as a finite expansion, the equivalent linear models are infinite. However, it might still be interesting to approximate such GPs by a finite linear model, which results in *degenerate* Gaussian Processes. In Sect. 3.3 we introduce degenerate GPs and explain that they often correspond to inappropriate priors over functions, implying counterintuitive predictive variances. We then show how to modify these degenerate GPs at test time to obtain more appropriate priors over functions.

3.1 From Linear Models to GPs

Consider the following extended linear model, where the model outputs are a linear combination of the response of a set of basis functions $\{\phi_j(\boldsymbol{x})|j = 1, \ldots, m\} \subset [\mathbb{R}^D \to \mathbb{R}]$:

$$f(\boldsymbol{x}_i) = \sum_{j=1}^{m} \phi_j(\boldsymbol{x}_i)\,\alpha_j = \boldsymbol{\phi}(\boldsymbol{x}_i)\,\boldsymbol{\alpha} \ , \qquad \mathbf{f} = \boldsymbol{\Phi}\boldsymbol{\alpha} \ , \qquad (9)$$

where as earlier $\mathbf{f} = [f(\boldsymbol{x}_1), \ldots, f(\boldsymbol{x}_n)]^\top$ are the function outputs. The weights are organized in a vector $\boldsymbol{\alpha} = [\alpha_1, \ldots, \alpha_M]^\top$, and $\phi_j(\boldsymbol{x}_i)$ is the response of the j-th basis function to input x_i. $\boldsymbol{\phi}(\boldsymbol{x}_i) = [\phi_1(\boldsymbol{x}_i), \ldots, \phi_m(\boldsymbol{x}_i)]$ is a row vector that contains the response of all m basis functions to input $\boldsymbol{x}_i$ and matrix $\boldsymbol{\Phi}$ (sometimes called *design matrix*) has as its i-th row vector $\boldsymbol{\phi}(\boldsymbol{x}_i)$. Let us define a Gaussian prior over the weights, of the form $p(\boldsymbol{\alpha}|A) \sim \mathcal{N}(0, A)$. Since $\mathbf{f}$ is a linear function of $\boldsymbol{\alpha}$ it has a Gaussian distribution under the prior on $\boldsymbol{\alpha}$, with mean zero. The prior distribution of $\mathbf{f}$ is:

$$p(\mathbf{f}|A, \boldsymbol{\Phi}) \sim \mathcal{N}(0, C) \ , \qquad C = \boldsymbol{\Phi} A \boldsymbol{\Phi}^\top \ . \qquad (10)$$

The model we have defined corresponds to a Gaussian Process. Now, if the number of basis functions m is smaller than the number of training points n, then C will not have full rank and the probability distribution of $\mathbf{f}$ will be an elliptical pancake confined to an m-dimensional subspace in the n-dimensional space where $\mathbf{f}$ lives (Mackay, 1997).

Let again $\mathbf{y}$ be the vector of observed training targets, and assume that the output noise is additive Gaussian i.i.d. of mean zero and variance σ^2. The likelihood of the weights is then Gaussian (in $\mathbf{y}$ and in $\boldsymbol{\alpha}$) given by

$p(\mathbf{y}|\boldsymbol{\alpha}, \boldsymbol{\Phi}, \sigma^2) \sim \mathcal{N}(\boldsymbol{\Phi}\boldsymbol{\alpha}, \sigma^2 I)$. The prior over the training targets is then given by

$$p(\mathbf{y}|A, \boldsymbol{\Phi}, \sigma^2) \sim (0, \sigma^2 I + C) , \qquad (11)$$

and has a full rank covariance, even if C is rank deficient.

To make predictions, one option is to build the joint distribution of the training targets and the new test function value and then condition on the targets. The other option is to compute the posterior distribution over the weights from the likelihood and the prior. Williams (1997b) refers to the first option as the "function-space view" and to the second as the "weight-space view". This distinction has inspired us for writing the next two sections.

The Parameter Space View. Using Bayes' rule, we find that the posterior is the product of two Gaussians in $\boldsymbol{\alpha}$, and is therefore a Gaussian distribution:

$$p(\boldsymbol{\alpha}|\mathbf{y}, A, \boldsymbol{\Phi}, \sigma^2) = \frac{p(\mathbf{y}|\boldsymbol{\alpha}, \boldsymbol{\Phi}, \sigma^2)\, p(\boldsymbol{\alpha}|A)}{p(\mathbf{y}|A, \boldsymbol{\Phi}, \sigma^2)} \sim \mathcal{N}(\boldsymbol{\mu}, \Sigma) ,$$

$$\boldsymbol{\mu} = \sigma^{-2}\, \Sigma\, \boldsymbol{\Phi}^\top \mathbf{y} , \qquad \Sigma = \left[\sigma^{-2}\, \boldsymbol{\Phi}^\top \boldsymbol{\Phi} + A^{-1} \right]^{-1} . \qquad (12)$$

The maximum a posteriori (MAP) estimate of the model weights is given by $\boldsymbol{\mu}$. If we rewrite this quantity as $\boldsymbol{\mu} = [\boldsymbol{\Phi}^\top \boldsymbol{\Phi} + \sigma^2 A]^{-1}\boldsymbol{\Phi}^\top \mathbf{y}$, we can see that the Gaussian assumption on the prior over the weights and on the output noise results in $\boldsymbol{\mu}$ being given by a regularized version of the normal equations. For a new test point $\boldsymbol{x}_*$, the corresponding function value is $f_* = \boldsymbol{\phi}(\boldsymbol{x}_*)\,\boldsymbol{\alpha}$; for making predictions the α's are drawn from the posterior. Since f_* is linear in $\boldsymbol{\alpha}$, it is quite clear that the predictive distribution $p(f_*|\mathbf{y}, A, \boldsymbol{\Phi}, \sigma^2)$ is Gaussian, with mean and variance given by:

$$m(\boldsymbol{x}_*) = \boldsymbol{\phi}(\boldsymbol{x}_*)^\top \boldsymbol{\mu} , \qquad v(\boldsymbol{x}_*) = \boldsymbol{\phi}(\boldsymbol{x}_*)^\top \Sigma\, \boldsymbol{\phi}(\boldsymbol{x}_*) . \qquad (13)$$

We can rewrite the posterior covariance using the matrix inversion lemma (see Appendix A.1) as $\Sigma = A - A[\sigma^2 I + \boldsymbol{\Phi} A \boldsymbol{\Phi}^\top]^{-1} A$. This expression allows us to rewrite the predictive mean and variance as:

$$m(\boldsymbol{x}_*) = \boldsymbol{\phi}(\boldsymbol{x}_*)^\top A \boldsymbol{\Phi}^\top [\sigma^2 I + \boldsymbol{\Phi} A \boldsymbol{\Phi}^\top]^{-1}\mathbf{y} ,$$

$$v(\boldsymbol{x}_*) = \boldsymbol{\phi}(\boldsymbol{x}_*)^\top A \boldsymbol{\phi}(\boldsymbol{x}_*) - \boldsymbol{\phi}(\boldsymbol{x}_*)^\top A \boldsymbol{\Phi}^\top [\sigma^2 I + \boldsymbol{\Phi} A \boldsymbol{\Phi}^\top]^{-1} \boldsymbol{\Phi} A \boldsymbol{\phi}(\boldsymbol{x}_*) , \qquad (14)$$

which will be useful for relating the parameter space view to the GP view.

The Gaussian Process View. There exists a Gaussian Process that is equivalent to our linear model with Gaussian priors on the weights given by (9). The covariance function of the equivalent GP is given by:

$$k(\boldsymbol{x}_i, \boldsymbol{x}_j) = \boldsymbol{\phi}(\boldsymbol{x}_i)^\top A \boldsymbol{\phi}(\boldsymbol{x}_j) = \sum_{k=1}^{m} \sum_{l=1}^{m} A_{kl}\, \phi_k(\boldsymbol{x}_i)\, \phi_l(\boldsymbol{x}_j) . \qquad (15)$$

The covariance matrix of the prior over training function values is given by $K = \Phi A \Phi^\mathsf{T}$ and we recover the same prior as in (10). Taking the same noise model as previously, the prior over targets is identical to (11).

Given a new test input $\boldsymbol{x}_*$, the vector of covariances between $\mathbf{f}_*$ and the training function values is given by $\mathbf{k}_* = \Phi A \phi(\boldsymbol{x}_*)$ and the prior variance of f_* is $k_{**} = \phi(\boldsymbol{x}_*) A \phi(\boldsymbol{x}_*)$. Plugging these expressions into the equations for the predictive mean and variance of a GP (6) one recovers the expressions given by (14) and (13). The predictive mean and variance of a GP with covariance function given by (15) are therefore identical to the predictive mean and variance of the linear model.

A fundamental property of the GP view of a linear model is that the set of m basis functions appear *exclusively* as inner products. Linear models where m is infinite are thus tractable under the GP view, provided that the basis functions and the prior over the weights are appropriately chosen. By appropriately chosen we mean such that a generalized dot product exists in feature space, that allows for the use of the *"kernel trick"*. Schölkopf and Smola (2002) provide with extensive background on kernels and the *"kernel trick"*.

Let us reproduce here an example given by Mackay (1997). Consider a one-dimensional input space, and let us use squared exponential basis functions $\phi_c(x_i) = \exp(-(x_i - c)/(2\lambda^2))$, where c is a given center in input space and λ is a known lengthscale. Let us also define an isotropic prior over the weights, of the form $A = \sigma_\alpha^2 I$. We want to make m go to infinity, and assume for simplicity uniformly spaced basis functions. To make sure that the integral converges, we set variance of the prior over the weights to $\sigma_\alpha^2 = s/\Delta m$, where Δm is the density of basis functions in the input space. The covariance function is given by:

$$
\begin{aligned}
k(x_i, x_j) &= s \int_{c_{min}}^{c_{max}} \phi_c(x_i)\, \phi_c(x_j)\, \mathrm{d}c \ , \\
&= s \int_{c_{min}}^{c_{max}} \exp\left[-\frac{(x_i - c)^2}{2\lambda^2}\right] \exp\left[-\frac{(x_j - c)^2}{2\lambda^2}\right] \mathrm{d}c \ .
\end{aligned}
\tag{16}
$$

Letting the limits of the integral go to infinity, we obtain the integral of the product of two Gaussians (but for a normalization factor), and we can use the algebra from Sect. A.2 to obtain:

$$
k(x_i, x_j) = s\sqrt{\pi\lambda^2} \exp\left[-\frac{(x_i - x_j)^2}{4\lambda^2}\right] ,
\tag{17}
$$

which is the squared exponential covariance function that we presented in (1). We now see that a GP with this particular covariance function is equivalent to a linear model with infinitely many squared exponential basis functions.

In the following we will show that for any valid covariance function, a GP has an equivalent linear model. The equivalent linear model will have infinitely many weights if the GP has a covariance function that has no finite expansion.

3.2   From GPs to Linear Models

We have just seen how to go from any linear model, finite or infinite, to an equivalent GP. We will now see how to go the opposite way, from an arbitrary GP to an equivalent linear model, which will in general be infinite and will be finite only for particular choices of the covariance function.

We start by building a linear model where all the function values considered (training *and* test inputs) are equal to a linear combination of the rows of the corresponding covariance matrix of the GP we wish to approximate, computed with the corresponding covariance function $K(\boldsymbol{x}_i, \boldsymbol{x}_j)$. As in Sect. 2, the covariance function is parametrized by the hyperparameters θ. A Gaussian prior distribution is defined on the model weights, with zero mean and covariance equal to the inverse of the covariance matrix:

$$\begin{bmatrix} \mathbf{f} \\ f_* \end{bmatrix} = \begin{bmatrix} K & \mathbf{k}_* \\ \mathbf{k}_*^\top & k_{**} \end{bmatrix} \cdot \begin{bmatrix} \boldsymbol{\alpha} \\ \alpha_* \end{bmatrix} \ , \qquad p\left(\begin{bmatrix} \boldsymbol{\alpha} \\ \alpha_* \end{bmatrix} \middle| \boldsymbol{x}_*, X, \theta \right) \sim \mathcal{N}\left(0, \begin{bmatrix} K & \mathbf{k}_* \\ \mathbf{k}_*^\top & k_{**} \end{bmatrix}^{-1} \right) \ . \quad (18)$$

To compute the corresponding prior over function values we need to integrate out the weights $[\boldsymbol{\alpha}, \alpha_*]^\top$ from the left expression in (18) by averaging over the prior (right expression in (18)):

$$p\left(\begin{bmatrix} \mathbf{f} \\ f_* \end{bmatrix} \middle| \boldsymbol{x}_*, X, \theta \right) = \int \delta\left(\begin{bmatrix} \mathbf{f} \\ f_* \end{bmatrix} - \begin{bmatrix} K & \mathbf{k}_* \\ \mathbf{k}_*^\top & k_{**} \end{bmatrix} \cdot \begin{bmatrix} \boldsymbol{\alpha} \\ \alpha_* \end{bmatrix} \right) p\left(\begin{bmatrix} \boldsymbol{\alpha} \\ \alpha_* \end{bmatrix} \middle| \boldsymbol{x}_*, X, \theta \right) d\boldsymbol{\alpha}$$

$$\sim \mathcal{N}\left(0, \begin{bmatrix} K & \mathbf{k}_* \\ \mathbf{k}_*^\top & k_{**} \end{bmatrix} \right) \ ,$$

$$(19)$$

and we recover exactly the same prior over function values as for the Gaussian Process, see (3).

Notice that for the linear model to correspond to the full GP two requirements need to be fulfilled:

1. There must be a weight associated to each training input.
2. There must be a weight associated to each possible test input.

Since there are as many weights as input instances, we consider that there is an infinite number of weights of which we only use as many as needed and qualify such a linear model of *infinite*.

Of course, for covariance functions that have a finite expansion in terms of m basis functions, the rank of the covariance matrix will never be greater than m and the equivalent linear model can be readily seen to be finite, with m basis functions. A trivial example is the case where the covariance function is built from a finite linear model with Gaussian priors on the weights. The linear model equivalent to a GP is only infinite if the covariance function of the GP has no finite expansion. In that case, independently of the number of training and test cases considered, the covariance matrix of the prior (independently of its size) will always have full rank.[7]

---

[7] The covariance matrix can always be made rank deficient by replicating a function value in the joint prior, but we do not see any reason to do this in practice.

It becomes evident how one should deal with GPs that have an equivalent finite linear model. If there are more training cases than basis functions, $n > m$, then the finite linear model should be used. In the case where there are less training cases than basis functions, $m > n$, it is computationally more interesting to use the GP.

One strong motivation for the use of Gaussian Processes is the freedom to directly specify the covariance function. In practice, common choices of GP priors imply covariance functions that do not have a finite expansion. For large datasets, this motivates the equivalent infinite linear model by a finite or sparse one. The approximated GP is called Reduced Rank GP since its covariance matrix has a maximum rank equal to the number of weights in the finite linear model.

We will see later in Sect. 4 that the finite linear approximation is built by relaxing the requirement of a weight being associated to each training input, resulting in training inputs with no associated weight. This relaxation should only be done at training time. In the next Section we show the importance of maintaining the requirement of having a weight associated to each test input.

3.3   "Can I Skip α_*?" or Degenerate Gaussian Processes

One may think that having just "as many weights as training cases" with no additional weight α_* associated to each test case gives the same prior as a full GP. It does only for the function evaluated at the training inputs, but it does not anymore for any additional function value considered. Indeed, if we posed $\mathbf{f} = K\,\alpha$ with a prior over the weights given by $p(\alpha|X,\theta) \sim \mathcal{N}(0, K^{-1})$, we would obtain that the corresponding prior over the training function values is $p(\mathbf{f}|X,\theta,\sigma^2) \sim \mathcal{N}(0, K)$. It is true that the linear model would be equivalent to the GP, but only when the function values considered are in $\mathbf{f}$. Without addition of α_*, the linear model and prior over function values are respectively given by:

$$\begin{bmatrix} \mathbf{f} \\ f_* \end{bmatrix} = \begin{bmatrix} K \\ \mathbf{k}_*^\top \end{bmatrix} \cdot \alpha \;, \qquad p\left(\begin{bmatrix} \mathbf{f} \\ f_* \end{bmatrix} \middle| \, \boldsymbol{x}_*, X, \theta\right) \sim \mathcal{N}\left(0, \begin{bmatrix} K & \mathbf{k}_* \\ \mathbf{k}_*^\top & \mathbf{k}_*^\top K^{-1}\mathbf{k}_* \end{bmatrix}\right) \;. \qquad (20)$$

The prior over the new function values f_* differs now from that of the full GP. Notice that the *prior* variance of f_* *depends* on the training inputs: for the common choice of an RBF-type covariance function, if $\boldsymbol{x}_*$ is far from the training inputs, then there is *a priori* no signal, that is f_* is zero without uncertainty! Furthermore, the distribution of $\mathbf{f}_*$ conditioned on the training function outputs, which for the full GP is given by (4), has now become:

$$p(f_*|\mathbf{f}, \boldsymbol{x}_*, X, \theta) \sim \mathcal{N}\left(\mathbf{k}_*^\top K^{-1}\mathbf{f}, 0\right) \;. \qquad (21)$$

Given $\mathbf{f}$, any additional function value f_* is not a random variable anymore, since its conditional distribution has zero variance: f_* is fully determined by $\mathbf{f}$.

If α has a fixed finite size, the prior over functions implied by the linear model ceases to correspond to the GP prior. The joint prior over sets of function values is still Gaussian, which raises the question "is this still a GP?". We choose to call such a *degenerate* process a "degenerate Gaussian Process".

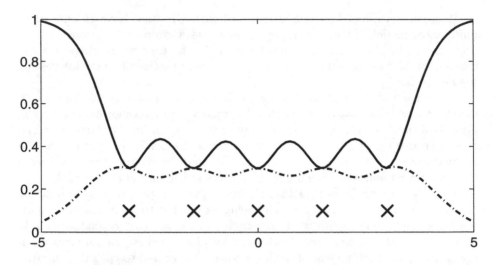

Fig. 1. Predictive standard deviation for a full GP (solid line) and for a degenerate GP (slash-dotted line). The hyperparameters θ_i are all set to 1. The crosses indicate the horizontal location of the 5 training inputs.

Degenerate GPs produce a predictive distribution that has maximal variability around the training inputs, while the predictive variance fades to the noise level as one moves away from them. We illustrate this effect on Fig. 1. We plot the predictive standard deviation of a full GP and its degenerate counterpart for various test points. The training set consists of 5 points: both models have thus 5 weights associated to the training set. The full GP has an additional weight, associated to each test point one at a time. Though it might be a reasonable prior in particular contexts, we believe that it is in general *inappropriate* to have smaller predictive variance far away from the observed data. We believe that appropriate priors are those under which the predictive variance is reduced when the test inputs approach training inputs. In the remaining of the paper we will consider that appropriate priors are desirable, and qualify the prior corresponding to a degenerate GP of inappropriate.

4   Finite Linear Approximations

As we have discussed in Sect. 3.1, a weight must be associated to each test case to avoid inappropriate priors that produce inappropriate predictive error bars. However, the requirement of each training case having a weight associated to it can be relaxed. For computational reasons it might be interesting to approximate, *at training time*, a GP by a finite linear model with less weights than training cases. The model and the prior on the weights are respectively given by:

$$\mathbf{f} = K_{nm}\,\boldsymbol{\alpha}_m \;, \qquad\qquad p(\boldsymbol{\alpha}_m|X,\theta) \sim \mathcal{N}(0, K_{mm}^{-1}) \;, \qquad (22)$$

The subscripts m and n are used to indicate the dimensions: α_m is of size $n \times 1$ and K_{mn} of size $m \times n$; in the following we will omit these subscripts where unnecessary or cumbersome. Sparseness arises when $m < n$: the induced prior over training function values is $p(\mathbf{f}|X, \theta) \sim \mathcal{N}\left(0, K_{nm}K_{mm}^{-1}K_{nm}^{\top}\right)$, and rank of the covariance matrix is at most m. We call such an approximation a Reduced Rank Gaussian Process (RRGP).

The m inputs associated to the weights in α_m do not need to correspond to training inputs. They can indeed be any set of arbitrary points in input space. We will call such points *support inputs* (in recognition to the large amount of work on sparse models done by the Support Vector Machines community). In this paper we will adopt the common restriction of selecting the support set from the training inputs. We discuss ways of selecting the support points in Sect. 4.4.

Learning an RRGP consists both in learning the hyperparameters of the covariance function and in selecting the support set. In practice however, it is hard to do both simultaneously. Besides the technical difficulties of the optimization process (observed for example by Csató (2002)), there is the fundamental issue of having an excessive amount of flexibility that may lead to overfitting (observed for example by Rasmussen (2002) and Seeger et al. (2003)). Smola and Bartlett (2001) address the issue of selecting the support set (Sect. 4.5), assuming that the covariance hyperparameters are given. However, we show that they do this in a way that does not guarantee generalization and we propose an alternative theoretically more sound approach in Sect. 4.4. In the next Section we show how to learn the hyperparameters of the covariance function for the RRGP for a fixed support set. We also show how to make predictions under a degenerate GP, that is, without an additional weight for the new test case, and with the inclusion of a new weight that ensures appropriate predictive variances.

4.1 Learning a Reduced Rank Gaussian Process

The likelihood of the weights is Gaussian in $\mathbf{y}$ and is a linear combination of α_m, given by $p(\mathbf{y}|X, \theta, \alpha_m, \sigma^2) \sim \mathcal{N}(K_{nm}\,\alpha_m, \sigma^2\,I)$, where σ^2 is again the white noise variance. The marginal likelihood of the hyperparameters of the full GP is given by (7). For the sparse finite linear approximation, the marginal likelihood is obtained by averaging the weights out of the likelihood over their prior:

$$p(\mathbf{y}|X, \theta, \sigma^2) = \int p(\mathbf{y}|X, \theta, \alpha_m, \sigma^2)\, p(\alpha_m|X, \theta)\, \mathrm{d}\alpha_m$$
$$\sim \mathcal{N}\left(0, \sigma^2\,I + K_{nm}\,K_{mm}^{-1}K_{nm}^{\top}\right) \ . \tag{23}$$

As expected, for the case where the support set comprises all training inputs and $m = n$, we recover the full Gaussian Process.

Let us define $\tilde{Q} \equiv \left[\sigma^2\,I + K_{nm}\,K_{mm}^{-1}K_{nm}^{\top}\right]$, the covariance of the RRGP evidence. Maximum likelihood learning of the hyperparameters can be achieved by minimizing the negative log evidence. The cost function and its derivatives are given by (8) where Q is replaced by $\tilde{Q}$. Since the simple linear algebra involved can be tedious, we give here the explicit expression of the different terms. For the terms involving $\log |\tilde{Q}|$ we have:

$$\log |\tilde{Q}| = (n - m)\, log(\sigma^2) + \log \left| K_{nm}^\top K_{nm} + \sigma^2 K_{mm} \right| \ ,$$

$$\frac{\partial \log |\tilde{Q}|}{\partial \theta_i} = \mathrm{Tr}\left[\tilde{Q}^{-1} \frac{\partial \tilde{Q}}{\partial \theta_i} \right] = 2\, \mathrm{Tr}\left[\frac{\partial K_{nm}}{\partial \theta_i} Z^\top \right] - \mathrm{Tr}\left[K_{mm}^{-1} K_{nm}^\top Z \frac{\partial K_{nm}}{\partial \theta_i} \right] \ ,$$

$$\frac{\partial \log |\tilde{Q}|}{\partial \sigma^2} = \frac{n - m}{\sigma^2} + \mathrm{Tr}\left[Z_{mm} \right] \ ,$$

$$(24)$$

where we have introduced $Z \equiv K_{nm} \left[K_{nm}^\top K_{nm} + \sigma^2 K_{mm} \right]^{-1}$. For the terms involving $\tilde{Q}^{-1}$ we have:

$$\mathbf{y}^\top \tilde{Q}^{-1} \mathbf{y} = \left(\mathbf{y}^\top \mathbf{y} - \mathbf{y}^\top Z K_{nm}^\top \mathbf{y} \right) /\sigma^2 \ ,$$

$$\frac{\partial \mathbf{y}^\top \tilde{Q}^{-1} \mathbf{y}}{\partial \theta_i} = \mathbf{y}^\top Z \frac{\partial K_{mm}}{\partial \theta_i} Z^\top \mathbf{y} - 2\, \mathbf{y}^\top \left(I - Z K_{nm}^\top \right) \frac{\partial K_{nm}}{\partial \theta_i} Z^\top \mathbf{y}/\sigma^2 \ , \quad (25)$$

$$\frac{\partial \mathbf{y}^\top \tilde{Q}^{-1} \mathbf{y}}{\partial \sigma^2} = -\mathbf{y}^\top \mathbf{y}/\sigma^4 + \mathbf{y}^\top Z K_{nm}^\top \mathbf{y}/\sigma^4 + \mathbf{y}^\top Z K_{mm} Z^\top \mathbf{y}/\sigma^2 \ .$$

The hyperparameters and the output noise variance can be learnt by using the expressions we have given for the negative log marginal likelihood and its derivatives in conjunction with some gradient descent algorithm. The computational complexity of evaluating the evidence and its derivatives is $\mathcal{O}(nm^2 + nDm)$, which is to be compared with the corresponding cost of $\mathcal{O}(n^3)$ for the full GP model.

4.2   Making Predictions Without α_*

The posterior over the weights associated to the training function values is $p(\boldsymbol{\alpha}_m | \mathbf{y}, K_{mn}, \sigma^2) \sim \mathcal{N}(\boldsymbol{\mu}, \boldsymbol{\Sigma})$ with:

$$\boldsymbol{\mu} = \sigma^{-2} \boldsymbol{\Sigma} K_{mn}^\top \mathbf{y} \ , \qquad \boldsymbol{\Sigma} = \left[\sigma^{-2} K_{mn}^\top K_{mn} + K_{mm} \right]^{-1} \ . \qquad (26)$$

At this point one can choose to make predictions right now, based on the posterior of $\boldsymbol{\alpha}_m$ and without adding an additional weight α_* associated to the new test point $\boldsymbol{x}_*$. As discussed in Sect. 3.3, this would correspond to a degenerate GP, leading to inappropriate predictive variance. The predictive mean on the other hand can still be a reasonable approximation to that of the GP: Smola and Bartlett (2001) approximate the predictive mean exactly in this way. The expressions for the predictive mean and variance, when not including α_*, are respectively given by:

$$m(\boldsymbol{x}_*) = \mathbf{k}(\boldsymbol{x}_*)^\top \boldsymbol{\mu}, \qquad v(\boldsymbol{x}_*) = \sigma^2 + \mathbf{k}(\boldsymbol{x}_*)^\top \boldsymbol{\Sigma}\, \mathbf{k}(\boldsymbol{x}_*). \qquad (27)$$

$\mathbf{k}(\boldsymbol{x}_*)$ denotes the $m \times 1$ vector $[K(\boldsymbol{x}_*, \boldsymbol{x}_1), \dots, K(\boldsymbol{x}_*, \boldsymbol{x}_m)]^\top$ of covariances between $\boldsymbol{x}_*$ and at the m support inputs (as opposed to $\mathbf{k}_*$ which is the $n \times 1$

vector of covariances between $\boldsymbol{x}_*$ and at the n training inputs). Note that if no sparseness is enforced, $(m = n)$, then $\boldsymbol{\mu} = (K_{nn} + \sigma^2 I)^{-1}\mathbf{y}$ and the predictive mean $m(\boldsymbol{x}_*)$ becomes identical to that of the full GP. Also, note that for decaying covariance functions,[8] if $\boldsymbol{x}_*$ is far away from the selected training inputs, the predictive variance collapses to the output noise level, which we have defined as an inappropriate prior.

The computational cost of predicting without α_* is an initial $\mathcal{O}(nm^2)$ to compute $\boldsymbol{\Sigma}$, and then an additional $\mathcal{O}(m)$ for the predictive mean and $\mathcal{O}(m^2)$ for the predictive variance per test case.

4.3   Making Predictions with α_*

To obtain a better approximation to the full GP, especially in terms of the predictive variance, we add an extra weight α_* to the model for each test input $\boldsymbol{x}_*$. Unless we are interested in the predictive covariance for a set of test inputs, it is enough to add one single α_* at a time. The total number of weights is therefore only augmented by one for any test case.

For a new test point, the mean and covariance matrix of the new posterior over the augmented weights vector are given by:

$$
\boldsymbol{\mu}_* = \sigma^{-2} \boldsymbol{\Sigma}_* \begin{bmatrix} K_{mn}^\top \\ \mathbf{k}_*^\top \end{bmatrix} \mathbf{y} \;,
$$

$$
\boldsymbol{\Sigma}_* = \begin{bmatrix} \boldsymbol{\Sigma}^{-1} & \mathbf{k}(\boldsymbol{x}_*) + \sigma^{-2} K_{nm}^\top \mathbf{k}_* \\ \mathbf{k}(\boldsymbol{x}_*)^\top + \sigma^{-2} \mathbf{k}_*^\top K_{nm} & k_{**} + \sigma^{-2} \mathbf{k}_*^\top \mathbf{k}_* \end{bmatrix}^{-1} \;. \tag{28}
$$

and the computational cost of updating the posterior and computing the predictive mean and variance is $\mathcal{O}(nm)$ for each test point. The most expensive operation is computing $K_{nm}^\top \mathbf{k}_*$ with $\mathcal{O}(nm)$ operations. Once this is done and given that we have previously computed $\boldsymbol{\Sigma}$, computing $\boldsymbol{\Sigma}_*$ can be efficiently done using inversion by partitioning in $\mathcal{O}(m^2)$ (see Sect. A.1 for the details). The predictive mean and variance can be computed by plugging the updated posterior parameters (28) into (27), or alternatively by building the updated joint prior over the training and new test function values. We describe in detail the algebra involved in the second option in App. A.5. The predictive mean and variance when including α_* are respectively given by:

$$
m_*(\boldsymbol{x}_*) = \mathbf{k}_*^\top \left[K_{nm} K_{mm}^{-1} K_{nm}^\top + \sigma^2 I + \mathbf{v}_* \mathbf{v}_*^\top / c_* \right]^{-1} \mathbf{y} \;,
$$

$$
v_*(\boldsymbol{x}_*) = \sigma^2 + k_{**} + \mathbf{k}_*^\top \left[K_{nm} K_{mm}^{-1} K_{nm}^\top + \sigma^2 I + \mathbf{v}_* \mathbf{v}_*^\top / c_* \right]^{-1} \mathbf{k}_* \;. \tag{29}
$$

where $\mathbf{v}_* \equiv \mathbf{k}_* - K_{nm} K_{mm}^{-1} \mathbf{k}(\boldsymbol{x}_*)$ is the difference between the actual and the approximated covariance of f_* and $\mathbf{f}$, and $c_* \equiv k_{**} - \mathbf{k}(\boldsymbol{x}_*)^\top K_{mm}^{-1} \mathbf{k}(\boldsymbol{x}_*)$ is the predictive variance at $\boldsymbol{x}_*$ of a full GP with the support inputs as training inputs.

---

[8] Covariance functions whose value decays with the distance between the two arguments. One example is the squared exponential covariance function described in Sect. 2. Decaying covariance functions are very commonly encountered in practice.

4.4   Selecting the Support Points

One way of addressing the problem of selecting the m support inputs is to select them from among the n training inputs. The number of possible sets of support inputs is combinatorial, C_n^m.[9] Since we will typically be interested in support sets much smaller than the training sets $(m < n)$, this implies that the number of possible support sets is roughly exponential in m. Ideally one would like to evaluate the evidence for the finite linear model approximation (23), for each possible support input set, and then select the set that yields a higher evidence. In most cases however, this is impractical due to computational limitations. One suboptimal solution is to opt for a greedy method: starting with an empty subset, one includes the input that results in a maximal increase in evidence. The greedy method exploits the fact that the evidence can be computed efficiently when a case is added (or deleted) to the support set.

Suppose that a candidate input x_i from the training set is considered for inclusion in the support set. The new marginal likelihood is given by:

$$\mathcal{L}_i = \frac{1}{2} \log |\tilde{Q}_i| + \frac{1}{2} \mathbf{y}^\top \tilde{Q}_i^{-1} \mathbf{y} \;, \qquad \tilde{Q}_i \equiv \sigma^2 I + K_{n\tilde{m}} K_{\tilde{m}\tilde{m}}^{-1} K_{n\tilde{m}}^\top \;, \qquad (30)$$

where $\tilde{m}$ is the set of $m + 1$ elements containing the m elements in the current support set plus the new case x_i. $\tilde{Q}_i$ is the updated covariance of the evidence of the RRGP augmented with x_i. Let us deal separately with the two terms in the evidence. The matrix inversion lemma allows us to rewrite $\tilde{Q}_i$ as:

$$\tilde{Q}_i = \sigma^{-2} I - \sigma^{-4} K_{n\tilde{m}} \Sigma_i K_{n\tilde{m}}^\top \;, \qquad \Sigma_i = \left[K_{n\tilde{m}}^\top K_{n\tilde{m}} / \sigma^2 + K_{\tilde{m}\tilde{m}} \right]^{-1} \;, \qquad (31)$$

where Σ_i is the covariance of the posterior over the weights augmented in α_i, the weight associated to x_i. Notice that Σ_i is the same expression as Σ_* in (28) if one replaces the index $*$ by i. In both cases we augment the posterior in the same way. Computing Σ_i from Σ costs therefore only $\mathcal{O}(nm)$.

The term of $\mathcal{L}$ quadratic in $\mathbf{y}$ can be rewritten as:

$$\mathcal{Q}_i = \frac{1}{2\sigma^2} \mathbf{y}^\top \mathbf{y} - \frac{1}{2\sigma^4} \mathbf{y}^\top K_{n\tilde{m}} \Sigma_i K_{n\tilde{m}}^\top \mathbf{y} \;, \qquad (32)$$

and can be computed efficiently in $\mathcal{O}(nm)$ if Σ and $K_{nm}^\top \mathbf{y}$ are known. In Sect. A.3 we provide the expressions necessary for computing $\mathcal{Q}_i$ incrementally in a robust manner from the Cholesky decomposition of Σ. In Sect. 4.5 we describe Smola and Bartlett's Sparse Greedy Gaussian Process (SGGP) Regression which uses $\mathcal{Q}_i$ solely as objective function for selecting the support set in a greedy manner.

The term of $\mathcal{L}$ that depends on $\log |\tilde{Q}_i|$ can be expressed as:

$$\mathcal{G}_i = \frac{1}{2} \left[\log |\Sigma_i| - \log |K_{\tilde{m}\tilde{m}}| + n \log \sigma^2 \right] \;, \qquad (33)$$

---

[9] C_n^m is "n choose m": the number of combinations of m elements out of n without replacement and where the order does not matter.

and computed at a cost of $\mathcal{O}(nm)$ (the cost of computing $K_{nm}^\top \mathbf{k}_i$). The algebra in Sect. A.3 can be used to update the determinants from the incremental Cholesky decompositions at no additional cost.

The overall cost of evaluating the evidence for each candidate point for the support set is $\mathcal{O}(nm)$. In practice, we may not want to explore the whole training set in search for the best candidate, since this would be too costly. We may restrict ourselves to exploring some reduced random subset.

4.5   Sparse Greedy Gaussian Process Regression

Smola and Bartlett (2001) and Schölkopf and Smola (2002) present a method to speed up the prediction stage for Gaussian processes. They propose a sparse greedy techniques to approximate the Maximum a Posteriori (MAP) predictions, treating separately the approximation of the predictive mean and that of the predictive variance.

For the predictive mean, Smola and Bartlett adopt a finite linear approximation of the form given by (22), where no extra weight α_* associated to the test input is added. Since this is a degenerate GP, it is understandable that they only use it for approximating the predictive mean: we now know that the predictive uncertainties of degenerate GPs are inappropriate.

The main contribution of their paper is to propose a method for selecting the m inputs in the support set from the n training inputs. Starting from a full posterior distribution (as many weights as training inputs), they aim at finding a sparse weight vector (with only m non-zero entries) with the requirement that *the posterior probability at the approximate solution be close to the maximum of the posterior probability* (quoted from (Schölkopf and Smola, 2002, Sect. 16.4.3)). Since the optimal strategy has again a prohibitive cost, they propose a greedy method where the objective function is the full posterior evaluated at the optimal weights vector with only m non-zeros weighs, those corresponding to the inputs in the support set.

The posterior on $\boldsymbol{\alpha}_n$ (full posterior) is given by (26), where $m = n$, i.e. matrix K_{nm} is replaced by the full $n \times n$ matrix K. The objective function used in (Smola and Bartlett, 2001; Schölkopf and Smola, 2002) is the part of the negative log posterior that depends on $\boldsymbol{\alpha}_n$, which is the following quadratic form:

$$-\frac{1}{2}\mathbf{y}^\top K_{nm}\,\boldsymbol{\alpha}_m + \frac{1}{2}\,\boldsymbol{\alpha}_m^\top \left[K_{nm}^\top K_{nm} + \sigma^2 K_{mm}\right]\boldsymbol{\alpha}_m \ , \qquad (34)$$

where as usual $\boldsymbol{\alpha}_m$ denotes the part of $\boldsymbol{\alpha}_n$ that hasn't been clamped to zero. Notice that it is essential for the objective function to be the full posterior evaluated at a sparse $\boldsymbol{\alpha}_n$, rather than the posterior on $\boldsymbol{\alpha}_m$ (given by (26) with indeed $m \neq n$). In the latter case, only the log determinant of the covariance would play a rôle in the posterior, since $\boldsymbol{\alpha}_m$ would have been made equal to the posterior mean, and we would have a completely different objective function from that in (Smola and Bartlett, 2001; Schölkopf and Smola, 2002).

Given two candidates to the support set, the one resulting in a support set for which the minimum of (34) is smaller is chosen. The minimum of (34) is given by:

$$-\frac{1}{2}\mathbf{y}^\top K_{nm} \left[K_{nm}^\top K_{nm} + \sigma^2 K_{mm}\right] K_{nm}^\top \mathbf{y} \ , \tag{35}$$

and it is in fact this quantity that is minimized with respect to the m elements in the support set in a greedy manner. The expression given in (35) with $m \neq n$ is in fact an upper bound to the same expression with $m = n$, which corresponds to selecting the whole training set as active set. Smola and Bartlett (2001); Schölkopf and Smola (2002) also provide a lower bound to the latter, which allows them to give a stop criterion to the greedy method based on the relative difference between upper and lower bound. The computational cost of evaluating the expression given in (35) for each candidate to the support set is $\mathcal{O}(nm)$, and use can be made of an incremental Cholesky factorization for numerical stability. The expressions in Sect. A.3 can be used. The computational cost is therefore the same for the SGGP method as for the greedy approach based on maximizing the evidence that we propose in Sect. 4.4.

Why Does It Work? One might at this point make abstraction from the algorithmic details, and ask oneself the fair question of why obtaining a sparse weight vector that evaluated under the posterior over the full weight vector yields a probability close to that of the non-sparse solution is a good approximation. Along the same lines, one may wonder whether the stopping criterion proposed relates in any way with good generalization.

It turns out that the method often works well in practice, in a very similar way as our proposed greedy criterion based on maximizing the evidence. One explanation for the SGGP method to select meaningful active sets is that it is in fact minimizing a part of the negative log evidence $\mathcal{L}_i$, given by (30). Indeed, notice that minimizing the objective function given by (35) is exactly equivalent to minimizing the part of the negative log evidence quadratic in $\mathbf{y}$ given by (32). So why would the method work if it only maximizes $\mathcal{Q}_i$ (32), the part of $\mathcal{L}_i$ that has to do with fitting the data, and ignores $\mathcal{G}_i$ (33), the part that enforces regularization? We believe that overfitting will seldom happen because m is typically significantly smaller than n, and that therefore we are selecting from a family of models that are all very simple. In other words, it is the sparsity itself that guarantees some amount of regularization, and therefore $\mathcal{G}_i$ can be often safely omitted from the negative log evidence. However, as we will see in what follows, the SGGP can fail and indeed overfit. The problem is that the SGGP fails to provide a valid stopping criterion for the process of adding elements to the support set.

But, How Much Sparsity? If sparsity seemingly ensures generalization, then it would also seem that a criterion is needed to know *the minimum sparsity level required*. In other words, we need to know how many inputs it is safe to include in the support set. (Smola and Bartlett, 2001; Schölkopf and Smola, 2002) use a measure they call the "gap", which is the relative difference between the upper and lower bound on the negative log posterior. They choose an arbitrary threshold below which they consider that the approximate posterior has been maximized to a value close enough to the maximum of the full posterior. Once again we fail to see what such a criterion has to do with ensuring generalization, and we are not the only ones: Schwaighofer and Tresp (2003) report "we did not observe any correlation between the gap and the generalization performance in our experiments". It might be that for well chosen hyperparameters of the covariance, or for datasets that do not lend themselves to sparse approximations, keeping on adding cases to the support set cannot be harmful. Yet the SGGP does not allow learning the hyperparameters, and those must be somehow guessed (at least not in a direct way).

We provide a simple toy example (Fig. 2) in which the value of minimizing the negative log evidence becomes apparent. We generate 100 one-dimensional training inputs, equally spaced from -10 to 10. We generate the corresponding training inputs by applying the function $\sin(x)/x$ to the inputs, and adding Gaussian noise of variance 0.01. We generate the test data from 1000 test inputs equally spaced between -12 and 12. We use a squared exponential covariance function as given by (1), and we set the hyperparameters in the following way: the lengthscale is $\theta_1 = 1$, the prior standard deviation of the output signal is $\theta_2 = 1$ and the noise variance is $\sigma^2 = \theta_3 = 0.01$. Note that we provide the model with the *actual* variance of the noise. We apply the greedy strategy for selecting the support set by minimizing in one case the negative log evidence and in the other case the negative log posterior. Interesting things happen. We plot the test squared error as a function of m, the size of the support set for both greedy strategies. Both have a minimum for support sets of size around 8 to 10 elements, and increase again as for larger support sets. Additionally, we compute the negative log evidence as a function of m, and we see that it has a minimum around the region where the test error is minimal. This means that we can actually use the evidence to determine good levels of sparsity. We also plot the "gap" as a function of m, and indicate the location of the arbitrary threshold of 0.025 used by Smola and Bartlett (2001); Schölkopf and Smola (2002). The gap cannot provide us with useful information in any case, since it is *always* a monotonically decreasing function of m! The threshold is absolutely arbitrary, and has no relation to the expected generalization of the model.

Approximating Predictive Variances. Obtaining the predictive variance based on the posterior of the weights associated to the support set is a bad idea, since those will be smaller the further away the test input is from the inputs in the support set. An explicit approximation to the predictive variance of a full GP, given in (6) is proposed instead. For a given test input x_*, Smola and Bartlett (2001); Schölkopf and Smola (2002) propose to approximate the term:

$$-\mathbf{k}_*^\top \left[K + \sigma^2 I \right]^{-1} \mathbf{k}_* , \tag{36}$$

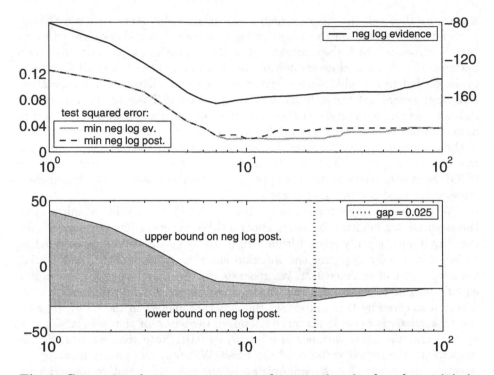

Fig. 2. Comparison between a sparse greedy approximation based on minimizing the negative log evidence, and one based on minimizing the negative log posterior. In both figures the horizontal axis indicates the size of the support set. *Top:* the solid black curve is the negative log evidence, with values given by the right vertical axis, the other two curves are the test squared error of the greedy methods based on minimizing the negative log evidence (solid gray) and the negative log posterior (dashed black), with values given on the left vertical axis. *Bottom:* for the SGGP approach the upper and lower bounds on the negative lower posterior are given, and the vertical dotted line shows the minimum size of the support set for which the "gap" is smaller that 0.025.

using the fact that it is the minimum (with respect to the $n \times 1$ weights vector β, one weight associated to each training input) of the quadratic form:

$$-2\,\mathbf{k}_*^\top \beta + \beta^\top \left[K + \sigma^2 I \right] \beta \ . \tag{37}$$

They then go on to propose finding a sparse version β_m of β with only m non-zero elements.[10] The method is again a greedy incremental minimization of the

---

[10] This m does not have anything to do with the number of inputs in the support set of our previous discussion. It corresponds to a *new* support set, this time for approximating the predictive variance at $\mathbf{x}_*$. We insist on using the same symbol though because it still corresponds to a support set with $m < n$.

expression in (37). For a given choice of active elements (non-zero) in β, the minimum of the objective function is given by:

$$-\mathbf{k}(\boldsymbol{x}_*)^\top \left[K_{mm} + \sigma^2 I\right]^{-1} \mathbf{k}(\boldsymbol{x}_*) \ , \tag{38}$$

where here again $\mathbf{k}(\boldsymbol{x}_*)$ represents an $m \times 1$ vector containing the covariance function evaluated at $\boldsymbol{x}_*$ and at the m inputs in the support set. Again, the support set yielding a minimal value of the expression in (38) will be chosen. The expression in (38) is also an upper bound on the (36), which means that bad approximations only mean an overestimate of the predictive variance, which is less bad than an underestimate. For each candidate to the support set, (38) can be evaluated in $\mathcal{O}(m^2)$ (this cost includes updating $\left[K_{mm} + \sigma^2 I\right]^{-1}$). Luckily, in practice the typical size of the support sets for approximating predictive variances is around one order of magnitude smaller than the size of the support set for approximating predictive means. Smola and Bartlett (2001); Schölkopf and Smola (2002) also provide a lower bound to (36), which allows to use a similar stop criterion as in the approximation of the predictive means.

Limitations Though it does work in practice and for the datasets on which we have tried it, there is no fundamental guarantee that SGGP will always work, since it does not maximize the whole of the evidence: it ignores the term in $\log |\tilde{Q}|$.

The hyperparameters of the covariance function need to be known: they cannot be learned by maximizing the posterior, since this would lead to overfitting.

While for approximating the predictive means one needs to find a unique support set, a specific support set needs to be estimated for *each* different test input if one wants to obtain good approximations to the predictive variance. The computational cost becomes then $\mathcal{O}(knm^2)$ per training case, where k is the size of a reduced random search set (Smola and Bartlett (2001) suggest using $k = 59$).

5   Experiments

We use the KIN40K dataset (for more details see Rasmussen, 1996, Chap. 5). This dataset represents the forward dynamics of an 8 link all-revolve robot arm. The dataset contains 40000 examples, the input space is 8-dimensional, and the 1-dimensional output represents the distance of an end-point of the robot arm from a fixed point. The mapping to be learned is low noise and highly nonlinear. This is of importance, since it means that the predictions can be improved by training on more data, and sparse solutions do not arise trivially.

We divide the dataset into 10 disjoint subsets of 4000 elements, that we then further split into training and test sets of 2000 elements each. The size of the support set is set to 512 elements in all cases. For each method we perform then 10 experiments, and compute the following losses: the Mean Absolute Error (MAE), the Mean Squared Error (MSE) and the Negative Test

Log-density (NTL). We also compute the training negative log likelihood per training case. Averaged results over the 10 disjoint sub-datasets are shown in the upper part of Table 1. SGGP is the sparse support set selection method proposed by Smola and Bartlett (2001); to compute predictive uncertainties, we do not use the sparse greedy approximation they suggest, since it has a too high computational cost of $\mathcal{O}(knm^2)$ per test case, with $k = 59$ and $m \approx 250$ in our case to reach $gap < 0.025$. As an alternative, they suggest to use the predictive uncertainties given by a reduced GP trained only on the support set obtained for approximating the predictive mean; the computational cost is low, $\mathcal{O}(m^2)$ per test case, but the performance is too poor to be worth reporting (NTL of the order of 0.3). To compute predictive uncertainties with the SGGP method we use the expressions given by (27) and (29). SGEV is our alternative greedy support set selection method based on maximizing the evidence. The HPEV-rand method selects a support set at random and learns the covariance hyperparameters by maximizing the evidence of the approximate model, as described in Sect. 4.1. The HPEV-SGEV and HPEV-SGGP methods select the support set for fixed hyperparameters according to the SGEV and and SGGP methods respectively, and then for that selected support set learn the hyperparameters by using HPEV. This procedure is iterated 10 times for both algorithms, which is enough for the likelihood to apparently converge. For all algorithms we present the results for the naïve non-augmented degenerate prediction model, and for the augmented non-degenerate one.

The experimental results show that the performance is systematically superior when using the augmented non-degenerate RRGP with an additional weight α_*. This superiority is expressed in all three losses, mean absolute, mean squared and negative test predictive density (which takes into account the predictive uncertainties). We believe that the relevant loss is the last one, since it reflects the fundamental theoretical improvement of the non-degenerate RRGP. The fact that the losses related to the predictive mean are also better can be explained by the model being slightly more flexible. We performed paired t-tests that confirmed that under all losses and algorithms considered, the augmented RRGP is significantly superior than the non-augmented one, with p-values always smaller than 1%. We found that for the dataset considered SGGP, SGEV and HPEV-rand are not significantly different. It would then seem that learning the hyperparameters for a random support set, or learning the support set for (carefully selected) hyperparameters by maximizing the posterior or the evidence are methods with equivalent performance. We found that both for the augmented and the non-augmented case, HPEV-SGEV and HPEV-SGGP are significantly superior to the other three methods, under all losses, again with p-values below 1%. On the other hand, HPEV-SGEV and HPEV-SGGP are not significantly different from each other under any of the losses.

The lower part of Table 1 shows the results of an additional experiment we made, where we compare SGEV to HPEV-rand on a larger training set. We generate this time 10 disjoint test sets of 4000 cases, and 10 corresponding training sets of 36000 elements. The size of the support sets remains 512. We

method	tr. neg log lik	_non-augmented_			_augmented_		
		MAE	MSE	NTL	MAE	MSE	NTL
SGGP	–	0.0481	0.0048	−0.3525	0.0460	0.0045	−0.4613
SGEV	−1.1555	0.0484	0.0049	−0.3446	0.0463	0.0045	−0.4562
HPEV-rand	−1.0978	0.0503	0.0047	−0.3694	0.0486	0.0045	−0.4269
HPEV-SGEV	−1.3234	0.0425	0.0036	−0.4218	0.0404	0.0033	−0.5918
HPEV-SGGP	−1.3274	0.0425	0.0036	−0.4217	0.0405	0.0033	−0.5920
		2000 _training_ - 2000 _test_					
SGEV	−1.4932	0.0371	0.0028	−0.6223	0.0346	0.0024	−0.6672
HPEV-rand	−1.5378	0.0363	0.0026	−0.6417	0.0340	0.0023	−0.7004
		36000 _training_ - 4000 _test_					

Table 1. Comparison of different learning methods for RRGPs on the KIN40K dataset, for 2000 training and test cases (*upper subtable*) and for 36000 training and 4000 test cases (*lower subtable*). The support set size is set to 512 for all methods. For each method the training negative log marginal likelihood per case is given, together with the Mean Absolute Error (MAE), Mean Squared Error (MSE) and Negative Test Log-likelihood (NTL) losses. SGGP (Smola and Bartlett, 2001) and SGEV (our alternative to SGGP based on maximizing the evidence) are based on learning the support set for fixed hyperparameters. HPEV-random learns the hyperparameters for a random subset, and HPEV-SGEV and HPEV-SGGP are methods where SGEV and SGGP are respectively interleaved with HPEV, for 10 repetitions.

compute the same losses as earlier, and consider also the augmented and the non-augmented RRGPs for making predictions. Paired t-tests[11] confirm once again the superiority of the augmented model to the non-augmented one for both models and all losses, with p-values below 1%.

6   Discussion

We have proposed to augment RRGPs at test time, by adding an additional weight α_* associated to the new test input x_*. The computational cost for the predictive mean increases to $\mathcal{O}(nm)$ per case, i.e. $\mathcal{O}(n)$ more expensive than the non-augmented case. It might seem surprising that this is more expensive than the $\mathcal{O}(n)$ cost per case of the full GP! Of course, the full GP has has an initial cost of $\mathcal{O}(n^2)$ provided that the covariance matrix has been inverted, which costs $\mathcal{O}(n^3)$. Computing predictive variances has an initial cost of $\mathcal{O}(nm^2)$ like for the non-augmented case, and then a cost per case of $\mathcal{O}(nm)$ which is more expensive than the $\mathcal{O}(m^2)$ for the non-augmented case, and below the $\mathcal{O}(n^2)$ of the full GP. It may be argued that the major improvement brought by augmenting the RRGP is in terms of the predictive variance, and that one might therefore

---

[11] Due to dependencies between the training sets, assumptions of independence needed for the t-test could be compromised, but this is probably not a major effect.

consider computing the predictive mean from the non-augmented model, and the predictive variance from the augmented. However, the experiments we have conducted show that the augmented RRGP is systematically superior to the non-augmented, for all losses and learning schemes considered. The mean predictions are also better, probably due to the gain in flexibility by having an additional basis function.

Which method should be used for computing predictive variances? We have shown that using the degenerate RRGP, (27), has a computational cost of $\mathcal{O}(m^2)$ per test case. Using the augmented non-degenerate RRGP is preferable though because it gives higher quality predictive uncertainties, but the cost augments to $\mathcal{O}(nm)$ per test case. Smola and Bartlett (2001) propose two possibilities. A cost efficient option, $\mathcal{O}(m^2)$ per test case, is to base the calculation of all test predictive variances on the support set selected by approximating the posterior, which is in fact equivalent to computing predictive variances from a small full GP trained only on the support set. They show that the predictive variances obtained will always be an upper bound on the ones given by the full GP, and argue that inaccuracy (over estimation) is for that reason benign. We found experimentally that the errorbars from a small full GP trained only on the support set are very poor. The more accurate, yet more costly option consists is selecting a new support set for each test point. While they argue that the typical size of such test sets is very small (of the order of 25 for reasonable hyperparameters for the abalone dataset, but of the order of 250 for the KIN40K dataset), the computational cost per test case rises to $\mathcal{O}(knm^2)$. As we have explained, k is the size of a reduced random search set that can be fixed to 59 (see Smola and Bartlett, 2001). For their method to be computationally cheaper than our augmented RRGP, the support set that our method selects should contain more than $59 \times 25^2 = 36875$ elements. This is two orders of magnitude above the reasonable size of support sets that we would choose. In the experiments, we ended up computing the predictive variances for the SGGP from our expressions (27) and (29).

We found that none of the two possible "one-shot" approaches to training a RRGP is significantly superior to the other. In other words, selecting support sets at random and optimizing the hyperparameters does not provide significantly different performance than fixing the hyperparameters and selecting the support set in a supervised manner. Furthermore, on the dataset we did our experiments SGGP and SGEV did not prove to be significantly different either. We expect SGEV to perform better than SGGP on datasets where for the given hyperparameters the learning curve saturates, or even deteriorates as the support set is increased, as is the case in the example we give in Fig. 2. Interleaving support set selection and hyperparameter learning schemes proves on the other hand to be promising. The experiments on KIN40K show that this scheme gives much superior performance to the two isolated learning schemes.

It is interesting to note the relation between the RRGP and the Nyström approximation proposed by Williams and Seeger (2001). In that approach the predictive mean and variance are respectively given by:

$$m(\boldsymbol{x}_*) = \mathbf{k}_*^\top \left[K_{nm} K_{mm}^{-1} K_{nm}^\top + \sigma^2 I \right]^{-1} \mathbf{y} \ ,$$

$$v(\boldsymbol{x}_*) = \sigma^2 + k_{**} + \mathbf{k}_*^\top \left[K_{nm} K_{mm}^{-1} K_{nm}^\top + \sigma^2 I \right]^{-1} \mathbf{k}_* \ . \tag{39}$$

These expressions are very similar to those obtained for the augmented RRGP, given by (29). However, the additional term in the approximate covariance for the augmented RRGP ensures that it is positive definite, see (Williams et al., 2002), and that therefore our approach does not suffer from negative predictive variances as is the case for the Nyström approximation for GPs.

Future work will involve the theoretical study of other sparse approximations to GPs that have been recently proposed, which we enumerate in Sect. 1, and the experimental comparison of these methods to those presented in this paper.

A Useful Algebra

A.1 Matrix Identities

The matrix inversion lemma, also known as the Woodbury, Sherman & Morrison formula states that:

$$(Z + UWV^\top)^{-1} = Z^{-1} - Z^{-1}U(W^{-1} + V^\top Z^{-1}U)^{-1}V^\top Z^{-1}, \tag{A-40}$$

assuming the relevant inverses all exist. Here Z is $n \times n$, W is $m \times m$ and U and V are both of size $n \times m$; consequently if Z^{-1} is known, and a low rank (ie. $m < n$) perturbation are made to Z as in left hand side of eq. (A-40), considerable speedup can be achieved. A similar equation exists for determinants:

$$|Z + UWV^\top| = |Z| \, |W| \, |W^{-1} + V^\top Z^{-1}U| \ . \tag{A-41}$$

Let the symmetric $n \times n$ matrix A and its inverse A^{-1} be partitioned into:

$$A = \begin{pmatrix} P & Q \\ Q^T & S \end{pmatrix} \ , \qquad A^{-1} = \begin{pmatrix} \tilde{P} & \tilde{Q} \\ \tilde{Q}^T & \tilde{S} \end{pmatrix} \ , \tag{A-42}$$

where P and $\tilde{P}$ are $n_1 \times n_1$ matrices and S and $\tilde{S}$ are $n_2 \times n_2$ matrices with $n = n_1 + n_2$. The submatrices in A^{-1} are given in Press et al. (1992, p. 77):

$$\begin{aligned} \tilde{P} &= P^{-1} + P^{-1}QM^{-1}Q^T P^{-1}, \\ \tilde{Q} &= -P^{-1}QM^{-1}, \qquad\qquad \text{where } M = S - Q^T P^{-1}Q \\ \tilde{S} &= M^{-1} \ . \end{aligned} \tag{A-43}$$

There are also equivalent formulae

$$\begin{aligned} \tilde{P} &= N^{-1}, \\ \tilde{Q} &= -N^{-1}QS^{-1}, \qquad\qquad \text{where } N = P - QS^{-1}Q^T \\ \tilde{S} &= S^{-1} + S^{-1}Q^T N^{-1}QS^{-1} \ . \end{aligned} \tag{A-44}$$

A.2 Product of Gaussians

When using linear models with Gaussian priors, the likelihood and the prior are both Gaussian. Their product is proportional to the posterior (also Gaussian), and their integral is equal to the marginal likelihood (or evidence). Consider the random vector $\boldsymbol{x}$ of size $n \times 1$ and the following product:

$$\mathcal{N}(\boldsymbol{x}|\mathbf{a}, A)\,\mathcal{N}(P\,\boldsymbol{x}|\mathbf{b}, B) = z_c\,\mathcal{N}(\boldsymbol{x}|\mathbf{c}, C) \; , \tag{A-45}$$

where $\mathcal{N}(\boldsymbol{x}|\mathbf{a}, A)$ denotes the probability of $\boldsymbol{x}$ under a Gaussian distribution centered on $\mathbf{a}$ (of size $n \times 1$) and with covariance matrix A (of size $n \times n$). P is a matrix of size $n \times m$ and vectors $\mathbf{b}$ and $\mathbf{c}$ are of size $m \times 1$, and matrices B and C of size $m \times m$. The product of two Gaussians is proportional to a new Gaussian with covariance and mean given by:

$$C = \left(A^{-1} + P\,B^{-1}P^{\top}\right)^{-1} \; , \qquad c = C\,\left(A^{-1}\mathbf{a} + P\,B\,\mathbf{b}\right) \; .$$

The normalizing constant z_c is gaussian in the means $\mathbf{a}$ and $\mathbf{b}$ of the two Gaussians that form the product on the right side of (A-45):

$$z_c = (2\,\pi)^{-\frac{m}{2}}|B + P^{\top}A^{-1}P|$$
$$\times \exp\left(-\frac{1}{2}(\mathbf{b} - P\,\mathbf{a})^{\top}\left(B + P^{\top}A^{-1}P\right)^{-1}(\mathbf{b} - P\,\mathbf{a})\right) \; .$$

A.3 Incremental Cholesky Factorization for SGQM

Consider the quadratic form:

$$Q(\boldsymbol{\alpha}) = -\mathbf{v}^{\top}\boldsymbol{\alpha} + \frac{1}{2}\boldsymbol{\alpha}^{\top}A\,\boldsymbol{\alpha} \; , \tag{A-46}$$

where A is a symmetric positive definite matrix of size $n \times n$ and $\mathbf{v}$ is a vector of size $n \times 1$. Suppose we have already obtained the minimum and the minimizer of $Q(\boldsymbol{\alpha})$, given by:

$$Q_{opt} = -\frac{1}{2}\mathbf{v}^{\top}A^{-1}\mathbf{v} \; , \qquad \boldsymbol{\alpha}_{opt} = A^{-1}\mathbf{v} \; . \tag{A-47}$$

We now want to minimize an augmented quadratic form $Q_i(\boldsymbol{\alpha})$, where $\boldsymbol{\alpha}$ is now of size $n \times 1$ and A and $\mathbf{v}$ are replaced by A_i and $\mathbf{v}_i$ of size $n+1 \times n+1$ and $n + 1 \times 1$ respectively, given by:

$$A_i = \begin{bmatrix} A & \mathbf{b}_i \\ \mathbf{b}_i^{\top} & c_i \end{bmatrix} \; , \qquad \mathbf{v}_i = \begin{bmatrix} \mathbf{v} \\ v_i \end{bmatrix} \; .$$

Assume that vector $\mathbf{b}_i$ of size $n \times 1$ and scalars c_i and v_i are somehow obtained. We want to exploit the incremental nature of A_i and $\mathbf{v}_i$ to reduce the number of operations necessary to minimize $Q_i(\boldsymbol{\alpha})$. One option would be to compute A_i^{-1} using inversion by partitioning, with cost $\mathcal{O}\left((n+1)^2\right)$ if A^{-1} is known.

For iterated incremental computations, using the Cholesky decomposition of A_i is numerically more stable. Knowing L, the Cholesky decomposition of A, the Cholesky decomposition L_i of A_i can be computed as:

$$L_i = \begin{bmatrix} L & 0 \\ \mathbf{z}_i^\top & d_i \end{bmatrix} , \qquad L\,\mathbf{z}_i = \mathbf{b}_i, \qquad d_i^2 = c_i - \mathbf{z}_i^\top \mathbf{z}_i . \tag{A-48}$$

The computational cost is $\mathcal{O}(n^2/2)$, corresponding to the computation of $\mathbf{z}_i$ by back-substitution. Q_i^{min} can be computed as:

$$Q_i^{min} = Q^{min} - \frac{1}{2}\,u_i^2 , \qquad u_i = \frac{1}{d_i}\,(v_i - \mathbf{z}_i^\top \mathbf{u}) , \qquad L\,\mathbf{u} = \mathbf{v} , \tag{A-49}$$

and the minimizer $\boldsymbol{\alpha}^{opt}$ is given by:

$$L^\top \boldsymbol{\alpha}^{opt} = \mathbf{u}_i , \qquad \mathbf{u}_i = \begin{bmatrix} \mathbf{u} \\ u_i \end{bmatrix} . \tag{A-50}$$

Notice that knowing $\mathbf{u}$ from the previous iteration, computing Q_i^{min} has a cost of $\mathcal{O}(n)$. This is interesting if many different i's need to be explored, for which only the minimum of Q_i is of interest, and not the minimizer. Once the optimal i has been found, computing the minimizer $\boldsymbol{\alpha}^{opt}$ requires a back-substitution, with a cost of $\mathcal{O}(n^2/2)$.

It is interesting to notice that as a result of computing L_i one obtains "for free" the determinant of A_i (an additional cost of $\mathcal{O}(m)$ to th e$\mathcal{O}(nm)$ cost of the incremental Cholesky). In Sect. A.4 we give a general expression of incremental determinants.

A.4 Incremental Determinant

Consider a square matrix A_i that has a row and a column more than square matrix A of size $n \times n$:

$$A_i = \begin{bmatrix} A & \mathbf{b}_i \\ \mathbf{c}_i^\top & d_i \end{bmatrix} . \tag{A-51}$$

The determinant of A_i is given by

$$|A_i| = |A| \cdot (d_i - \mathbf{b}_i^\top A^{-1} \mathbf{c}_i) . \tag{A-52}$$

In the interesting situation where A^{-1} is known, the new determinant is computed at a cost of $\mathcal{O}(m^2)$.

A.5 Derivation of (29)

We give here details of the needed algebra for computing the predictive distribution of the Reduced Rank Gaussian Process. Recall that at training time we use a finite linear model approximation, with less weights than training inputs.

Each weight has an associated support input possibly selected from the training inputs. The linear model and prior on the weights are:

$$\begin{bmatrix} \mathbf{f} \\ f_* \end{bmatrix} = \varPhi_{nm} \cdot \begin{bmatrix} \boldsymbol{\alpha} \\ \alpha_* \end{bmatrix} \;,\qquad p\left(\begin{bmatrix} \boldsymbol{\alpha} \\ \alpha_* \end{bmatrix}\middle|\, \boldsymbol{x}_*, X, \theta\right) \sim \mathcal{N}\left(0, A^{-1}\right)\;.$$

where we have defined

$$\varPhi_{nm} = \begin{bmatrix} K_{nm} & \mathbf{k}_* \\ \mathbf{k}(\boldsymbol{x}_*)^\top & k_{**} \end{bmatrix}\;,\qquad A = \begin{bmatrix} K_{mm} & \mathbf{k}(\boldsymbol{x}_*) \\ \mathbf{k}(\boldsymbol{x}_*)^\top & k_{**} \end{bmatrix}\;. \tag{A-53}$$

The induced prior over functions is Gaussian with mean zero and covariance matrix C:

$$p\left(\begin{bmatrix} \mathbf{f} \\ f_* \end{bmatrix}\middle|\, \boldsymbol{x}_*, X, \theta\right) \sim \mathcal{N}\left(0, C\right)\;,\qquad C = \varPhi_{nm}\, A^{-1}\, \varPhi_{nm}^\top\;. \tag{A-54}$$

We use inversion by partitioning to compute A^{-1}:

$$A^{-1} = \begin{bmatrix} K_{mm}^{-1} + K_{mm}^{-1}\mathbf{k}(\boldsymbol{x}_*)\,\mathbf{k}(\boldsymbol{x}_*)^\top K_{mm}^{-1} & -K_{mm}^{-1}\mathbf{k}(\boldsymbol{x}_*)/c_* \\ -\mathbf{k}(\boldsymbol{x}_*)^\top K_{mm}^{-1}/c_* & 1/c_* \end{bmatrix}\;,$$

$$c_* = k_{**} - \mathbf{k}(\boldsymbol{x}_*)^\top K_{mm}^{-1}\mathbf{k}(\boldsymbol{x}_*)\;,$$

which allows to obtain C:

$$C = \begin{bmatrix} C_{nn} & \mathbf{k}_* \\ \mathbf{k}_*^\top & k_{**} \end{bmatrix}\;,\qquad C_{nn} \equiv K_{nm}\,K_{mm}^{-1}\,K_{nm}^\top + \mathbf{v}_*\mathbf{v}_*^\top/c_*\;, \tag{A-55}$$

where $\mathbf{v}_* \equiv \mathbf{k}_* - K_{nm}\,K_{mm}^{-1}\,\mathbf{k}(\boldsymbol{x}_*)$. We can now compute the distribution of f_* conditioned $\mathbf{f}$:

$$p(f_*|\mathbf{f}, \boldsymbol{x}_*, X, \theta) \sim \mathcal{N}\left(\mathbf{k}_*^\top C_{nn}^{-1}\mathbf{f}, k_{**} - \mathbf{k}_*^\top C_{nn}^{-1}\mathbf{k}_*\right)\;. \tag{A-56}$$

The predictive distribution, obtained as in (5), is Gaussian with mean and variance given by (29). We repeat their expressions here for convenience:

$$m_*(\boldsymbol{x}_*) = \mathbf{k}_*^\top \left[K_{nm}\,K_{mm}^{-1}\,K_{nm}^\top + \sigma^2\,I + \mathbf{v}_*\mathbf{v}_*^\top/c_*\right]^{-1} \mathbf{y}\;,$$

$$v_*(\boldsymbol{x}_*) = \sigma^2 + k_{**} + \mathbf{k}_*^\top \left[K_{nm}\,K_{mm}^{-1}\,K_{nm}^\top + \sigma^2\,I + \mathbf{v}_*\mathbf{v}_*^\top/c_*\right]^{-1} \mathbf{k}_*\;.$$

B   Matlab Code for the RRGP

We believe that one very exciting part of looking at a new algorithm is "trying it out"! We would like the interested reader to be able to train our Reduced Rank Gaussian Process (RRGP) algorithm. Training consists in finding the value of the hyperparameters that minimizes the negative log evidence of the RRGP (we give it in Sect. 4.1). To do this we first need to be able to compute the negative log evidence and its derivatives with respect to the hyperparameters. Then we can plug this to a gradient descent algorithm to perform the actual learning.

We give a Matlab function, rrgp_nle, that computes the negative log evidence of the RRGP and its derivatives for the squared exponential covariance function (given in (1)). The hyperparameters of the squared exponential covariance function are all positive. To be able to use unconstrained optimization, we optimize with respect to the logarithm of the hyperparameters.

An auxiliary Matlab function sq_dist is needed to compute squared distances. Given to input matrices of sizes $d \times n$ and $d \times m$, the function returns the $n \times m$ matrix of squared distances between all pairs of columns from the inputs matrices. The authors would be happy to provide their own Matlab MEX implementation of this function upon request.

Inputs to the Function rrgp_nle:

- X: $D + 2 \times 1$ vector of log hyperparameters, $X = [\log \theta_1, \ldots \log \theta_{D+1}, \log \sigma]^\top$, see (1)
- input: $n \times D$ matrix of training inputs
- target: $n \times 1$ matrix of training targets
- m: scalar, size of the support set

Outputs of the Function rrgp_nle:

- f: scalar, evaluation of the negative log evidence at X
- f: $D + 2 \times 1$ vector of derivatives of the negative log evidence evaluated at X

Matlab Code of the Function rrgp_nle:

```
function [f,df] = rrgp_nle(X,input,target,m)

% number of examples and dimension of input space
[n, D] = size(input);
input = input ./ repmat(exp(X(1:D))',n,1);

% write the noise-free covariance of size n x m
Knm = exp(2*X(D+1))*exp(-0.5*sq_dist(input',input(1:m,:)'));
% add little jitter to Kmm part
Knm(1:m,:) = Knm(1:m,:)+1e-8*eye(m);

Cnm = Knm/Knm(1:m,:);
Smm = Knm'*Cnm + exp(2*X(D+2))*eye(m);
Pnm = Cnm/Smm;
wm = Pnm'*target;

% compute function evaluation
invQt = (target-Pnm*(Knm'*target))/exp(2*X(D+2));
logdetQ = (n-m)*2*X(D+2) + sum(log(abs(diag(lu(Smm)))));
f = 0.5*logdetQ + 0.5*target'*invQt + 0.5*n*log(2*pi);
```

```
% compute derivatives
df = zeros(D+2,1);

for d=1:D
  Vnm = -sq_dist(input(:,d)',input(1:m,d)').*Knm;
  df(d) = (invQt'*Vnm)*wm - 0.5*wm'*Vnm(1:m,:)*wm+...
      -sum(sum(Vnm.*Pnm))+0.5*sum(sum((Cnm*Vnm(1:m,:)).*Pnm));
end
aux = sum(sum(Pnm.*Knm));
df(D+1) = -(invQt'*Knm)*wm+aux;
df(D+2) = (n-aux) - exp(2*X(D+2))*invQt'*invQt;
```

References

Cressie, N. A. C. (1993). *Statistics for Spatial Data*. John Wiley and Sons, Hoboken, New Jersey.

Csató, L. (2002). *Gaussian Processes – Iterative Sparse Approximation*. PhD thesis, Aston University, Birmingham, United Kingdom.

Csató, L. and Opper, M. (2002). Sparse online gaussian processes. *Neural Computation*, 14(3):641–669.

Gibbs, M. and MacKay, D. J. C. (1997). Efficient implementation of gaussian processes. Technical report, Cavendish Laboratory, Cambridge University, Cambridge, United Kingdom.

Lawrence, N., Seeger, M., and Herbrich, R. (2003). Fast sparse gaussian process methods: The informative vector machine. In Becker, S., Thrun, S., and Obermayer, K., editors, *Neural Information Processing Systems 15*, pages 609–616, Cambridge, Massachussetts. MIT Press.

MacKay, D. J. C. (1994). Bayesian non-linear modelling for the energy prediction competition. *ASHRAE Transactions*, 100(2):1053–1062.

Mackay, D. J. C. (1997). Gaussian Processes: A replacement for supervised Neural Networks? Technical report, Cavendish Laboratory, Cambridge University, Cambridge, United Kingdom. Lecture notes for a tutorial at NIPS 1997.

Neal, R. M. (1996). *Bayesian Learning for Neural Networks*, volume 118 of *Lecture Notes in Statistics*. Springer, Heidelberg, Germany.

Press, W., Flannery, B., Teukolsky, S. A., and Vetterling, W. T. (1992). *Numerical Recipes in C*. Cambridge University Press, Cambridge, United Kingdom, second edition.

Rasmussen, C. E. (1996). *Evaluation of Gaussian Processes and Other Methods for Non-linear Regression*. PhD thesis, Department of Computer Science, University of Toronto, Toronto, Ontario.

Rasmussen, C. E. (2002). Reduced rank gaussian process learning. Unpublished Manuscript.

Schölkopf, B. and Smola, A. J. (2002). *Learning with Kernels*. MIT Press, Cambridge, Massachussetts.

Schwaighofer, A. and Tresp, V. (2003). Transductive and inductive methods for approximate gaussian process regression. In Becker, S., Thrun, S., and Obermayer, K., editors, *Advances in Neural Information Processing Systems 15*, pages 953–960, Cambridge, Massachussetts. MIT Press.

Seeger, M. (2003). *Bayesian Gaussian Process Models: PAC-Bayesian Generalisation Error Bounds and Sparse Approximations*. PhD thesis, University of Edinburgh, Edinburgh, Scotland.

Seeger, M., Williams, C., and Lawrence, N. (2003). Fast forward selection to speed up sparse gaussian process regression. In Bishop, C. M. and Frey, B. J., editors, *Ninth International Workshop on Artificial Intelligence and Statistics*. Society for Artificial Intelligence and Statistics.

Smola, A. J. and Bartlett, P. L. (2001). Sparse greedy Gaussian process regression. In Leen, T. K., Dietterich, T. G., and Tresp, V., editors, *Advances in Neural Information Processing Systems 13*, pages 619–625, Cambridge, Massachussetts. MIT Press.

Smola, A. J. and Schölkopf, B. (2000). Sparse greedy matrix approximation for machine learning. In Langley, P., editor, *International Conference on Machine Learning 17*, pages 911–918, San Francisco, California. Morgan Kaufmann Publishers.

Tipping, M. E. (2001). Sparse bayesian learning and the relevance vector machine. *Journal of Machine Learning Research*, 1:211–244.

Tresp, V. (2000). A bayesian committee machine. *Neural Computation*, 12(11):2719–2741.

Wahba, G., Lin, X., Gao, F., Xiang, D., Klein, R., and Klein, B. (1999). The bias-variance tradeoff and the randomized GACV. In Kerns, M. S., Solla, S. A., and Cohn, D. A., editors, *Advances in Neural Information Processing Systems 11*, pages 620–626, Cambridge, Massachussetts. MIT Press.

Williams, C. (1997a). Computation with infinite neural networks. In Mozer, M. C., Jordan, M. I., and Petsche, T., editors, *Advances in Neural Information Processing Systems 9*, pages 295–301, Cambridge, Massachussetts. MIT Press.

Williams, C. (1997b). Prediction with gaussian processes: From linear regression to linear prediction and beyond. Technical Report NCRG/97/012, Dept of Computer Science and Applied Mathematics, Aston University, Birmingham, United Kingdom.

Williams, C., Rasmussen, C. E., Schwaighofer, A., and Tresp, V. (2002). Observations of the nyström method for gaussiam process prediction. Technical report, University of Edinburgh, Edinburgh, Scotland.

Williams, C. and Seeger, M. (2001). Using the Nyström method to speed up kernel machines. In Leen, T. K., Dietterich, T. G., and Tresp, V., editors, *Advances in Neural Information Processing Systems 13*, pages 682–688, Cambridge, Massachussetts. MIT Press.

Acknowledgements

The authors would like to thank Lehel Csató, Alex Zien and Olivier Chapelle for useful discussions.

This work was supported by the Multi-Agent Control Research Training Network - EC TMR grant HPRN-CT-1999-00107, and the German Research Council (DFG) through grant RA 1030/1.

Filtered Gaussian Processes for Learning with Large Data-Sets

Jian Qing Shi[1], Roderick Murray-Smith[2,3], D. Mike Titterington[4], and
Barak A. Pearlmutter[3]

[1] School of Mathematics and Statistics, University of Newcastle, UK
j.q.shi@ncl.ac.uk
[2] Department of Computing Science, University of Glasgow, Scotland
rod@dcs.gla.ac.uk
[3] Hamilton Institute, NUI Maynooth, Co. Kildare, Ireland
barak@cs.may.ie
[4] Department of Statistics, University of Glasgow, Scotland
mike@stats.gla.ac.uk

Abstract. Kernel-based non-parametric models have been applied widely
over recent years. However, the associated computational complexity imposes limitations on the applicability of those methods to problems with
large data-sets. In this paper we develop a filtering approach based on
a Gaussian process regression model. The idea is to generate a small-
dimensional set of filtered data that keeps a high proportion of the information contained in the original large data-set. Model learning and
prediction are based on the filtered data, thereby decreasing the computational burden dramatically.

Keywords: Filtering transformation, Gaussian process regression model,
Karhunen-Loeve expansion, Kernel-based non-parametric models, Principal component analysis.

1 Introduction

Kernel-based non-parametric models such as Splines (Wahba, 1990), Support
Vector Machines (Vapnik, 1995) and Gaussian process regression models (see
for example O'Hagan (1978), and Williams and Rasmussen, (1996)) have become very popular in recent years. A major limiting factor with such methods
is the computational effort associated with dealing with large training data-sets,
as the complexity grows at rate $O(N^3)$, where N is the number of observations
in the training set. A number of methods have been developed to overcome this
problem. So far as the Gaussian process (GP) regression model is concerned,
such methods include the use of mixtures of GPs (Shi, Murray-Smith and Titterington, 2002) for a large data-set with repeated measurements, and the use
of approximation methods such as the Subset of Regressors method (Poggio and
Girosi, 1990; Luo and Wahba, 1997), the iterative Lanczos method (Gibbs and
Mackay, 1997), the Bayesian Committee Machine (Tresp, 2000), the Nyström
Method (Williams and Seeger, 2001) and Selection Mechanisms (Seeger *et al.*,
2003).

R. Murray-Smith, R. Shorten (Eds.): Switching and Learning, LNCS 3355, pp. 128–139, 2005.
© Springer-Verlag Berlin Heidelberg 2005

Gaussian process prior systems generally consist of noisy measurements of samples of the putatively Gaussian process of interest, where the samples serve to constrain the posterior estimate. In Murray-Smith and Pearlmutter (2003), the case was considered where the measurements are instead noisy weighted sums of samples. Adapting the idea of the transformation of GPs described in Murray-Smith and Pearlmutter (2003), we describe here a specific filtering approach to deal with the modelling of large data-sets. The approach involves two stages. In the first stage, a set of filtered data of dimension n is generated, where usually $n \ll N$, the dimension of the original training data-set. The value of n can be selected such that the filtered data can represent a proportion of the information of the original whole data-set. This therefore amounts to a question of experiment design, involving specification of how to design physical filters to generate filtered data. In the second stage, we carry out model learning and prediction based on the filtered data. The approach is also extended to online learning where data arrive sequentially and training must be performed sequentially as well.

The paper is organized as follows. Section 2 discusses the details of the filtering approach. We first discuss an orthogonal expansion of a kernel covariance function of a GP model based on its eigenfunctions and eigenvalues in Section 2.1. Using the results, we develop a filtering approach, the details of which are given in Section 2.2. Section 2.3 discusses statistical inference based on the filtered data, including model learning and prediction. Section 2.4 extends the approach to online learning. A simulation study is given in Section 3 to illustrate the performance of the method, and some discussion is given in Section 4.

2   Filtering Approach for Large Data-Sets

2.1   Expansion of a Gaussian Process and Its Transformations

Consider a Gaussian process $y(\boldsymbol{x})$, which has a normal distribution with zero mean and kernel covariance function $k(\boldsymbol{x}, \boldsymbol{u})$, where $\boldsymbol{x}$ is a vector of input variables. The related observation is $t(\boldsymbol{x}) = y(\boldsymbol{x}) + \epsilon(\boldsymbol{x})$, where $\epsilon(\boldsymbol{x}) \sim N(0, \sigma^2)$ and $\epsilon(\boldsymbol{x})$'s for different $\boldsymbol{x}$'s are assumed independent. The Gaussian process $y(\boldsymbol{x})$ can be decomposed, according to the Karhunen-Loève orthogonal expansion, as

$$y(\boldsymbol{x}) = \sum_{i=1}^{\infty} \phi_i(\boldsymbol{x})\xi_i, \tag{1}$$

and the covariance kernel function $k(\boldsymbol{x}, \boldsymbol{u})$ can be expanded as

$$k(\boldsymbol{x}, \boldsymbol{u}) = \sum_{i=1}^{\infty} \lambda_i \phi_i(\boldsymbol{x})\phi_i(\boldsymbol{u}), \tag{2}$$

where $\lambda_1 \geq \lambda_2 \geq \cdots \geq 0$ denote the eigenvalues and $\phi_1, \phi_2, \cdots$ are the related eigenfunctions of the operator whose kernel is $k(\boldsymbol{x}, \boldsymbol{u})$, so that

$$\int k(\boldsymbol{u}, \boldsymbol{x})\phi_i(\boldsymbol{x})p(\boldsymbol{x})d\boldsymbol{x} = \lambda_i \phi_i(\boldsymbol{u}), \tag{3}$$

where $p(\boldsymbol{x})$ is the density function of the input vector $\boldsymbol{x}$. The eigenfunctions are p-orthogonal, i.e.

$$\int \phi_i(\boldsymbol{x})\phi_j(\boldsymbol{x})p(\boldsymbol{x})d\boldsymbol{x} = \delta_{ij}.$$

In (1) ξ_i is given by

$$\xi_i = \int \phi_i(\boldsymbol{x})y(\boldsymbol{x})p(\boldsymbol{x})d\boldsymbol{x}. \tag{4}$$

Given a random sample $\{\boldsymbol{x}_i, i = 1, \cdots, N\}$ of inputs, independent and identically distributed according to $p(\boldsymbol{x})$, we have the discrete form of $y(\boldsymbol{x})$; that is, $\boldsymbol{Y}' = (y(\boldsymbol{x}_1), \cdots, y(\boldsymbol{x}_N))$. From (2), the covariance kernel $k(\boldsymbol{x}, \boldsymbol{u}; \boldsymbol{\theta})$ can be expanded into a feature space of dimension N as

$$k(\boldsymbol{x}, \boldsymbol{u}; \boldsymbol{\theta}) = \sum_{i=1}^{N} \lambda_i \phi_i(\boldsymbol{x})\phi_i(\boldsymbol{u}), \tag{5}$$

where $\boldsymbol{\theta}$ is a vector of unknown parameters of interest. Typically N is very large, so that the above expansion is a good approximation to (2). The discrete form of (4) is

$$\xi_i \approx \frac{1}{N} \sum_{j=1}^{N} \phi_i(\boldsymbol{x}_j)y(\boldsymbol{x}_j). \tag{6}$$

Let $\boldsymbol{\Sigma}^{(N)}$ be the covariance matrix of $\boldsymbol{Y}$, $\lambda_i^{(N)}$ be an eigenvalue of $\boldsymbol{\Sigma}^{(N)}$ and $\phi_i^{(N)}$ be the related N-dimensional eigenvector, where $\lambda_1^{(N)} \geq \lambda_2^{(N)} \geq \cdots \geq 0$. Then $(\phi_i(\boldsymbol{x}_1), \cdots, \phi_i(\boldsymbol{x}_N)) \approx \sqrt{N}\phi_i^{(N)}$ and $\lambda_i \approx \lambda_i^{(N)}/N$ for $i = 1, \cdots, N$; for details see Williams and Seeger (2001).

We will now assume that instead of observing the $\boldsymbol{Y}$'s directly, we observe a transformation z of the latent vector $\boldsymbol{Y}$, given by

$$z_k = \sum_{j=1}^{N} \boldsymbol{K}_{kj} y(\boldsymbol{x}_j) = \boldsymbol{K}_k \boldsymbol{Y} \tag{7}$$

for $k = 1, \cdots, n$. In other words, for the vector of latent variables $\boldsymbol{Y}$ we observe outputs $\boldsymbol{Z} = \boldsymbol{K}\boldsymbol{Y}$, where $\boldsymbol{K}$ is an $n \times N$ known matrix and $\boldsymbol{Z}^T = (z_1, \cdots, z_n)$. The above transformations define n data filters, and usually $n \ll N$. Each of n physical filters can be designed by the values of each row of $\boldsymbol{K}$.

A special case corresponds to constructing $\boldsymbol{K}$ from the first n eigenvectors of $\boldsymbol{\Sigma}^{(N)}$. When the kth row of $\boldsymbol{K}$ consists of the eigenvector $\phi_k^{(N)}$, z_k is calculated by (7). Comparing (7) with (6), we have that $\xi_k \approx z_k/\sqrt{N}$. The n filtered observations z correspond to the n largest eigenvalues. Therefore, if we use the n-dimensional filtered data, we approximate the covariance kernel in (5) by

$$k(\boldsymbol{x}, \boldsymbol{u}) \approx \sum_{i=1}^{n} \lambda_i \phi_i(\boldsymbol{x})\phi_i(\boldsymbol{u}). \tag{8}$$

Then the subspace spanned by the n-dimensional transformed data contains the 'best' n-dimensional view of the original N-dimensional data. If the remaining eigenvalues are very small in comparison, (8) should be a good approximation to (5). This idea is used to develop a filtering approach, the details of which are given in the next subsection.

2.2   Filtering Approach

If we have N observations, the related $N \times N$ covariance matrix is calculated by the covariance kernel function $\boldsymbol{\Sigma}^{(N)} = (k(\boldsymbol{x}_i, \boldsymbol{x}_j; \boldsymbol{\theta}))$. Following the discussion in the above section, $\boldsymbol{K}$ is constructed from the n eigenvectors of $\boldsymbol{\Sigma}^{(N)}(\boldsymbol{\theta})$ which are associated with the first n largest eigenvalues. Since $\boldsymbol{\theta}$ is unknown, we need to use an estimate $\hat{\boldsymbol{\theta}}$ based on those N observations. A standard method (see for example Williams and Rasmussen, 1996, and Shi, Murray-Smith and Titterington, 2002) can be used to calculate $\hat{\boldsymbol{\theta}}$. Then $\boldsymbol{\Sigma}^{(N)}$ is approximated by $\boldsymbol{\Sigma}^{(N)}(\hat{\boldsymbol{\theta}})$. The related eigenvalues and eigenvectors are calculated from $\boldsymbol{\Sigma}^{(N)}$ and are used to construct filtered data. Since the complexity of obtaining the estimate $\hat{\boldsymbol{\theta}}$ and calculating eigenvalues is $O(N^3)$, it is very time consuming for large N. Fortunately, the Nyström method (Williams and Seeger, 2001) can be used to calculate the n largest eigenvalues and the associated eigenvectors approximately. It is a very efficient approach, especially when $n \ll N$.

The procedure for generating a filtered data-set is as follows.

Step 1. Select a subset of training data of size m at random from the N observations. This m may be much less than N. We use a standard method to calculate an estimate $\hat{\boldsymbol{\theta}}^{(m)}$ using those m observations. The covariance matrix of those m observations is estimated by the covariance kernel function $\boldsymbol{\Sigma}^{(m)} = \left(k(\boldsymbol{x}_i, \boldsymbol{x}_j; \hat{\boldsymbol{\theta}}^{(m)}) \right)$, which is an $m \times m$ matrix. We calculate its eigenvalues $\lambda_1 \geq \cdots \geq \lambda_m \geq 0$ and the related eigenvectors $\boldsymbol{v}_1, \cdots, \boldsymbol{v}_m$.

Step 2. By the Nyström method, the first m largest eigenvalues of $\boldsymbol{\Sigma}^{(N)}$ can be approximated by

$$\frac{N}{m}\lambda_i,$$

for $i = 1, \cdots, m$, and their associated eigenvectors are

$$\sqrt{\frac{m}{N}} \frac{1}{\lambda_i} \boldsymbol{\Sigma}_{N,m} \boldsymbol{v}_i, \tag{9}$$

where $\boldsymbol{\Sigma}_{N,m}$ is the appropriate $N \times m$ submatrix of $\boldsymbol{\Sigma}^{(N)}$.

Step 3. We select the first n ($\leq m$) eigenvectors in order to construct the transformation matrix $\boldsymbol{K}$ in (7), and thereby generate an n-dimensional filtered data-set.

In the above procedure, we need to select m and n. We first discuss how to select n. The basic idea of the filtering approach is to use (8) to approximate (5). In the extreme case where $\lambda_i = 0$ for all $i > n$, the filtered data are equivalent to the original data, in terms of the covariance kernel. This typically does not happen in practice. However, if the values of λ_i for all $i > n$ are very small compared to the first n eigenvalues, (8) is a good approximation of (5). Though it is difficult to compare (8) and (5) directly, we can compare the values of eigenvalues and choose n such that the remaining eigenvalues are very small in comparison to the largest eigenvalue. Alternatively, we might select n such that

$$\frac{\sum_{i=1}^{n} \lambda_i}{\sum_{i=1}^{m} \lambda_i} \geq c,$$

where c is a constant, such as $c = 0.99$. More discussion will be given in Section 3 in the context of the simulation study.

The other problem is how to select m. In Step 1, we select m observations and use them to learn the eigen-structure of the covariance kernel $k(\boldsymbol{x}, \boldsymbol{u})$. It is obvious that a larger value of m should lead to a more accurate approximation of eigenvalues and eigenvectors. However, we usually just need to learn the eigen-structure once. It can then be used repeatedly in similar systems. It will not increase the computational burden very much if we select a relatively large value of m. On the other hand, since the eigenvectors are used to generate a 'best' n-dimensional view of the original data, the accuracy of the 'design' in the first stage will not have much influence on carried out in the second stage. Some numerical results will be presented in the next section.

2.3   Model Learning and Prediction Using Filtered Data

The procedure proposed in the last subsection is used to generate a filtered data-set. Here we discuss how to carry out inference based on the filtered data.

The filtered data are defined via a linear transformation $\boldsymbol{Z} = \boldsymbol{KY}$, which can be used to design a set of filters. The observed filtered data may be obtained through those filters, so for generality we can consider observed errors. The observed filtered data are assumed to be

$$s_k = z_k + e_i, \quad \text{for } i = 1, \cdots, n,$$

where $\boldsymbol{Z} = (z_k) = \boldsymbol{KY}$ and the e_i are independent and identically distributed as $N(0, \sigma_s^2)$ which is the random error when the filtered data are observed. In matrix form, the filtered data $\boldsymbol{S} = (s_1, \cdots, s_n)^T$ are distributed as

$$\boldsymbol{S} \sim N(0, \ \boldsymbol{\Sigma}_s),$$

where $\boldsymbol{\Sigma}_s = \boldsymbol{K\Sigma K}^T + \sigma_s^2 \boldsymbol{I}_n$, and $\boldsymbol{K}$ is a known matrix which is designed in the first stage. We still use $\boldsymbol{\theta}$ to denote the unknown parameters involved in $\boldsymbol{\Sigma}_s$, which includes σ_s^2 and the unknown parameters in kernel covariance function. Then the log-likelihood of $\boldsymbol{\theta}$ is

$$L(\boldsymbol{\theta}) = -\frac{1}{2} \log |\boldsymbol{\Sigma}_s| - \frac{1}{2} \boldsymbol{S}^T \boldsymbol{\Sigma}_s^{-1} \boldsymbol{S} - \frac{n}{2} \log(2\pi).$$

Maximizing $L(\boldsymbol{\theta})$ leads to a maximum likelihood estimate of $\boldsymbol{\theta}$.

Suppose that we wish to predict $z^* = K^{*T} Y^*$, where $Y^* = Y(X^*)$ is q-dimensional, and X^* represents q test data points of input. K^* is a known q-dimensional vector, which can be thought of as a filter. Given the filtered data S, the conditional mean and variance of z^* are

$$\hat{\mu}^* = K^{*T} \Sigma_{X^* X} K^T \Sigma_s^{-1} S$$
$$\hat{\sigma}^{*2} = K^{*T} \Sigma_{X^* X^*} K^* - K^{*T} \Sigma_{X^* X} K^T \Sigma_s^{-1} K \Sigma_{XX^*} K^*,$$

where $\Sigma_{X^* X} = \left(k(x_i^*, x_j; \theta) \right)$ is the $q \times N$ covariance matrix between X^* and X evaluated at $\hat{\theta}$ which is an estimate using S, and so are the other similar notations.

If we want to make a single prediction at a new $y^* = y(x^*)$, we just need to take $q = 1$ and $K^* = 1$. Bayesian inference can also be used. The implementation is similar to the methods discussed in Rasmussen (1996) and Shi, Murray-Smith and Titterington (2002).

2.4   Online Filtering Approach

We assume that data arrive sequentially. Let $D_a = (Y_a, X_a)$ denote the data collected between time $t(a)$ and $t(a-1)$. We can apply the filtering approach online and adapt the predictive distribution for test data point. For each subset of data D_a, we have a set of filtered data sets,

$$S_a = Z_a + e_a, \quad Z_a = K_a Y_a$$

for $a = 1, 2, \cdots, A$, where A is the number of data sets up to time $t(A)$. The transformation matrix K_a can be constructed by Step 2 of the filtering approach discussed in Subsection 2.2. We assume that the eigenstructure of the covariance kernel for the new data is similar to the previous data, so we just need to learn the eigenstructure once. It can be used repeatedly for new data sets, so the computation to generate a new filtered data-set is therefore very efficient. An estimate of θ based on filtered data S_a is obtained by the method discussed in the last subsection, and is denoted by $\hat{\theta}_a$. If we are interested in prediction at a new $y^* = y(x^*)$, the predictive mean and variance based on the filtered data are

$$\hat{\mu}_a^* = \Sigma_a^* K_a^T \Sigma_{s,a}^{-1} S_a \qquad (10)$$
$$\hat{\sigma}_a^{*2} = k(x^*, x^*; \hat{\theta}_a) - \Sigma_a^* K_a^T \Sigma_{s,a}^{-1} K_a \Sigma_a^{*T} \qquad (11)$$

where $\Sigma_a^* = \left(k(x^*, x_{j,a}; \hat{\theta}_a) \right)$, and all the other quantities are defined in the last subsection but evaluated at $\hat{\theta}_a$.

Therefore, we have

$$\hat{\mu}_a^* \sim N(\mu^*, \hat{\sigma}_a^{*2})$$

for $a = 1, \cdots, A$. Here, $\hat{\mu}_a^*$'s are correlated with each other with covariance

$$\hat{\sigma}_{ab}^* = \text{cov}(\hat{\mu}_a^*, \hat{\mu}_b^*)$$
$$= \Sigma_a^* K_a^T \Sigma_{s,a}^{-1} (K_a \Sigma_s^{ab} K_b^T) \Sigma_{s,b}^{-1} K_a \Sigma_b^{*T}, \qquad (12)$$

where $\boldsymbol{\Sigma}_s^{ab}$ are the covariance matrix between $\boldsymbol{Y}_a$ and $\boldsymbol{Y}_b$. The correlation is calculated by $\rho_{ab}^* = \hat{\sigma}_{ab}^*/\hat{\sigma}_a^*\hat{\sigma}_b^*$. The overall mean of prediction based on A data-sets can be calculated by

$$\mu_* = \mathbf{1}^T \Omega^{-1} \hat{\boldsymbol{\mu}}_* / (\mathbf{1}^T \Omega^{-1} \mathbf{1})$$

and the variance is

$$\sigma_*^2 = (\mathbf{1}^T \Omega^{-1} \mathbf{1})^{-1},$$

where Ω is the covariance matrix of $\hat{\boldsymbol{\mu}}^* = (\hat{\mu}_1^*, \cdots, \hat{\mu}_A^*)^T$, with the diagonal element $\hat{\sigma}_a^{*2}$ and off-diagonal element $\hat{\sigma}_{ab}^*$, and $\mathbf{1} = (1, \cdots, 1)^T$.

If the correlation ρ_{ab}^* is not very large, we may approximate the predictive mean by

$$\mu_* = \frac{\sum_a \hat{\mu}_a^*/\hat{\sigma}_a^{*2}}{\sum_a 1/\hat{\sigma}_a^{*2}},$$

and the variance is

$$\sigma_*^2 = \frac{\sum_a 1/\hat{\sigma}_a^{*2} + 2\sum_{a \neq b} \rho_{ab}^*/(\hat{\sigma}_a^*\hat{\sigma}_b^*)}{(\sum_a 1/\hat{\sigma}_a^{*2})^2}.$$

This approximate method can be replaced by an iterative method. Each time we have a new data-set, we calculate the predictive mean (10), variance (11) and the covariance (12). Then, we can define the following iterative method:

$$u^{(a)} = u^{(a-1)} + \hat{\mu}_a^*/\hat{\sigma}_a^{*2},$$
$$v^{(a)} = v^{(a-1)} + 1/\hat{\sigma}_a^{*2},$$
$$w^{(a)} = w^{(a-1)} + 2\sum_{j=1}^{a-1} \rho_{aj}^*/(\hat{\sigma}_a^*\hat{\sigma}_j^*),$$
$$\hat{\mu}^{*(a)} = u^{(a)}/v^{(a)},$$
$$\hat{\sigma}^{*(a)2} = (v^{(a)} + w^{(a)})/(v^{(a)})^2.$$

We can therefore maintain a much smaller set of training data, and can subsequently update the predictions online, as new data becomes available.

3   Applications

3.1   Learning with Large Data-Sets

As we have discussed in Section 1, a major limiting factor in the use of Gaussian process models is the heavy computational burden associated with large training data-sets, as the complexity grows at rate $O(N^3)$. Some methods have been proposed for overcoming this. Murray-Smith and Pearlmutter (2003) argued that the complexity is $O(n^3) + O(N^2n)$ for the model learning and prediction based on the filtered data, which corresponding to the second stage in this paper. In our first stage, the complexity associated with generating filtered data is $O(m^3)$, and therefore the overall complexity is $O(m^3) + O(n^3) + O(N^2n)$. Since $n \leq m$ and usually $m \ll N$, the complexity is generally dominated by $O(N^2n)$, and thus the filtering approach results in substantially decreased computational burden.

An example is used here to illustrate the filtering approach discussed in this paper. The original 500 training data (dots) and the $m = 50$ randomly selected data points (circles) are presented in Figure 1(a). The true model used to generate the data is $y_i = \sin((0.5x_i)^3) + \epsilon_i$, where the ϵ_i's are independent and identical distributed as $N(0, 0.01^2)$ and $x_i \in (-5, 5)$. The 50 selected data points are used to calculate the eigenvalues, and the related eigenfunctions and eigenvectors using the method described in Step 1 in Section 2.2. We take the values of c as 0.99, 0.999 and 0.9999, obtaining the values of n as 27, 33 and 39 respectively. The predictions and 95% confidence intervals for a test data set are presented in Figures 1(d) to 1(f).

There are several interesting findings from these figures. Figure 1(e) gives the best results in terms of the value of root of mean squared error between the true test values and the predictions. Though it involves just $n = 33$ filtered data, the results are better than the results in Figure 1(b), obtained from 50 randomly selected data points. Figure 1(f) gives the results obtained from $n = 39$ filtered data points. The performance is slightly worse than Figure 1(e), based on only 33 filtered data points, though the difference is very small. This in fact coincides with the theory we discussed in Section 2.1. The filtering approach always chooses the largest eigenvalues and the related transformed data. It will not add much information to add more filtered data associated with relatively small eigenvalues. Comparing Figures 1(e) and 1(f), six more filtered data points are added. The associated eigenvalues are range from $\lambda_{34} = 0.0056$ to $\lambda_{39} = 0.0009$ and, relative to the largest eigenvalue $\lambda_1 = 2.5365$, the valuess ranged from 0.0022 to 0.0003, which are extremely small. In contrast, the numerical error may increase because the covariance matrix deteriorates due to those small eigenvalues. Thus it is not surprising that the performance of 1(f) is slightly worse than Figure 1(e). It shows that only a certain small number of filtered data points are needed to provide a good representation of the whole data set of N observations.

Figure 1(d) gives the results based on only $n = 27$ filtered data points. If we just use a randomly selected subset of 27 data points, the results are presented in Figure 1(c). The former is obviously much better than the latter. The other problem of using subset of data points is the sensitivity of the method to the choice of the data points. Our simulation study shows that the performance may be improved if those 27 data points are advantageously distributed over the whole range. For this, the training set must be located in all parts of the range, and there must be enough training data in regions where the mean response changes rapidly, such as near the two ends of the range in this example. Obviously, it is not easy to guarantee this. The performance will be poor when the data points are concentrated in certain areas. However, the filtering approach is quite robust.

In our Step 1, an m-dimensional subset is selected for the calculation of the eigenvalues and eigenvectors. The accuracy depends on the value of m. For more

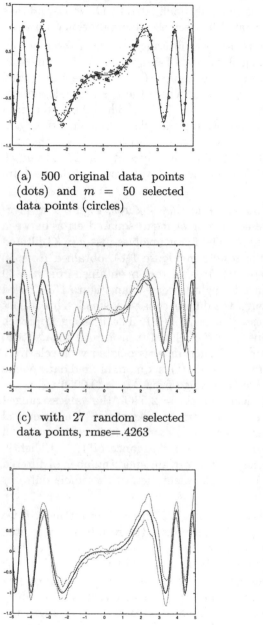
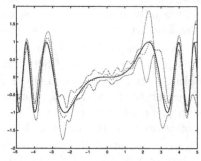
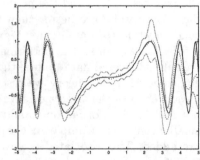
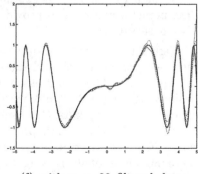

(a) 500 original data points (dots) and $m = 50$ selected data points (circles)

(b) with 50 random selected points; rmse=.1438

(c) with 27 random selected data points, rmse=.4263

(d) with $n = 27$ filtered data, rmse=.2063

(e) with $n = 33$ filtered data, rmse=.0945

(f) with $n = 39$ filtered data, rmse=.1019

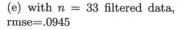

Fig. 1. Simulation study with $m = 50$: plot of true curve (solid line), prediction (dotted line) and 95% confidence intervals.

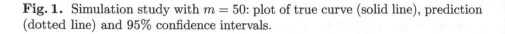

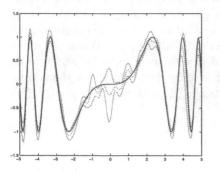

(a) with 100 random selected points; rmse=.0695

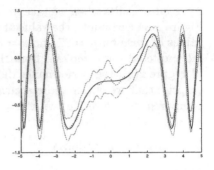

(b) with $n = 30$ filtered data, rmse=.1139

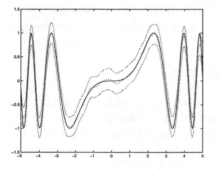

(c) with $n = 39$ filtered data rmse=.0611

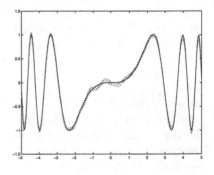

(d) with $n = 46$ filtered data, rmse=.0293

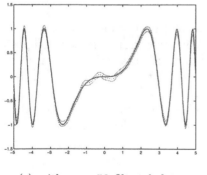

(e) with $n = 56$ filtered data, rmse=.0291

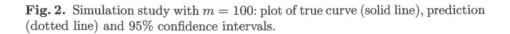

Fig. 2. Simulation study with $m = 100$: plot of true curve (solid line), prediction (dotted line) and 95% confidence intervals.

accurate results, we should obviously select a relatively larger value of m. Figure 2 presents results when we take $m = 100$. We get quite similar results to those in Figure 1. For example, the value of rmse for $n = 30$ with $m = 100$ is between the values of rmse for $n = 27$ and $n = 33$ with $m = 50$ in Figure 1. When we added more filtered data, moving from the case of $n = 46$ in Figure 2(d) to $n = 56$ in Figure 2(e), the performance did not improve further. Of course, there is no surprise that we obtain more accurate results in 2(d) with $n = 46$ and $m = 100$, compared to Figure 1(e) with $m = 50$.

3.2   Inverse Problems

Suppose we want to transfer an image, which typically corresponds to a very large data-set, across a communication channel. One method is to compress the image into a data-set of much smaller size. On receipt of the compressed data-set, the original image is estimated. We can use the method discussed in Murray-Smith and Pearlmutter (2003). If the filtered data are represented by Z, the transformation matrix used to construct the filtered data is K, and the the original data-set Y can be estimated by

$$Y = \Sigma K^T (K \Sigma K^T)^{-1} Z.$$

However, here we construct K in advance by selecting m data points from the original N data points using the method discussed in Section 2.2. There are two distinguishing features of this method. First, the filtered data provide approximately the 'best' n-dimensional view of the original N-dimensional data-set. Secondly, $K \Sigma K^T$ is approximately a diagonal matrix $\text{diag}(\lambda_1, \cdots, \lambda_n)$, so that numerically the inversion of $K \Sigma K^T$ is well conditioned.

4   Conclusions

In this paper we have developed the work in Murray-Smith & Pearlmutter (2003), and have proposed a filtering approach based on approximate eigendecompositions of the covariance matrix, for dealing with large data-sets. There are two stages. The first stage is to generate a small-sized filtered data-set, which is a good representation of the original data-set so far as the covariance kernel is concerned. The second stage carries out model learning and prediction based on the filtered data. The method can be used in multi-scale learning, the solution of inverse problems and other areas.

References

1. Gibbs, Mark and MacKay, D. J. C. (1997). Efficient implementation of Gaussian processes.
2. Luo, Z and Wahba, G. (1997). Hybrid adaptive splines. *J. Amer. Statist. Assoc.*, **92**, 107-116.

3. Murray-Smith, R. and Pearlmutter, B. A. (2003). Transformations of Gaussian process priors. TR-2003-149, Department of Computing Science, University of Glasgow, Scotland, June.
4. O'Hagan, A. (1978). On curve fitting and optimal design for regression (with discussion). *Journal of the Royal Statistical Society B*, **40**, 1-42.
5. Poggio, T. and Girosi, F. (1990). Networks for approximation and learning. *Proceedings of IEEE*, **78**, 1481-1497.
6. Seeger, M., Willians, C. K. I., and Lawrence, N. D. (2003). Fast forward selection to speed up sparse Gaussian process regression. In Bishop, C. M. and Frey, B. J., editors, *Proceedings of the Ninth International Workshop on AI and Statistics*.
7. Rasmussen, C. E. (1996). *Evaluation of Gaussian Processes and Other Methods for Non-linear Regression*. PhD Thesis. University of Toronto. (Available from http://bayes.imm.dtu.dk)
8. Shi, J. Q., Murray-Smith, R. and Titterington, D. M. (2002). Hierarchical Gaussian Process Mixtures for Regression, DCS Technical Report TR-2002-107/Dept. Statistics Tech. Report 02-7, University of Glasgow, Scotland.
9. Tresp, V. (2000). The Bayesian committee machine. *Neural Computation*, **12**, 2719-2741.
10. Vapnik, V. N. (1995). *The Nature of Statistical Learning Theory*. Springer Verlag, New York.
11. Wahba, G. (1990). *Spline Models for Observational Data*. SIAM, Philadelphia, PA. CBMS-NSF Regional Conference series in applied mathematics.
12. Williams, C. K. I. and Rasmussen, C. E. (1996). Gaussian process for regression. in D.S. Touretzky *et al* (eds), *Advances in Neural Information Processing Systems 8*, MIT Press.
13. Williams, C. K. I. and Seeger, M. (2001). Using the Nyström method to speed up kernel machines. *Advances in Neural Information Processing Systems*, **13**. Eds T. K. Leen, T. G. Diettrich and V. Tresp. MIT Press.

Self-tuning Control of Non-linear Systems Using Gaussian Process Prior Models

Daniel Sbarbaro[1] and Roderick Murray-Smith[2,3]

[1] Departamento de Ingeniería Eléctrica, Universidad de Concepción, Chile
dsbarbar@die.udec.cl
[2] Department of Computing Science, University of Glasgow, Scotland
rod@dcs.gla.ac.uk
[3] Hamilton Institute, NUI Maynooth, Co. Kildare, Ireland

Abstract. Gaussian Process prior models, as used in Bayesian non-parametric statistical models methodology are applied to implement a nonlinear adaptive control law. The expected value of a quadratic cost function is minimised, without ignoring the variance of the model predictions. This leads to implicit regularisation of the control signal (caution) in areas of high uncertainty. As a consequence, the controller has dual features, since it both tracks a reference signal and learns a model of the system from observed responses. The general method and its unique features are illustrated on simulation examples.

1 Introduction

Linear control algorithms have been successfully applied to control nonlinear systems, since they can adapt their parameters to cope the nonlinear characteristics of real systems. However, their performance degrades as the system undergoes rapid and larger changes in its operating point. Several authors have proposed the use of non-linear models as a base to build nonlinear adaptive controllers. Agarwal and Seborg [1], for instance, have proposed the use of known nonlinearities, capturing the main characteristic of the process, to design a Generalized Minimum Variance type of self-tuning controller. In many applications, however, these nonlinearities are not known, and non-linear parameterisation must be used instead. A popular choice has been the use of Artificial Neural Networks for estimating the nonlinearities of the system [2, 3, 4, 5]. All these works have adopted the certainty equivalence principle for designing the controllers, where the model is used in the control law as if it were the true system. In order to improve the performance of nonlinear adaptive controllers based on nonlinear models, the accuracy of the model predictions should also be taken into account. A common approach to consider the uncertainty in the parameters, is to add an extra term in the cost function of a Minimum Variance controller, which penalizes the uncertainty in the parameters of the nonlinear approximation [6]. Another similar approach based on the minimization of two separate cost functions, has been proposed in [7], the first one is used to improve the parameter estimation and the second one to drive the system output to follow a given reference signal. This approach is called bicriterial, and it has also beeen extended to deal with nonlinear systems [8].

R. Murray-Smith, R. Shorten (Eds.): Switching and Learning, LNCS 3355, pp. 140–157, 2005.

Most of these engineering applications are still based on parametric models, where the functional form is fully described by a finite number of parameters, often a linear function of the parameters. Even in the cases where flexible parametric models are used, such as neural networks, spline-based models, multiple models etc, the uncertainty is usually expressed as uncertainty of parameters (even though the parameters often have no physical interpretation), and do not take into account uncertainty about model structure, or distance of current prediction point from training data used to estimate parameters.

Non-parametric models retain the available data and perform inference conditional on the current state and local data (called 'smoothing' in some frameworks). As the data are used directly in prediction, unlike the parametric methods more commonly used in control contexts, non-parametric methods have advantages for off-equilibrium regions, since normally in these regions the amount of data available for identification is much smaller than that available in steady state. The uncertainty of model predictions can be made dependent on local data density, and the model complexity automatically related to the amount and distribution of available data (more complex models need more evidence to make them likely). Both aspects are very useful in sparsely-populated transient regimes. Moreover, since weaker prior assumptions are typically applied in a non-parametric model, the bias is typically less than in parametric models.

Non-parametric models are also well-suited to initial data analysis and exploration, as they are powerful models of the data, with robust behaviour despite few prior structural assumptions. This paper describes an approach based on Gaussian process priors, as an example of a non-parametric model with particularly nice analytic properties. This allow us to analytically obtain a control law which perfectly minimises the expected value of a quadratic cost function, which does not disregard the variance of the model prediction as an element to be minimised. This leads naturally, and automatically to a suitable combination of regularising *caution* in control behaviour in following the reference trajectory, depending on model accuracy. This paper expands on previous work [9] by making the cost function more flexible, introducing priors and investigating modelling and control performance for nonlinear systems affine in control inputs.

The above ideas are closely related to the work done on dual adaptive control, where the main effort has been concentrated on the analysis and design of adaptive controllers based on the use of the uncertainty associated with parameters of models with fixed structure [10, 11].

The paper is organised as follows: section 2 describes the characteristics of non-parametric models. Section 3 introduces Gaussian Process priors. Section 4 illustrates how to design controllers based on the above representation. In section 5, we illustrate the control behaviour via simulation. Finally, some conclusions and future directions are outlined.

2 Controller Design

The objective of this paper is to control a multi-input, single-output, affine nonlinear system of the form,

$$y(t+1) = f(\mathbf{x}(t)) + g(\mathbf{x}(t))u(t) + \varepsilon(t+1),\tag{1}$$

where $\mathbf{x}(t)$ is the state vector at a discrete time t, which in this paper will be defined as $\mathbf{x}(t) = [y(t), \ldots, y(t-n), u(t-1), \ldots, u(t-m)]$, $y(t+1)$ the output, $u(t)$ the current control vector, f and g are unknown smooth nonlinear functions. We also assume that g is bounded away from zero. In addition, it is also assumed that the system is minimum phase, as defined in [4]. For notational simplicity we consider single control input systems, but extending the presentation to vector $u(t)$ is trivial. The noise term $\varepsilon(t)$ is assumed zero mean Gaussian, but with unknown variance σ_n^2. The control strategy consists in choosing a control variable $u(t)$ so as to minimize the following cost function:

$$J = E\{(y_d(t+1) - y(t+1))^2\} + (R(q^{-1})u(t))^2, \qquad (2)$$

where $y_d(t)$ is a bounded reference signal, the polynomial $R(q^{-1})$ is defined as:

$$R(q^{-1}) = r_0 + r_1 q^{-1} + \ldots + r_{n_r} q^{-n_r} \qquad (3)$$

where q^{-1} is a unit backward shift operator. The polynomial coefficients can be used as tuning parameters.

Using the fact that $\text{Var}\{y\} = E\{y^2\} - \mu_y^2$, where $\mu_y = E\{y\}$, the cost function can be written as:

$$J = (y_d(t+1) - E\{y(t+1)\})^2 + \text{Var}\{y(t+1)\} + (R(q^{-1})u(t))^2. \qquad (4)$$

Note that we have not 'added' the model uncertainty term, $\text{Var}\{y(t+1)\}$, to the classical quadratic cost function – most conventional work has 'ignored' it, or have added extra terms to the cost function [10, 11].

Since f and g are unknown, it will be necessary to use a model to predict the output of the system.

3   Non-parametric Models: Gaussian Process Priors

In a Bayesian framework the model must be based on a prior distribution over the infinite-dimensional space of functions. As illustrated in [12], such priors can be defined as Gaussian processes. These models have attracted a great deal of interest recently, in for example reviews such as [13]. Rasmussen [14] showed empirically that Gaussian processes were extremely competitive with leading nonlinear identification methods on a range of benchmark examples. The further advantage that they provide analytic predictions of model uncertainty makes them very interesting for control applications. Use of GPs in a control systems context is discussed in [15, 16]. A variation which can include ARMA noise models is described in [17]. k-step ahead prediction with GP's is described in [18] and integration of prior information in the form of state or control linearisations is presented in [19].

Let's assume a model $y(i) = h(\phi(i)) + \epsilon(i)$, where $\phi(i) \in \mathbb{R}^p$ is the input vector, $\epsilon(i)$ is a noise term, and $y(i) \in \mathbb{R}$ is the corresponding output. Instead of parameterizing $h(\phi(i))$ as a parametric model, we obtain an inference of function $h(\phi(i))$ by computing the distribution $P(h(\phi(i))|\mathcal{D}, \phi(i))$ of the scalar output $h(\phi(i))$, given the input vector

$\phi(i)$ and a set of N training data points $\mathcal{D} = \{(\phi(i), y(i))\ \ i = 1, 2, .., N\}$. The given N data pairs used for identification are stacked in matrices Φ_N and $\mathbf{y}_N$. The vector with the stacked values of the function is defined as $\mathbf{h}_N$. A Gaussian process represents the simplest form of prior over functions and introduces a set of N stochastic variables $\mathbf{H}(1)...\mathbf{H}(N)$, for modelling the function at the corresponding inputs $\phi(1)...\phi(N)$ [14]. Then, a multivariable prior distribution with zero mean[4] and covariance function $\mathbf{K}_N$ is assumed for these variables:

$$P(h(1)...h(N)|\Phi_N) \propto \exp[-\frac{1}{2}(\mathbf{h}_N^T \mathbf{K}_N^{-1} \mathbf{h}_N)], \tag{5}$$

this prior specifies the joint distribution of the function values given the inputs. On the other hand, the likelihood relates the underlying function which is modeled by the function $h(\phi(i))$ to the observed outputs $y(i)$, $i = 1, .., N$. If the noise is assumed to be Gaussian with some unknown variance σ_n^2, then by combining the prior and the likelihood the distribution of the observed data will simply be:

$$P(\mathbf{y}_N|\Phi_N) \propto \exp[-\frac{1}{2}(\mathbf{y}_N^T \mathbf{C}_N^{-1} \mathbf{y}_N)], \tag{6}$$

where $\mathbf{C}_N = \mathbf{K}_N + \sigma_n^2 \mathbf{I}$. The prediction $y(N + 1)$ given the data $\mathcal{D}$ and a new input vector $\phi(N + 1)$ can be calculated by obtaining the following conditioned Gaussian distribution :

$$P(y(N + 1)|\mathcal{D}, \phi(N + 1)) = \frac{P(\mathbf{y}_{N+1}|\Phi_N, \phi(N + 1))}{P(\mathbf{y}_N|\Phi_N)}$$

$$\propto \exp[-\frac{1}{2}(\mathbf{y}_{N+1}^T \mathbf{C}_{N+1} \mathbf{y}_{N+1} - \mathbf{y}_N^T \mathbf{C}_N \mathbf{y}_N)] \tag{7}$$

where $\mathbf{C}_{N+1}$ can be partitioned as:

$$\mathbf{C}_{N+1} = \begin{bmatrix} \mathbf{C}_N & \mathbf{k} \\ \mathbf{k}^T & \kappa \end{bmatrix}. \tag{8}$$

The partitioned form (8) can be used, as it is illustrated in [21], to obtain the parameters of the conditioned Gaussian distribution:

$$P(y(N + 1)|\mathcal{D}, \phi(N + 1)) = \frac{1}{(2\pi\hat{\sigma}_y^2)^{\frac{1}{2}}} \exp[-\frac{(y(t + 1) - \hat{\mu}_y)^2}{2\hat{\sigma}_y^2}],$$

where the mean and variance are:

$$\hat{\mu}_y = \mathbf{k}^T \mathbf{C}_N^{-1} \mathbf{y}_N, \tag{9}$$
$$\mathrm{Var}\{y\} = \hat{\sigma}_y^2 = \kappa - \mathbf{k}^T \mathbf{C}_N^{-1} \mathbf{k}. \tag{10}$$

We can use $\hat{\mu}_y(\phi(N + 1))$ as the expected model output, with a variance of $\hat{\sigma}(\phi(N + 1))^2$. Thus the dynamical system (1) can be modelled under this framework by considering the input vector as $\phi(i) = [\mathbf{x}(t)\ u(t)]$ and the corresponding output $y(i) = y(t+1)$.

---

[4] Note, as explained in [20] the zero mean assumption does not mean that we expect the regression function to be spread equally on either side of zero. If a covariance function had a large constant term the actual function could be always positive or always negative over the range of interest. The zero mean reflects or ignorance as to what that sign will be. There are good numerical computational reasons for transforming data to be zero mean.

3.1   The Covariance Function

The Normal assumption may seem strangely restrictive initially, but represents a powerful tool since the model's prior expectations can be adapted to a given application by altering the covariance function. The choice of this function is only constrained in that it must always generate a non-negative definite covariance matrix for any inputs Φ, so we can represent a spectrum of systems from very local nonlinear models, to standard linear models using the same framework. The covariance function will also often be viewed as being the combination of a covariance function due to the underlying model K and one due to measurement noise C_n. The entries ij of this matrix are then:

$$\mathbf{C}_{N_{ij}} = K(\Phi_{N_i}, \Phi_{N_j}; \Theta) + C_n(\Phi_{N_i}, \Phi_{N_j}; \Theta_n) \tag{11}$$

where Θ denotes a set of parameters, which in the GP framework are also called hyperparameters. As pointed out in [14] it is convenient to specify priors in terms of the hyperparameters, which then can be adapted as the model is fit to the identification data. The covariance associated to the noise $C_n()$ could be $\delta_{ij}\mathcal{N}(\Phi_N; \Theta_n)$, which would be adding a noise model $\mathcal{N}$ to the diagonal entries of $\mathbf{C}_N$. This framework allows the use of different noise models, as discussed in [17], where ARMA noise models were used.

Since the output is an affine function of the control input, it is reasonable to propose a covariance function with a contribution from the control inputs as an affine function as well:

$$K(\phi(i), \phi(j); \Theta) = C_x(\mathbf{x}(i), \mathbf{x}(j); \Theta_x) + u(i)C_x(\mathbf{x}(i), \mathbf{x}(j); \Theta_u)u(j) \tag{12}$$

where the first term represents the contribution of the state vector and the second one the contribution of the input signal. The covariance function C_u can be parameterised in any suitable way. Here, we use the same structure as in C_x above, but with different set of hyperparameters, Θ_u, to those used in C_x.

The covariance function for C_x represents a straightforward covariance function proposed by [14], which has demonstrated to work well in practice:

$$C_x(\mathbf{x}(i), \mathbf{x}(j); \Theta) = v_0 \rho(|\mathbf{x}(i) - \mathbf{x}(j)|, \alpha) + \sum_{k=1}^{p} a_k x_k(i) x_k(j) + a_0, \tag{13}$$

so that the parameter vector $\Theta = log[v_0, \alpha_{1,..p}, a_0]^T$ (the log is applied elementwise) and p is the dimension of vector $\mathbf{x}$. The parameters are defined to be the log of the variable in equation (13) since these are positive scale-parameters. The function $\rho(d)$ is a function of a distance measure d, which should be one at $d = 0$ and which should be a monotonically decreasing function of d. The one used here was

$$\rho(|\mathbf{x}(i) - \mathbf{x}(j)|, \alpha) = e^{-\frac{1}{2}\sum_{k=1}^{p} \alpha_k (x_k(i) - x_k(j))^2}. \tag{14}$$

The α_k's determine how quickly the function varies in dimension k. This estimates the relative smoothness of different input dimensions, and can therefore be viewed as an automatic relevance detection (ARD) tool [22], which helps weight the importance of different input dimensions. Other bases which included a nonlinear transformation of x, like the RBF neural networks used in [6], could be put into this framework. The prior associated with this covariance function states that outputs associated with ϕ's closer together should have higher covariance than points further apart.

The Gaussian Process approach to regression is simple and elegant, and can model nonlinear problems in a probabilistic framework. There tend also to be far fewer parameters to identify in the Gaussian Process approach than for competing approaches (such as e.g. artificial neural networks). The disadvantage is its computational complexity, as estimating the mean $\hat{\mu}_y$ requires a matrix inversion of the $N \times N$ covariance matrix, which becomes problematic for identification data where $N > 1000$. In transient regimes, however, we have very few data points and we wish to make robust estimates of model behaviour, which are now possible. This suggests that a multiple-model style partitioning of the state-space could make GPs more feasible in many applications [23].

Adapting the Covariance Function Parameters The hyperparameter vector Θ provides flexibility to define a family of covariance functions which provide suitable prior distributions over functions. In most cases we will only have uncertain knowledge of Θ. Given unknown hyperparameters we can use numerical methods such as Markov-Chain Monte Carlo (MCMC) to integrate over hyperparameters, or use maximum likelihood methods, with standard gradient-based optimisation tools to optimise hyperparameters. The log-likelihood l of the training data can be calculated analytically as [13]:

$$l = -\frac{1}{2} \log \det \mathbf{C}_N - \frac{1}{2} \mathbf{y}_N^T \mathbf{C}_N^{-1} \mathbf{y}_N - \frac{n}{2} \log 2\pi. \tag{15}$$

The partial derivative of the log likelihood with respect to the hyperparameters is:

$$\frac{\partial l}{\partial \theta_i} = -\frac{1}{2} \text{tr} \left[\mathbf{C}_N^{-1} \frac{\partial \mathbf{C}_N}{\partial \theta_i} \right] + \frac{1}{2} \mathbf{y}_N^T \mathbf{C}_N^{-1} \frac{\partial \mathbf{C}_N}{\partial \theta_i} \mathbf{C}_N^{-1} \mathbf{y}_N. \tag{16}$$

Given l and its derivative with respect to θ_i it is straightforward to use an efficient optimization program in order to obtain a local maximum of the likelihood.

In parametric models the parameters must to be updated each sampling time, but in the nonparametric framework this is not necessary, as it will be illustrated in section 5, since the model also relies on the data contained in the identification data set.

Hierarchical Priors The hyperparameters of the covariance function will rarely be known exactly in advance, so they are usually given a vague prior distribution, such as a gamma prior [20].

$$p(\phi) = \frac{(a/2\omega)^{a/2}}{\Gamma(a/2)} \psi^{((a/2)-1)} \exp\left(-\frac{\psi a}{2\omega}\right) \tag{17}$$

where $\psi = \theta^{-2}$ for a hyperparameter θ. a is a positive shape parameter and ω is the mean of ψ. Large values of a produce priors for θ concentrated near ω^{-2} and small values lead to vague priors. Each hyperparameter of the covariance function can be given an independent prior distribution. If prior distributions on the hyperparameters, such as equation (17) are used then obviously these are included in the likelihood equations and the derivative terms. The use of gamma priors does not add significant complexity to the optimisation, and if used appropriately makes the model behaviour more robust with small numbers of training data.

4   Derivation of Control Law

Given the cost function (4), and observations to time t, if we wish to find the optimal $u(t)$, we need the derivative of J,

$$\frac{\partial J}{\partial u(t)} = -2\left(y_d(t+1) - \mu_y\right)\frac{\partial \mu_y}{\partial u(t)} + \frac{\partial \mathrm{Var}\{y(t+1)\}}{\partial u(t)} + 2r_0 R(q^{-1})u(t). \quad (18)$$

With most models, estimation of $\mathrm{Var}\{y\}$, or $\frac{\partial \mathrm{Var}\{y\}}{\partial u(t)}$ would be difficult, but with the Gaussian process prior (assuming smooth, differentiable covariance functions – see [24]) the following straightforward analytic solutions can be obtained:

$$\frac{\partial \mu_y}{\partial u(t)} = \frac{\partial \mathbf{k}^T}{\partial u(t)}\mathbf{C}_N^{-1}\mathbf{y}_N \quad (19)$$

$$\frac{\partial \mathrm{Var}\{y\}}{\partial u(t)} = \frac{\partial \kappa}{\partial u(t)} - 2\mathbf{k}^T\mathbf{C}_N^{-1}\frac{\partial \mathbf{k}}{\partial u(t)}. \quad (20)$$

The covariance matrix $\mathbf{k}$ and κ can be expressed in terms of the independent control variable $u(t)$ as follows:

$$\mathbf{k} = \Omega_1 + u(t)\Omega_2 \quad (21)$$

$$\kappa = \Omega_3 + \Omega_4 u(t)^2, \quad (22)$$

where $\Omega_1 = C_x(\mathbf{x}(t), \Phi_N, \Theta_x)$, $\Omega_2 = C_x(\mathbf{x}(t), \Phi_N, \Theta_u). * U_N$, where U_N is a vector with all the values of $u(t)$ contained in the identification data set, and $.*$ indicates elementwise multiplication of two matrices. $\Omega_3 = C_x(\mathbf{x}(t), \mathbf{x}(t), \Theta_x)$, and $\Omega_4 = C_x(\mathbf{x}(t), \mathbf{x}(t), \Theta_u)$. The final expressions for μ_y and $\mathrm{Var}\{y\}$ are:

$$\mu_y = \Omega_1 \mathbf{C}_N^{-1}\mathbf{y}_N + \Omega_2 \mathbf{C}_N^{-1}\mathbf{y}_N u(t), \quad (23)$$

$$\mathrm{Var}\{y\} = \Omega_3 + \Omega_4 u(t)^2 - (\Omega_1 + u(t)\Omega_2)\mathbf{C}_N^{-1}(\Omega_1 + u(t)\Omega_2)^T. \quad (24)$$

Taking the partial derivatives of the variance and the mean expressions and replacing their values in (18), it follows:

$$\frac{\partial J}{\partial u(t)} = -2\left(y_d(t+1) - \Omega_1\mathbf{C}_N^{-1}\mathbf{y}_N - \Omega_2\mathbf{C}_N^{-1}\mathbf{y}_N u(t)\right)(\Omega_2\mathbf{C}_N^{-1}\mathbf{y}_N)^T$$
$$+2\Omega_4 u(t) - 2\Omega_1\mathbf{C}_N^{-1}\Omega_2^T - 2u(t)\Omega_2\mathbf{C}_N^{-1}\Omega_2^T) + 2r_0 R(q^{-1})u(t). \quad (25)$$

At $\frac{\partial J}{\partial u(t)} = 0$, the optimal control signal is obtained as:

$$u(t) = \frac{(y_d(t+1) - \Omega_1\mathbf{C}_N^{-1}\mathbf{y}_N)(\Omega_2\mathbf{C}_N^{-1}\mathbf{y}_N)^T + \Omega_1\mathbf{C}_N^{-1}\Omega_2^T - r_0(R(q^{-1}) - r_0)u(t)}{r_0^2 + \Omega_4 - \Omega_2\mathbf{C}_N^{-1}\Omega_2^T + \Omega_2\mathbf{C}_N^{-1}\mathbf{y}_N(\Omega_2\mathbf{C}_N^{-1}\mathbf{y}_N)^T}. \quad (26)$$

Note that equation (26) can also be presented as

$$u(t) = \frac{(y_d(t+1) - \Omega_1 \mathbf{C}_N^{-1} \mathbf{y}_N)(\Omega_2 \mathbf{C}_N^{-1} \mathbf{y}_N)^T + \alpha(t) - r_0(R(q^{-1}) - r_0)u(t)}{r_0^2 + \beta(t) + \Omega_2 \mathbf{C}_N^{-1} \mathbf{y}_N (\Omega_2 \mathbf{C}_N^{-1} \mathbf{y}_N)^T}. \quad (27)$$

where $\alpha(t) = \Omega_1 \mathbf{C}_N^{-1} \Omega_2^T$, and $\beta(t) = \Omega_4 - \Omega_2 \mathbf{C}_N^{-1} \Omega_2^T$. If we had not included the variance term in cost function (2), or if we were in a region of the state-space where the variance was zero, the optimal control law would be equation (27) with $\alpha = \beta = 0$. We can therefore see that analysing the values of α and β is a promising approach to gaining insight into the behaviour of the new form of controller. These terms make a control effort penalty constant, or regulariser unnecessary in many applications.

4.1 Adapting Control Behaviour with New Data

After $u(t)$ has been calculated, applied, and the output observed, we add the information $\mathbf{x}(t), u(t), y(t+1)$ to the training set, and the new $\mathbf{C}_N$ increases in size to $N+1 \times N+1$. Obviously, given the expense of inverting $\mathbf{C}_N$ for large N, this naive approach will only work for relatively small data sets. For a more general solution, we can potentially incorporate elements of Relevance Vector Machines [25], or use heuristics for selection of data for use in an active training set, as in e.g. [26].

We can then choose to optimise the hyperparameters of the covariance function to further refine the model, or keep the covariance function fixed, and just use the extra data points to improve model performance. In the next section, will be illustrated the performance obtained with both strategies.

5   Simulation Results

To illustrate the feasibility of the approach we used it to control several target plants based on noisy observed responses. We start off with only *two* training points, and add subsequent data to the model during operation. The model has had no prior adaptation to the system before the experiment. A gamma distribution was used for all hyperparameters, with ω set equal to the initial condition for each variable and shape parameter $a = 3$, indicating vague knowledge about the variable. The noise term σ_n, was given a tighter distribution, with $a = 5$. The covariance functions chosen are the same for all the experiments.

5.1   Non-linear System 1

Let non-linear system 1 be:

$$y(t+1) = f(\mathbf{x}(t)) + g(\mathbf{x}(t))u(t) + \varepsilon(t+1)$$

where $\mathbf{x}(t) = y(t)$, $f(\mathbf{x}(t)) = \sin(y(t)) + \cos(3y(t))$ and $g(\mathbf{x}(t)) = 2 + \cos(y(t))$, subject to noise with variance $\sigma_n^2 = 0.001$ [6]. Model hyperparameters are adapted after each iteration using conjugate gradient descent optmisation algorithm.

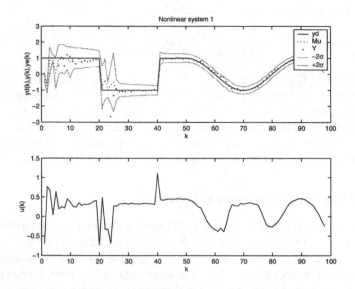

(a) Simulation of nonlinear GP-based controller

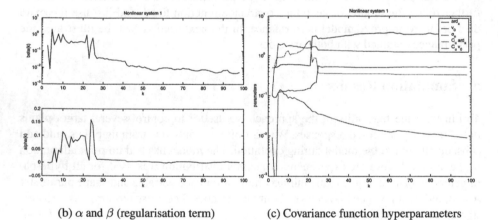

(b) α and β (regularisation term)   (c) Covariance function hyperparameters

Fig. 1. Simulation results for nonlinear system 1, showing modelling accuracy, control signals, tracking behaviour and levels of α and β at each stage.

Note how in Figure 1 β is large in the early stages of learning, but decreasing with the decrease in variance, showing how the regularising effect enforces caution in the face of uncertainty, but reduces caution as the model accuracy increases. In terms of the hyperparameters, most hyperparameters have converged by about 30 data points. The noise parameter σ_n decreases with increasing levels of training data. After this point the control signal u is also fairly smooth, despite the noisy nature of the data.

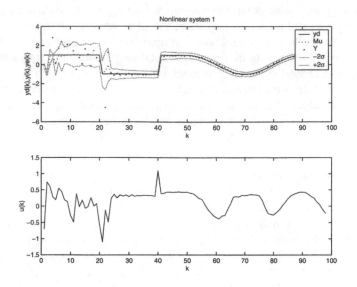

(a) Simulation of nonlinear GP-based controller

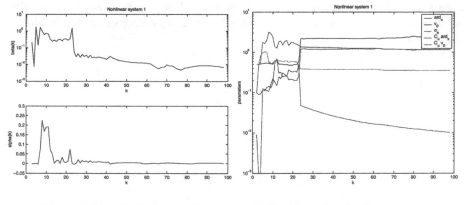

(b) α and β (regularisation term)   (c) Covariance function hyperparameters

Fig. 2. Simulation results on nonlinear system 1, *without including the α and β terms linked to the model variance in the control law*. Data shows modelling accuracy, control signals and tracking behaviour.

α can be seen to be larger in higher variance regions, essentially adding an excitatory component which decreases with the decrease in model uncertainty, and in this example plays almost no role after about iteration 30.

Figure 2 shows control performance on the same system where the variance part of the cost function is ignored (i.e. α and β are removed from the control law. In order to achieve any reasonable control behaviour, we set $r_0 = 0.5$. As can be seen in the figure, the system still tracks the trajectory, and after iteration 30 there is little visible difference between the two control laws, but ignoring the variance does lead to the use of greater control effort, with larger model uncertainty in the early stages of learning. The hyperparameter estimates also fluctuate much more in the early stages of learning, when variance is not considered, although both systems converge on similar values after the initial stages. The constant nature of r_0 as a regularising term, as opposed to the dynamically changing $\alpha(t), \beta(t)$ makes controller design more difficult, as we can see that in early stages of learning it tends to be too small, reducing robustness, while later it is larger than $\alpha(t), \beta(t)$ damaging performance.

We now plot the nonlinear mappings involved in non-linear system 1, to give the reader a clearer impression of the adaptation of the system. The surfaces in figure 4 show the development in the mean mapping from $x(t), u(t)$ to $y(t + 1)$ as the system acquires data, taken from the simulation shown in figure 1 at $t = 3, 20, 99$. For comparison, figure 5.1 shows the true mapping. Examining the surfaces in figure 4 we can see how the nonlinear mapping adapts gradually given increasing numbers of data points, but we also see that the standard deviation of the mapping also evolves in an appropriate manner, indicating clearly at each stage of adaptation the model uncertainty over the state-space. In the final plot, Figure 4(f) we can see a uniformly low uncertainty in the areas covered by data, but a rapid increase in uncertainty as we move beyond that in the x-axis. Note that the uncertainty grows much more slowly in the u-axis because of the affine assumption inherent in the covariance function, which constrains the freedom of the model.

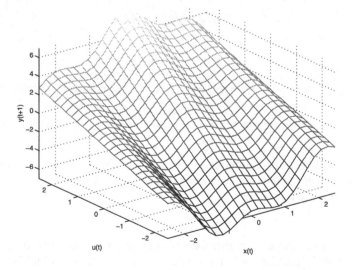

Fig. 3. True surface (mesh) of nonlinear system 1, $y(t+1) = \sin(y(t)) + \cos(3y(t)) + (2 + \cos(y(t)))u(t)$, over the space $y \times u$.

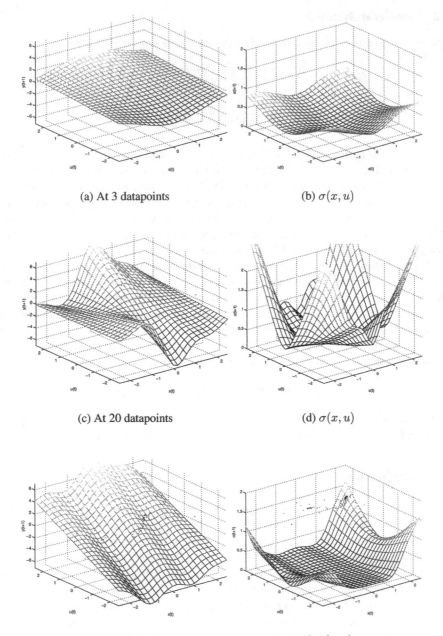

(a) At 3 datapoints

(b) $\sigma(x, u)$

(c) At 20 datapoints

(d) $\sigma(x, u)$

(e) At 99 datapoints

(f) $\sigma(x, u)$

Fig. 4. Left-hand figures show mean surface (mesh) $y(t + 1) = f(x(t), u(t))$ of non-linear system 1 over the space $x \times u$ during the learning process. These can be compared to the true mapping in Figure 5.1. Right-hand figures show condition standard deviation $\sigma(x(t), u(t))$ surfaces. Each figure also shows the available data at that point in the learning process.

5.2 Non-linear System 2

The second nonlinear example considers the following non-linear functions:

$$f(\mathbf{x}(t)) = \frac{y(t)y(t-1)y(t-2)u(t-1)(y(t-2)-1)}{1+y(t-1)^2+y(t-2)^2}$$

$$g(\mathbf{x}(t)) = \frac{1}{1+y(t-1)^2+y(t-2)^2},$$

where $\mathbf{x} = \begin{bmatrix} y(t) & y(t-1) & y(t-2) & u(t-1) \end{bmatrix}^T$ [2]. The system noise has a variance $\sigma_n^2 = 0.001$, and we had 6 initial data points. The results are shown in Figure 5. Again, the trend of decreasing α and β can be seen, although they do increase in magnitude following changes in the system state towards higher uncertainty regions, showing that the control signal will be appropriately damped when the system moves to a less well-modelled area of the state-space. The hyperparameters in Figure 5(c) make few rapid changes, seeming well-behaved during learning.

Figure 6 illustrates the effect of keeping constant the initial set of hyperparameters. As it can be seen in the figure, even with this set of paramaters, which have values very far away from the optimal ones, the system is capable of controlling the system.

The next example illustrates the use of the polynomial $R(q^{-1})$ for shaping the closed loop response. In this example, it was selected as $R(q^{-1}) = r_o - r_o q^{-1}$, so as to weight the control signal deviations for tuning the speed of response without introducing steady state errors, the associated cost function is:

$$J = E\{(y_d(t+1) - y(t+1))^2\} + (r_0(u(t) - u(t-1))^2. \qquad (28)$$

The response obtained for a $r_0 = 0.6$ is illustrated in Figure 7, where as we expected the speed of response is much slower that the case of having no control weighting.

6 Conclusions

This work has presented a novel adaptive controller based on non-parametric models. The control design is based on the expected value of a quadratic cost function, leading to a controller that not only will minimise the squared difference between the reference signal and the expected value of the output, but will also try to minimise the variance of the output, based on analytical estimates of model uncertainty. This leads to a robust control action during adaptation, and when extended to multi-step ahead prediction, forms the basis of full dual control with implicit excitatory components. Simulation results, considering linear and non-linear systems, demonstrate the interesting characteristics of this type of adaptive control algorithm.

The GP models are capable of high performance, with or without priors being placed on their hyperparameters. Use of gamma prior distributions led to increased robustness and higher performance in the early stages of adaptation with very few data points, but the relative advantage decreases with the amount of initial data available, as would be expected. Since the predictions are do not only rely on the hyperparameters,

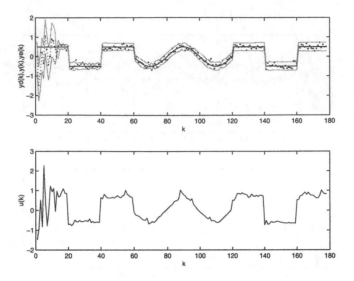

(a) Simulation of nonlinear GP-based controller

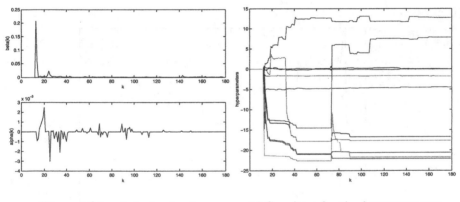

(b) α and β (regularisation term)   (c) Covariance function hyperparameters

Fig. 5. Simulation results for nonlinear system 2, showing modelling accuracy, control signals, tracking behaviour and levels of α and β at each stage.

but also on the training data set, their on-line adaptation can be carried out at a sampling interval much bigger than the one used for controlling the system.

The additional polynomial term in the cost function can be used to shape the closed loop response without introducing steady state error.

GP's have been successfully adopted from their statistics origins by the neural network community [13]. This paper is intended to bring the GP approach to the attention

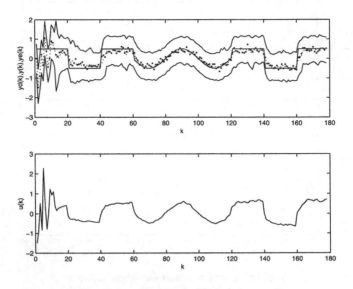

(a) Simulation of nonlinear GP-based controller

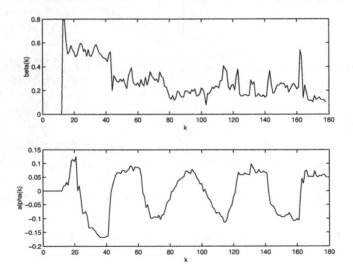

(b) α and β (regularisation term)

Fig. 6. Simulation results for nonlinear system 2 without adapting the hyperparameters, showing modelling accuracy, control signals, tracking behaviour and levels of α and β at each stage.

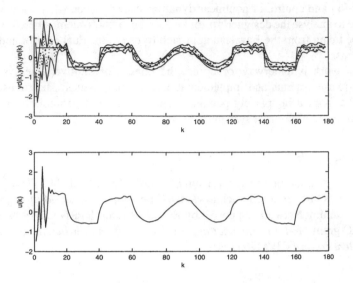

(a) Simulation of nonlinear GP-based controller

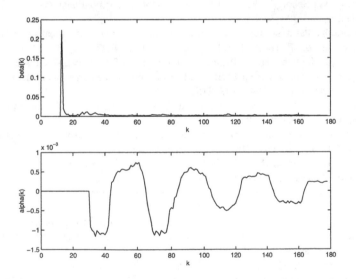

(b) α and β (regularisation term)

Fig. 7. Simulation results for nonlinear system 2 with $r_o = 0.6, r_1 = -0.6$, showing modelling accuracy, control signals, tracking behaviour and levels of α and β at each stage.

of the control community, and to show that the basic approach is a competitive approach for modelling and control of nonlinear dynamic systems, even when little attempt has been made to analyse the designer's prior knowledge of the system – there is much more that can be taken from the Bayesian approach to use in the dual control and nonlinear control areas.

Further work is underway to address the control of multivariable systems, non-minimum-phase systems and implementation efficiency issues. The robust inference of the GP approach in sparsely populated spaces makes it particularly promising in multivariable and high-order systems.

Acknowledgements

Both authors are grateful for support from FONDECYT Project 700397 and the Hamilton Institute. RM-S gratefully acknowledges the support of the *Multi-Agent Control Research Training Network* supported by EC TMR grant HPRN-CT-1999-00107, and the EPSRC grant *Modern statistical approaches to off-equilibrium modelling for nonlinear system control* GR/M76379/01.

References

[1] Agarwal, M., Seborg, D.E.: Self-tuning controllers for nonlinear systems. Automatica (1987) 209–214

[2] Narendra, K., Parthasarathy, P.: Identification and control of dynamical systems using neural networks. IEEE. Trans. Neural Networks **1** (1990) 4–27

[3] Liang, F., ElMargahy, H.: Self-tuning neurocontrol of nonlinear systems using localized polynomial networks with CLI cells. In: Proceedings of the American Control Conference, Baltimore, Maryland (1994) 2148–2152

[4] Chen, F., Khalil, H.: Adaptive control of a class of nonlinear discrete-time systems. IEEE. Trans. Automatic Control **40** (1995) 791–801

[5] Bittanti, S., Piroddi, L.: Neural implementation of GMV control shemes based on affine input/output models. Proc. IEE Control Theory Appl. (1997) 521–530

[6] Fabri, S., Kadirkamanathan, V.: Dual adaptive control of stochastic systems using neural networks. Automatica **14** (1998) 245–253

[7] Filatov, N., Unbehauen, H., Keuchel, U.: Dual pole placement controller with direct adaptation. Automatica **33** (1997) 113–117

[8] Sbarbaro, D., Filatov, N., Unbehauen, H.: Adaptive dual controller for a class of nonlinear systems. In: Proceedings of the IFAC Workshop on Adaptive systems in Control and Signal Processing, Glasgow, U.K. (1998) 28–33

[9] Murray-Smith, R., Sbarbaro, D.: Nonlinear adaptive control using non-paramtric gaussian process prior models. In: Proceedings of the 15th IFAC world congress, Barcelona, Spain (2002)

[10] Wittenmark, R.: Adaptive dual control methods: An overview. In: Proceedings of the 5th IFAC Symposium on Adaptive systems in Control and Signal Processing, Budapest (1995) 67–92

[11] Filatov, N., Unbehauen, H.: Survey of adaptive dual control methods. Proc. IEE Control Theory Appl. (2000) 119–128

[12] O'Hagan, A.: On curve fitting and optimal design for regression (with discussion). Journal of the Royal Statistical Society **B** (1978) 1–42

[13] Williams, C.: Prediction with Gaussian process: From linear regression to linear prediction and beyond. In Jordan, M., ed.: Learning and Inference in Graphical Models, Kluwer (1998) 599–621

[14] Rasmussen, C.: Evaluation of Gaussian Process and other Methods for non-linear regression. PhD thesis, Department of Computer Science, University of Toronto (1996)

[15] Murray-Smith, R., Johansen, T., Shorten, R.: On transient dynamics, off-equilibrium behaviour and identification in blended multiple model structures. In: Proceedings of the European Control Conference, Karlsruhe, Germany (1999) BA–14

[16] Leith, D., Murray-smith, R., Leithhead, W.: Nonlinear structure identification: A gaussian process prior/velocity-based approach. In: Proceedings of Control 2000, Cambridge, U.K. (2000)

[17] Murray-Smith, R., Girard, A.: Gaussian process priors with ARMA noise models. In: Proceedings of the Irish Sgnals and Systems Conference, Maynooth, Ireland (2001) 147–152

[18] Girard, A., Rasmussen, C.E., Quinonero-Candela, J., Murray-Smith, R.: Gaussian process priors with uncertain inputs – application to multiple-step ahead time series forecasting. In Becker, S., Thrun, S., Obermayer, K., eds.: Advances in Neural Information processing Systems 15. (2003)

[19] Solak, E., Murray-Smith, R., Leithead, W.E., Leith, D.J., Rasmussen, C.E.: Derivative observations in Gaussian process models of dynamic systems. In Becker, S., Thrun, S., Obermayer, K., eds.: Advances in Neural Information processing Systems 15. (2003)

[20] Neal, R.: Monte Carlo implementation of Gaussian process models for Bayesian regression and classification. Technical Report 9702, Department of Statistics, University of Toronto (1997)

[21] Gibbs, M.: Bayesian Gaussian processes for regression and classification. PhD thesis, Cavendish Laboratory, University of Cambridge (1998)

[22] MacKay, D.: Gaussian processes: A replacement for supervised neural networks? In: Lectures notes for the NIPS 1997, Denver, Colorado (1997)

[23] Shi, J., Murray-Smith, R., Titterington, D.M.: Bayesian regression and classification using mixtures of multiple Gaussian processes. International Journal of Adaptive Control and Signal Processing 17 (2003) 149–161

[24] O'Hagan, A.: Some Bayesian numerical analysis. In Bernardo, J., Berger, J., Dawid, A., Smith, F., eds.: Bayesian Statistics 4, Oxford University Press (2001) 345–363

[25] Tipping, M.E.: Sparse Bayesian learning and the relevance vector machine. Journal of Machine Learning Research 1 (2001) 211–244

[26] Seeger, M., Williams, C., Lawrence, D.: Fast forward selection to speed up sparse Gaussian process regression. In: Proceedings of the Ninth International Workshop on AI and Statistics. (2003)

Gaussian Processes: Prediction at a Noisy Input and Application to Iterative Multiple-Step Ahead Forecasting of Time-Series

Agathe Girard[1] and Roderick Murray-Smith[1,2]

[1] Department of Computing Science, University of Glasgow, 17 Lilybank Gardens, Glasgow G12 8QQ, UK,
agathe@dcs.gla.ac.uk
[2] Hamilton Institute, Maynooth, Ireland

Abstract. With the Gaussian Process model, the predictive distribution of the output corresponding to a new given input is Gaussian. But if this input is uncertain or noisy, the predictive distribution becomes non-Gaussian. We present an analytical approach that consists of computing only the mean and variance of this new distribution (*Gaussian approximation*). We show how, depending on the form of the covariance function of the process, we can evaluate these moments exactly or approximately (within a Taylor approximation of the covariance function). We apply our results to the iterative multiple-step ahead prediction of non-linear dynamic systems with propagation of the uncertainty as we predict ahead in time. Finally, using numerical examples, we compare the *Gaussian approximation* to the numerical approximation of the true predictive distribution by simple Monte-Carlo.

1 Background

Given a set of observed data $\mathcal{D} = \{\mathbf{x}_i, t_i\}_{i=1}^{N}$, where $\mathbf{x}_i \in \mathcal{R}^D$ and $t_i = f(\mathbf{x}_i) + \epsilon, \in \mathcal{R}$ (ϵ is a white noise with variance v_t), we model the input/output relationship using a zero-mean Gaussian Process (GP) with covariance function $C(\mathbf{x}_i, \mathbf{x}_j)$. For the moment, we do not specify the form of the covariance function and simply assume it is a valid one, generating a positive definite covariance matrix. We refer to [1, 2, 3, 4] for a review of GPs.

1.1 Prediction at a New x

With this model, given a new 'test' input $\mathbf{x}$, and based on the observed data, the predictive distribution of the corresponding output $y = f(\mathbf{x})$ is readily obtained. This distribution is Gaussian, $p(y|\mathcal{D}, \mathbf{x}) = \mathcal{N}(\mu(\mathbf{x}), \sigma^2(\mathbf{x}))$, with mean and variance respectively given by

R. Murray-Smith, R. Shorten (Eds.): Switching and Learning, LNCS 3355, pp. 158–184, 2005.
© Springer-Verlag Berlin Heidelberg 2005

$$\begin{cases} \mu(\mathbf{x}) = \sum_{i=1}^{N} \beta_i C(\mathbf{x}, \mathbf{x}_i) \\ \sigma^2(\mathbf{x}) = C(\mathbf{x}, \mathbf{x}) - \sum_{i,j=1}^{N} K_{ij}^{-1} C(\mathbf{x}, \mathbf{x}_i) C(\mathbf{x}, \mathbf{x}_j) \end{cases} \tag{1}$$

with $\boldsymbol{\beta} = \mathbf{K}^{-1}\mathbf{t}$, where $\mathbf{t}$ is the $N \times 1$ vector of observed noisy targets and $\mathbf{K}$ is the $N \times N$ data covariance matrix, such that $K_{ij} = C(\mathbf{x}_i, \mathbf{x}_j) + v_t \delta_{ij}$. The covariances between the new point and the training cases are given by $C(\mathbf{x}, \mathbf{x}_i)$, for $i = 1 \ldots N$, and $C(\mathbf{x}, \mathbf{x})$ is the covariance between the test point and itself.

In practice, the predictive mean $\mu(\mathbf{x})$ is used as a point estimate for the function output, while the variance $\sigma^2(\mathbf{x})$ can be translated into uncertainty bounds (error-bars) on this estimate. Although this variance corresponds to the model's uncertainty (and therefore depends on the prior and on the local data complexity), it represents valuable information as it enables us to quantify the uncertainty attached to the prediction. Figure 1 shows the predictive means and their 2σ error-bars computed for 81 test inputs. A Gaussian Process with zero-mean and Gaussian covariance function (Eq. (22)) was trained using only $N = 10$ points. Near the data points, the predictive variance is small, increasing as the test inputs are far away from the training ones.

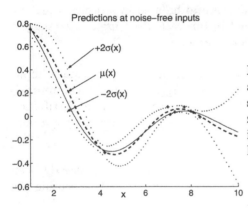

Fig. 1. Predictive means (dashed line) and 2σ error-bars (dotted lines) corresponding to 81 noise-free test inputs. A zero-mean GP was trained on 10 training points (crosses) to learn the underlying function (continuous line).

1.2   Motivation

We first motivate the necessity of being able to make a prediction at an uncertain or noisy input using a dynamic example.

Dynamic Case Let a time-series be known up to time t and assume a simple auto-regressive generative model of the form $y_{t+1} = f(y_t)$ where the input now corresponds to a delayed value of the time-series. Having formed a set of input/output pairs and trained a GP, we wish to predict the value of the time-series at, say, time $t+k$. With our one-step ahead model, we need to iterate predictions up to the desired horizon, i.e. we have $y_{t+k} = f(y_{t+k-1})$, $y_{t+k-1} = f(y_{t+k-2})$, so on, down to $y_{t+1} = f(y_t)$. Since y_t is known, the predictive distribution of y_{t+1} is simply Gaussian, $p(y_{t+1}|\mathcal{D}, y_t) = \mathcal{N}(\mu(y_t), \sigma^2(y_t))$, with mean and variance given by (1) evaluated at $\mathbf{x} = y_t$. For the next time-step, a naive approach consists in only using $\mu(y_t)$ as an estimate for y_{t+1}, $\hat{y}_{t+1} = \mu(y_t)$, and evaluate $p(y_{t+2}|\mathcal{D}, \hat{y}_{t+1}) = \mathcal{N}(\mu(\hat{y}_{t+1}), \sigma^2(\hat{y}_{t+1}))$. As we will see in our numerical examples, this approach is not advisable for two reasons: it is over-confident about the estimate (the variance $\sigma^2(\hat{y}_{t+1})$ will typically be very small) and it is also throwing away valuable information, namely, the uncertainty attached to the estimate $\hat{y}_{t+1}$, $\sigma(y_t)$. If we wish to account for this uncertainty, and thus *propagate* it as we predict ahead in time, we need to be able to evaluate $p(y_{t+2}|\mathcal{D}, y_{t+1})$ where $y_{t+1} \sim \mathcal{N}(\mu(y_t), \sigma^2(y_t))$. This means being able to evaluate the predictive distribution corresponding to an uncertain or noisy input, y_{t+1} here.

Static Case In real experiments and applications, we use sensors and detectors that can be corrupted by many different sources of disturbances. We might then only observe a noise corrupted version of the true input and the system senses the new input imperfectly. Again, if the model does not account for this 'extra' uncertainty (as opposed to the uncertainty usually acknowledged on the observed outputs), the model is too confident, which is misleading and could potentially be dangerous if, say, the model's output were to be used in a decision-making process of a critical application. Note that in this case, the approach we suggest assumes prior knowledge of the input noise variance.

In the next section, we present the problem of predicting at a noisy input when using a Gaussian Process model. We then suggest an analytical approximation and compute the mean and variance of the new predictive distribution (sections 3 and 4). In section 5, we return to the iterative forecasting of a nonlinear time-series to which we apply our results.

Although most of the material presented in this chapter has already been published [5, 6, 7], the present document aims at unifying and presenting the different results in a more principled manner.

2   Prediction at an Uncertain Input

Let the new test input be corrupted by some noise, $\epsilon_{\mathbf{x}} \sim \mathcal{N}_{\epsilon_{\mathbf{x}}}(0, \Sigma_x)$, such that $\mathbf{x} = \mathbf{u} + \epsilon_{\mathbf{x}}$. That is, we wish to make a prediction at $\mathbf{x} \sim \mathcal{N}_{\mathbf{x}}(\mathbf{u}, \Sigma_x)$ and to do so, we need to integrate the predictive distribution $p(y|\mathcal{D}, \mathbf{x})$ over the input distribution[3]

$$p(y|\mathcal{D}, \mathbf{u}, \Sigma_x) = \int p(y|\mathcal{D}, \mathbf{x}) p(\mathbf{x}|\mathbf{u}, \Sigma_x) d\mathbf{x} . \tag{2}$$

---

[3] When the bounds are not indicated, it means that the integrals are evaluated from $-\infty$ to $+\infty$.

For the GP, we have $p(y|\mathcal{D}, \mathbf{x}) = \frac{1}{\sqrt{2\pi\sigma^2(\mathbf{x})}} \exp\left[-\frac{1}{2}\frac{(y-\mu(\mathbf{x}))^2}{\sigma^2(\mathbf{x})}\right]$, which is a nonlinear function of $\mathbf{x}$, such that this integral cannot be solved without resorting to approximations.

2.1 Possible Approximations

Many techniques are available to approximate intractable integrals of this kind. Approximation methods are divided into deterministic approximations and Monte-Carlo numerical methods. The most popular deterministic approaches are variational methods,[4] Laplace's method and Gaussian quadrature that consist of analytical approximations of the integral. Refer to [4] for a review of these methods.

Numerical methods relying on Markov-Chain Monte-Carlo sampling techniques evaluate the integral numerically, thus approximating the true distribution. In our case, the numerical approximation by simple Monte-Carlo is straightforward since we simply need to sample from a Gaussian distribution $\mathcal{N}_{\mathbf{x}}(\mathbf{u}, \boldsymbol{\Sigma}_x)$. For each sample $\mathbf{x}^t$ from this distribution, $p(y|\mathcal{D}, \mathbf{x}^t)$ is normal, with mean and variance given by Eqs. (1):

$$p(y|\mathcal{D}, \mathbf{u}, \boldsymbol{\Sigma}_x) \simeq \frac{1}{T}\sum_{t=1}^{T} p(y|\mathcal{D}, \mathbf{x}^t) = \frac{1}{T}\sum_{t=1}^{T} \mathcal{N}_y(\mu(\mathbf{x}^t), \sigma^2(\mathbf{x}^t)) . \qquad (3)$$

The numerical approximation of $p(y|\mathcal{D}, \mathbf{u}, \boldsymbol{\Sigma}_x)$ is then a mixture of T Gaussians with identical mixing proportions. As the number of samples T grows, the approximate distribution will tend to the true distribution.

On Fig. 2, 100 predictive means with their corresponding uncertainties are plotted, corresponding to 100 samples x^t from $p(x)$, centered at the noisy observed input x (asterisks), with variance $v_x = 1$. The 'true' test inputs are $u = 2$ (left) and $u = 6$ (right). The histograms of the samples at which predictions are made are shown on Fig. 3. The circle and asterisk indicate the noise-free and noisy inputs (u and x respectively). After having computed the loss associated to each x^t[5], we find that for which the loss is minimum (triangle), which is close to the true value.

In the remaining of this document, we focus on an analytical approximation which consists of computing only the first two moments, the mean and variance, of $p(y|\mathcal{D}, \mathbf{u}, \boldsymbol{\Sigma}_x)$. As we will now see, approximate or exact moments are computed, depending on the form of the covariance function.

2.2 Analytical Approximation

To distinguish from $\mu(\mathbf{u})$ and $\sigma^2(\mathbf{u})$, the mean and variance of the Gaussian predictive distribution $p(y|\mathcal{D}, \mathbf{u})$ corresponding to a noise-free $\mathbf{u}$, we denote by

---

[4] Many references can be found at http://www.gatsby.ucl.ac.uk/vbayes/

[5] We compute the squared error and the minus log-predictive density, see section 6.

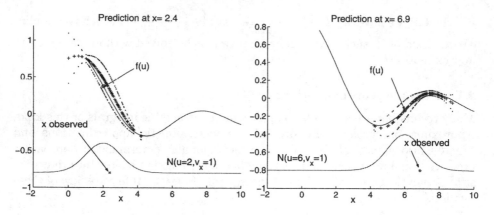

Fig. 2. Monte-Carlo approximation to the prediction at an observed noisy input x (asterisk). Predictive means $\mu(x^t)$ (crosses) with $2\sigma(x^t)$ error-bars (dots), computed for 100 samples x^t from $p(x)$, with mean x and variance v_x. The true input distribution is $x \sim \mathcal{N}_x(u, v_x)$, for $u = 2$ (left), $u = 6$ (right) and $v_x = 1$. The circle indicates the output corresponding to the noise-free input u.

$m(\mathbf{u}, \boldsymbol{\Sigma}_x)$ the mean and by $v(\mathbf{u}, \boldsymbol{\Sigma}_x)$ the variance of the non-Gaussian predictive distribution $p(y|\mathcal{D}, \mathbf{u}, \boldsymbol{\Sigma}_x)$, corresponding to $\mathbf{x} \sim \mathcal{N}_\mathbf{x}(\mathbf{u}, \boldsymbol{\Sigma}_x)$. This can be interpreted as a *Gaussian approximation*, such that

$$p(y|\mathcal{D}, \mathbf{u}, \boldsymbol{\Sigma}_x) \approx \mathcal{N}(m(\mathbf{u}, \boldsymbol{\Sigma}_x), v(\mathbf{u}, \boldsymbol{\Sigma}_x)) .$$

This mean and variance are respectively given by

$$m(\mathbf{u}, \boldsymbol{\Sigma}_x) = \int y \left\{ \int p(y|\mathcal{D}, \mathbf{x}) p(\mathbf{x}|\mathbf{u}, \boldsymbol{\Sigma}_x) d\mathbf{x} \right\} dy$$

$$v(\mathbf{u}, \boldsymbol{\Sigma}_x) = \int y^2 \left\{ \int p(y|\mathcal{D}, \mathbf{x}) p(\mathbf{x}|\mathbf{u}, \boldsymbol{\Sigma}_x) d\mathbf{x} \right\} dy - m(\mathbf{u}, \boldsymbol{\Sigma}_x)^2 .$$

Using the law of iterated expectations and that of conditional variances,[6] we directly have

$$m(\mathbf{u}, \boldsymbol{\Sigma}_x) = E_\mathbf{x}[\mu(\mathbf{x})] \tag{4}$$

$$v(\mathbf{u}, \boldsymbol{\Sigma}_x) = E_\mathbf{x}[\sigma^2(\mathbf{x})] + \mathrm{Var}_\mathbf{x}[\mu(\mathbf{x})] , \tag{5}$$

---

[6] Recall that $E[X] = E[E[X|Y]]$ and $\mathrm{Var}[X] = E[\mathrm{Var}[X|Y]] + \mathrm{Var}[E[X|Y]]$.

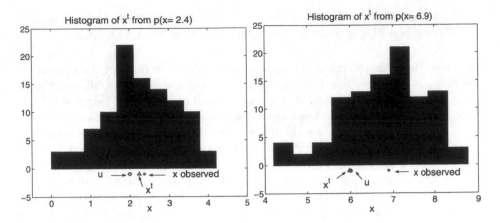

Fig. 3. Histogram of the samples x^t from $p(x)$ at which predictions were made, when the true input (circle) is $u = 2$ (left) and $u = 6$ (right). Also plotted, the observed noisy input (asterisk), taken as the mean of $p(x)$, and the sample x^t that leads to the minimum loss (triangle).

where $\mathrm{Var}_{\mathbf{x}}[\mu(\mathbf{x})] = E_{\mathbf{x}}[\mu(\mathbf{x})^2] - m(\mathbf{u}, \boldsymbol{\Sigma}_x)^2$. Replacing $\mu(\mathbf{x})$ and $\sigma^2(\mathbf{x})$ by their expressions (Eqs. (1)), we finally have

$$
\begin{cases}
m(\mathbf{u}, \boldsymbol{\Sigma}_x) = \sum_{i=1}^{N} \beta_i E_{\mathbf{x}}[C(\mathbf{x}, \mathbf{x}_i)] \\[2mm]
v(\mathbf{u}, \boldsymbol{\Sigma}_x) = E_{\mathbf{x}}[C(\mathbf{x}, \mathbf{x})] - \sum_{i,j=1}^{N} (K_{ij}^{-1} - \beta_i \beta_j) E_{\mathbf{x}}[C(\mathbf{x}, \mathbf{x}_i) C(\mathbf{x}, \mathbf{x}_j)] - m(\mathbf{u}, \boldsymbol{\Sigma}_x)^2 \, .
\end{cases}
$$

$$(6)$$

Let

$$l = \int C(\mathbf{x}, \mathbf{x}) p(\mathbf{x}) d\mathbf{x} \qquad (7)$$

$$l_i = \int C(\mathbf{x}, \mathbf{x}_i) p(\mathbf{x}) d\mathbf{x} \qquad (8)$$

$$l_{ij} = \int C(\mathbf{x}, \mathbf{x}_i) C(\mathbf{x}, \mathbf{x}_j) p(\mathbf{x}) d\mathbf{x} \, . \qquad (9)$$

How solvable integrals (7)-(9) are basically depends on the form of the covariance function.

1. If the covariance function is e.g. linear, Gaussian, polynomial (or a mixture of those), we can compute the integrals exactly and obtain the *exact* mean and variance. In section 4, we derive the 'exact' moments for the linear and Gaussian covariance functions.
2. Otherwise, we can again approximate (7)-(9) in a number of ways. Since we are mostly interested in closed form approximate solutions, we evaluate the

integrals within a Taylor approximation of the covariance function around the mean **u** of **x** and obtain the 'approximate' mean and variance.

Note that this second case might be required, if the form of the covariance function is definitely one for which one cannot solve the integrals exactly, or simply preferable, if the integrals are tractable but at the cost of long and tedious calculations (assuming one has access to software like Mathematica or Matlab's symbolic toolbox to compute the derivatives, the solutions obtained using the proposed approximations provide a suitable performance/implementation trade-off).

Figure 4 summarizes the different possible approximations and highlights the analytical one we take. We now turn to the evaluation of the mean and variance in the case of a 'general' the covariance function, that is when further approximations are needed to evaluate integrals (7)-(9) analytically.

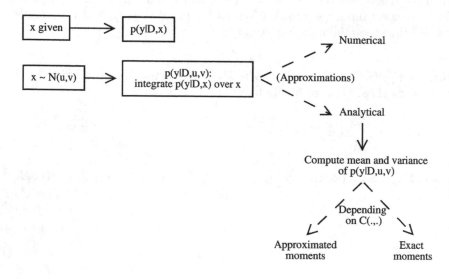

Fig. 4. Dealing with a noisy test input: With the GP model, the predictive distribution of the output corresponding to a new test input x is readily obtained, by conditioning on the training data D and on the new x. If x is noisy, such that $x \sim \mathcal{N}(u, v)$, the new predictive distribution is now obtained by integrating over the input distribution. Since $p(y|D, x)$ is nonlinear in x, the integral is analytically intractable. Although a numerical approximation of the integral is possible, we concentrate on an analytical approximation. We suggest to compute the mean and the variance of the new predictive distribution, which is done exactly or approximately, depending on the parametric form of the covariance function $C(.,.)$.

3   *Gaussian Approximation*: **Approximate Moments**

We use the Delta method (also called Moment Approximation), which consists of approximating the integrand by a Taylor polynomial. In the one-dimensional case, the Delta method is as follows [8, 9]: Let x be a random variable with mean $E_x[x] = u$ and variance $\text{Var}_x[x] = v_x$, and $y = \phi(x)$. For sufficiently small $\sigma_x = \sqrt{v_x}$ and well-behaved ϕ we can write

$$E_x[y] \simeq \phi(u) + \frac{1}{2}v_x\phi''(u) \tag{10}$$

$$\text{Var}_x[y] \simeq \phi'(u)^2 v_x \tag{11}$$

where ϕ' and ϕ'' are the first and second derivatives of ϕ evaluated at u.

These results are simply obtained by considering the expansion of $\phi(x)$ in Taylor series about u, up to the second order:

$$y = \phi(x) = \phi(u) + (x - u)\phi'(u) + \frac{1}{2}(x - u)^2\phi''(u) + O([(x - u)^3]) . \tag{12}$$

By taking the expectation on both sides, we directly find the approximation (10). For the variance, we have $\text{Var}[y] = E[y^2] - E[y]^2$ and the estimate given by (11) corresponds to an approximation of the second order estimate: Neglecting the term in v_x^2 for both $E[y^2]$ and $E[y]^2$, we have

$$E[y^2] \approx \phi(u)^2 + v_x\phi'(u)^2 + \phi(u)\phi''(u)v_x$$
$$E[y]^2 \approx \phi(u)^2 + \phi(u)\phi''(u)v_x$$

leading to (11). This approximation is motivated by the fact that the Taylor approximation is useful for small standard deviations (if σ_x is small, by Chebychev's inequality $P(|x - u| > k\sigma_x) < \frac{1}{k^2}$), such that x will depart only a little from u except on rare occasions and therefore $(x - u)$ will be small.

There are obviously conditions which $\phi(x)$ should fulfill to make the Taylor series possible (in the neighborhood of u) and to avoid anomalies of behavior away from u. As in [8], we do not state such conditions and assume the covariance function to be such that the expressions are valid.

3.1   Approximate Mean

Let $m^{ap}(\mathbf{u}, \Sigma_x)$ be the approximate mean, such that

$$m^{ap}(\mathbf{u}, \Sigma_x) = \sum_{i=1}^{N} \beta_i l_i^{ap}$$

with $l_i^{ap} = E_{\mathbf{x}}[C^{ap}(\mathbf{x}, \mathbf{x}_i)]$ and where $C^{ap}(\mathbf{x}, \mathbf{x}_i)$ corresponds to the second order Taylor polynomial of $C(\mathbf{x}, \mathbf{x}_i)$ around the mean $\mathbf{u}$ of $\mathbf{x}$,

$$C^{ap}(\mathbf{x}, \mathbf{x}_i) = C(\mathbf{u}, \mathbf{x}_i) + (\mathbf{x} - \mathbf{u})^T \mathbf{C}'(\mathbf{u}, \mathbf{x}_i) + \frac{1}{2}(\mathbf{x} - \mathbf{u})^T \mathbf{C}''(\mathbf{u}, \mathbf{x}_i)(\mathbf{x} - \mathbf{u}) .$$

We directly have

$$l_i^{ap} = C(\mathbf{u}, \mathbf{x}_i) + \frac{1}{2}\text{Tr}[\mathbf{C}''(\mathbf{u}, \mathbf{x}_i)\boldsymbol{\Sigma}_x]$$

so that the approximate mean is

$$m^{ap}(\mathbf{u}, \boldsymbol{\Sigma}_x) = \mu(\mathbf{u}) + \frac{1}{2}\sum_{i=1}^N \beta_i \text{Tr}[\mathbf{C}''(\mathbf{u}, \mathbf{x}_i)\boldsymbol{\Sigma}_x] \tag{13}$$

where $\mu(\mathbf{u}) = \sum_{i=1}^N \beta_i C(\mathbf{u}, \mathbf{x}_i)$ is the noise-free predictive mean computed at $\mathbf{u}$.

3.2   Approximate Variance

Similarly, the approximate variance is

$$v^{ap}(\mathbf{u}, \boldsymbol{\Sigma}_x) = l^{ap} - \sum_{i,j=1}^N (K_{ij}^{-1} - \beta_i\beta_j)l_{ij}^{ap} - m^{ap}(\mathbf{u}, \boldsymbol{\Sigma}_x)^2$$

with $l^{ap} = E_{\mathbf{x}}[C^{ap}(\mathbf{x}, \mathbf{x})]$ and $l_{ij}^{ap} = E_{\mathbf{x}}[C^{ap}(\mathbf{x}, \mathbf{x}_i)C^{ap}(\mathbf{x}, \mathbf{x}_j)]$, where $C^{ap}(.,.)$ is again the second order Taylor approximation of $C(.,.)$. We have

$$l^{ap} = C(\mathbf{u}, \mathbf{u}) + \frac{1}{2}\text{Tr}[\mathbf{C}''(\mathbf{u}, \mathbf{u})\boldsymbol{\Sigma}_x]$$

and

$$l_{ij}^{ap} \approx C(\mathbf{u}, \mathbf{x}_i)C(\mathbf{u}, \mathbf{x}_j) + \text{Tr}[\mathbf{C}'(\mathbf{u}, \mathbf{x}_i)\mathbf{C}'(\mathbf{u}, \mathbf{x}_j)^T\boldsymbol{\Sigma}_x] + \frac{1}{2}C(\mathbf{u}, \mathbf{x}_i)\text{Tr}[\mathbf{C}''(\mathbf{u}, \mathbf{x}_j)\boldsymbol{\Sigma}_x]$$
$$+ \frac{1}{2}C(\mathbf{u}, \mathbf{x}_j)\text{Tr}[\mathbf{C}''(\mathbf{u}, \mathbf{x}_i)\boldsymbol{\Sigma}_x]$$

where the approximation comes from discarding terms of higher order than $\boldsymbol{\Sigma}_x$ in $C^{ap}(\mathbf{x}, \mathbf{x}_i)C^{ap}(\mathbf{x}, \mathbf{x}_j)$, as discussed in the previous section. Similarly, approximating $m^{ap}(\mathbf{u}, \boldsymbol{\Sigma}_x)^2$ by

$$m^{ap}(\mathbf{u}, \boldsymbol{\Sigma}_x)^2 \approx \sum_{i,j=1}^N \beta_i\beta_j \left(C(\mathbf{u}, \mathbf{x}_i)C(\mathbf{u}, \mathbf{x}_j) + \frac{1}{2}C(\mathbf{u}, \mathbf{x}_i)\text{Tr}[\mathbf{C}''(\mathbf{u}, \mathbf{x}_j)\boldsymbol{\Sigma}_x] \right.$$
$$\left. + \frac{1}{2}C(\mathbf{u}, \mathbf{x}_j)\text{Tr}[\mathbf{C}''(\mathbf{u}, \mathbf{x}_i)\boldsymbol{\Sigma}_x] \right),$$

we find, after simplifications,

$$v^{ap}(\mathbf{u}, \boldsymbol{\Sigma}_x) = \sigma^2(\mathbf{u}) + \frac{1}{2}\text{Tr}[\mathbf{C}''(\mathbf{u}, \mathbf{u})\boldsymbol{\Sigma}_x] - \sum_{i,j=1}^N (K_{ij}^{-1} - \beta_i\beta_j)\text{Tr}[\mathbf{C}'(\mathbf{u}, \mathbf{x}_i)\mathbf{C}'(\mathbf{u}, \mathbf{x}_j)^T\boldsymbol{\Sigma}_x]$$
$$- \frac{1}{2}\sum_{i,j=1}^N K_{ij}^{-1}(C(\mathbf{u}, \mathbf{x}_i)\text{Tr}[\mathbf{C}''(\mathbf{u}, \mathbf{x}_j)\boldsymbol{\Sigma}_x] + C(\mathbf{u}, \mathbf{x}_j)\text{Tr}[\mathbf{C}''(\mathbf{u}, \mathbf{x}_i)\boldsymbol{\Sigma}_x])$$

$$\tag{14}$$

where $\sigma^2(\mathbf{u}) = C(\mathbf{u}, \mathbf{u}) - \sum_{i,j=1}^{N} K_{ij}^{-1} C(\mathbf{u}, \mathbf{x}_i) C(\mathbf{u}, \mathbf{x}_j)$ is the noise-free predictive variance.

Note that these results might be more easily derived by approximating $\mu(\mathbf{x})$ and $\sigma^2(\mathbf{x})$ directly in Eqs. (4) and (5), as done in [5, 7].[7] Applying (10) to $\mu(\mathbf{x})$, we have $E[\mu(\mathbf{x})] \simeq \mu(\mathbf{u}) + \frac{1}{2}\mathrm{Tr}[\boldsymbol{\mu}''(\mathbf{u})\boldsymbol{\Sigma}_x]$, and replacing into (4) gives

$$m^{ap}(\mathbf{u}, \boldsymbol{\Sigma}_x) = \mu(\mathbf{u}) + \frac{1}{2}\mathrm{Tr}[\boldsymbol{\mu}''(\mathbf{u})\boldsymbol{\Sigma}_x] \ .$$

Similarly, $E[\sigma^2(\mathbf{x})] \simeq \sigma^2(\mathbf{u}) + \frac{1}{2}\mathrm{Tr}[\boldsymbol{\sigma}^{2''}(\mathbf{u})\boldsymbol{\Sigma}_x]$ and, using (11), $\mathrm{Var}[\mu(\mathbf{x})] \simeq \mathrm{Tr}[\boldsymbol{\mu}'(\mathbf{u})\boldsymbol{\mu}'(\mathbf{u})^T\boldsymbol{\Sigma}_x]$. Replacing into (5) we obtain

$$v^{ap}(\mathbf{u}, \boldsymbol{\Sigma}_x) = \sigma^2(\mathbf{u}) + \mathrm{Tr}\left[\left(\frac{1}{2}\boldsymbol{\sigma}^{2''}(\mathbf{u}) + \boldsymbol{\mu}'(\mathbf{u})\boldsymbol{\mu}'(\mathbf{u})^T\right)\boldsymbol{\Sigma}_x\right] \ .$$

Although, replacing the derivatives by their expressions, these results are obviously the same as those obtained when working with the covariance function, working directly with $\mu(\mathbf{x})$ and $\sigma^2(\mathbf{x})$ lacks flexibility in that it is not clear that exact moments can be computed.

Both approximate mean and variance are composed of the noise-free predictive moments plus correction terms. Assuming $\boldsymbol{\Sigma}_x$ is diagonal, these correction terms consist of the sum of the derivatives of the covariance function in each input dimension, weighted by the variance of the new test input in the same direction. Figure 5 illustrates these results. On the x-axis, the asterisks indicate the observed noisy inputs and the distribution they come from ($p(x) = \mathcal{N}_x(u, v_x)$, for $u = 2, 6, 9.5$ and $v_x = 1$). The circles indicate the function output corresponding to the noise-free u's. The approximate means $m^{ap}(u, v_x)$ and associated uncertainties, $\pm 2\sqrt{v^{ap}}(u, v_x)$ are plotted as triangles and dotted lines. We can compare them to the *naive* (noise-free) means $\mu(u)$ with error-bars $\pm 2\sigma(u)$, which do not account for the noise on the input.

4   *Gaussian Approximation*: **Exact Moments**

We are now going to show that in the special cases of the linear and the Gaussian (squared exponential) covariance functions, we can evaluate integrals (7)-(9) exactly.

4.1   **Case of the Linear Covariance Function**

Let us write the linear covariance function as $C_L(\mathbf{x}_i, \mathbf{x}_j) = \mathbf{x}_i^T \mathbf{L} \mathbf{x}_j$ where $\mathbf{L} = \mathrm{diag}[\alpha_1 \ldots \alpha_D]$. In the noise-free case, the prediction at $\mathbf{u}$ leads to a Gaussian distribution with mean and variance

---

[7] In [5, 7], we only considered a first order approximation for the mean $\mu(\mathbf{x})$.

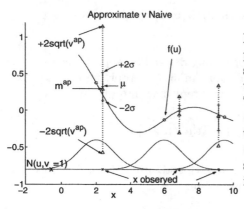

Fig. 5. *Gaussian approximation* to the prediction at x (asterisk), noisy version of the true $u = 2, 6, 9.5$, where the noise has variance and $v_x = 1$. The approximate mean and uncertainty $(m^{ap}(x) \pm 2\sqrt{v^{ap}(x)})$ are indicated by triangles and the noise-free moments $(\mu(x) \pm 2\sigma(x))$ by crosses. The circles show the function outputs corresponding to the noise-free u's.

$$\begin{cases} \mu_L(\mathbf{u}) = \sum_{i=1}^{N} \beta_i C_L(\mathbf{u}, \mathbf{x}_i) \\ \sigma_L^2(\mathbf{u}) = C_L(\mathbf{u}, \mathbf{u}) - \sum_{i,j=1}^{N} K_{ij}^{-1} C_L(\mathbf{u}, \mathbf{x}_i) C_L(\mathbf{u}, \mathbf{x}_j) . \end{cases} \tag{15}$$

When predicting at a noisy input, the predictive mean and variance are now given by

$$m^{exL}(\mathbf{u}, \boldsymbol{\Sigma}_x) = \sum_{i=1}^{N} \beta_i l_i^{exL} \tag{16}$$

$$v^{exL}(\mathbf{u}, \boldsymbol{\Sigma}_x) = l^{exL} - \sum_{i,j=1}^{N} (K_{ij}^{-1} - \beta_i \beta_j) l_{ij}^{exL} - m^{exL}(\mathbf{u}, \boldsymbol{\Sigma}_x)^2 \tag{17}$$

so that we need to evaluate

$$l^{exL} = E_{\mathbf{x}}[C_L(\mathbf{x}, \mathbf{x})] = \int \mathbf{x}^T \mathbf{L} \mathbf{x} \mathcal{N}_{\mathbf{x}}(\mathbf{u}, \boldsymbol{\Sigma}_x) d\mathbf{x}$$

$$l_i^{exL} = E_{\mathbf{x}}[C_L(\mathbf{x}, \mathbf{x}_i)] = \int \mathbf{x}^T \mathbf{L} \mathbf{x}_i \mathcal{N}_{\mathbf{x}}(\mathbf{u}, \boldsymbol{\Sigma}_x) d\mathbf{x}$$

$$l_{ij}^{exL} = E_{\mathbf{x}}[C_L(\mathbf{x}, \mathbf{x}_i) C_L(\mathbf{x}, \mathbf{x}_j)] = \int \mathbf{x}^T \mathbf{L} \mathbf{x}_i \mathbf{x}^T \mathbf{L} \mathbf{x}_j \mathcal{N}_{\mathbf{x}}(\mathbf{u}, \boldsymbol{\Sigma}_x) d\mathbf{x} .$$

Using the formula giving the expectation of a quadratic form under a Gaussian[8] we directly obtain

---
[8]

$$\int_{\mathbf{x}} (\mathbf{x} - \mathbf{m})^T \mathbf{M}^{-1} (\mathbf{x} - \mathbf{m}) \mathcal{N}_{\mathbf{x}}(\mathbf{u}, \boldsymbol{\Sigma}_x) d\mathbf{x} = (\mathbf{m} - \mathbf{u})^T \mathbf{M}^{-1}(\mathbf{m} - \mathbf{u}) + \text{Tr}[\mathbf{M}^{-1} \boldsymbol{\Sigma}_x]$$

$$l^{ex_L} = \mathbf{u}^T L\mathbf{u} + \text{Tr}[\mathbf{L}\boldsymbol{\Sigma}_x] = C_L(\mathbf{u}, \mathbf{u}) + \text{Tr}[\mathbf{L}\boldsymbol{\Sigma}_x]$$

$$l_i^{ex_L} = \mathbf{u}^T \mathbf{L}\mathbf{x}_i = C_L(\mathbf{u}, \mathbf{x}_i)$$

$$l_{ij}^{ex_L} = \mathbf{u}^T (\mathbf{L}\mathbf{x}_i\mathbf{x}_j^T \mathbf{L})\mathbf{u} + \text{Tr}[\mathbf{L}\mathbf{x}_i\mathbf{x}_j^T \mathbf{L}\boldsymbol{\Sigma}_x] = C_L(\mathbf{u}, \mathbf{x}_i)C_L(\mathbf{x}_j, \mathbf{u}) + \text{Tr}[\mathbf{L}\mathbf{x}_i\mathbf{x}_j^T \mathbf{L}\boldsymbol{\Sigma}_x] \ .$$

Therefore, the predictive mean is the same as the noise-free one, as we have

$$m^{ex_L}(\mathbf{u}, \boldsymbol{\Sigma}_x) = \sum_{i=1}^{N} \beta_i C_L(\mathbf{u}, \mathbf{x}_i) \ . \tag{18}$$

On the other hand, the variance becomes

$$v^{ex_L}(\mathbf{u}, \boldsymbol{\Sigma}_x) = C_L(\mathbf{u}, \mathbf{u}) + \text{Tr}[\mathbf{L}\boldsymbol{\Sigma}_x] - \sum_{i,j=1}^{N} (K_{ij}^{-1} - \beta_i\beta_j)\text{Tr}[\mathbf{L}\mathbf{x}_i\mathbf{x}_j^T \mathbf{L}\boldsymbol{\Sigma}_x])$$
$$- \sum_{i,j=1}^{N} K_{ij}^{-1} C_L(\mathbf{u}, \mathbf{x}_i)C_L(\mathbf{x}_j, \mathbf{u}) \tag{19}$$

after simplification of the $\beta_i\beta_j$ terms. Or, in terms of the noise-free variance,

$$v^{ex_L}(\mathbf{u}, \boldsymbol{\Sigma}_x) = \sigma_L^2(\mathbf{u}) + \text{Tr}[\mathbf{L}\boldsymbol{\Sigma}_x] - \sum_{i,j=1}^{N} (K_{ij}^{-1} - \beta_i\beta_j)\text{Tr}[\mathbf{L}\mathbf{x}_i\mathbf{x}_j^T \mathbf{L}\boldsymbol{\Sigma}_x]) \ . \tag{20}$$

If we note that $\mathbf{C}'_L(\mathbf{u}, \mathbf{x}_i) = \frac{\partial C_L(\mathbf{u}, \mathbf{x}_i)}{\partial \mathbf{u}} = \mathbf{L}\mathbf{x}_i$ and $\mathbf{C}''_L(\mathbf{u}, \mathbf{u}) = \frac{\partial^2 C_L(\mathbf{u}, \mathbf{u})}{\partial \mathbf{u} \partial \mathbf{u}^T} = 2\mathbf{L}$, we can also write it as

$$v^{ex_L}(\mathbf{u}, \boldsymbol{\Sigma}_x) = \sigma_L^2(\mathbf{u}) + \frac{1}{2}\text{Tr}[\mathbf{C}''_L(\mathbf{u}, \mathbf{u})\boldsymbol{\Sigma}_x]$$
$$- \sum_{i,j=1}^{N} (K_{ij}^{-1} - \beta_i\beta_j)\text{Tr}[\mathbf{C}'_L(\mathbf{x}, \mathbf{x}_i)\mathbf{C}'_L(\mathbf{x}, \mathbf{x}_j)^T \boldsymbol{\Sigma}_x]) \ . \tag{21}$$

As we would expect, the predictive mean and variance in the case of the linear covariance function correspond to the approximate moments we would obtain within a first order approximation of the covariance function.

4.2   Case of the Gaussian Covariance Function

The Gaussian (or squared exponential) covariance function became a popular choice especially after Rasmussen demonstrated that a GP with such a covariance function performed as well, if not better, than other models like neural networks [10]. It is usually expressed as

$$C_G(\mathbf{x}_i, \mathbf{x}_j) = v \exp\left[-\frac{1}{2}(\mathbf{x}_i - \mathbf{x}_j)^T \mathbf{W}^{-1}(\mathbf{x}_i - \mathbf{x}_j)\right] \tag{22}$$

with $\mathbf{W}^{-1} = \text{diag}[w_1 \ldots w_D]$, where w_d is a roughness parameter, inversely proportional to the square of the correlation length in direction d ($w_d = 1/\lambda_d^2$), which represents the length along which successive values are strongly correlated (with a role similar to the Automatic Relevance Determination tool of Mackay and Neal [11, 12]). The parameter v controls the overall vertical scale relative to the zero mean of the process in the output space (the vertical amplitude of variation of a typical function).

We now denote by $\mu_G(\mathbf{u})$ and $\sigma_G^2(\mathbf{u})$ the corresponding noise-free predictive mean and variance,

$$
\begin{cases}
\mu_G(\mathbf{u}) = \sum_{i=1}^{N} \beta_i C_G(\mathbf{u}, \mathbf{x}_i) \\[2mm]
\sigma_G^2(\mathbf{u}) = C_G(\mathbf{u}, \mathbf{u}) - \sum_{i,j=1}^{N} K_{ij}^{-1} C_G(\mathbf{u}, \mathbf{x}_i) C_G(\mathbf{u}, \mathbf{x}_j)
\end{cases}
\tag{23}
$$

where $C_G(\mathbf{x}, \mathbf{x}) = v$. In this case, the predictive mean and variance, obtained for a prediction at $\mathbf{x} \sim \mathcal{N}_{\mathbf{x}}(\mathbf{u}, \boldsymbol{\Sigma}_x)$, are given by

$$
m^{exG}(\mathbf{u}, \boldsymbol{\Sigma}_x) = \sum_{i=1}^{N} \beta_i l_i^{exG}
\tag{24}
$$

$$
v^{exG}(\mathbf{u}, \boldsymbol{\Sigma}_x) = l^{exG} - \sum_{i,j=1}^{N} (K_{ij}^{-1} - \beta_i \beta_j) l_{ij}^{exG} - m^{exG}(\mathbf{u}, \boldsymbol{\Sigma}_x)^2 .
\tag{25}
$$

We directly have $l^{exG} = E_{\mathbf{x}}[C_G(\mathbf{x}, \mathbf{x})] = v = C_G(\mathbf{u}, \mathbf{u})$, and we need to evaluate

$$
l_i^{exG} = E_{\mathbf{x}}[C_G(\mathbf{x}, \mathbf{x}_i)] = c \int N_{\mathbf{x}}(\mathbf{x}_i, \mathbf{W}) \mathcal{N}_{\mathbf{x}}(\mathbf{u}, \boldsymbol{\Sigma}_x) d\mathbf{x}
$$

$$
l_{ij}^{exG} = E_{\mathbf{x}}[C_G(\mathbf{x}, \mathbf{x}_i) C_G(\mathbf{x}, \mathbf{x}_j)] = c^2 \int N_{\mathbf{x}}(\mathbf{x}_i, \mathbf{W}) \mathcal{N}_{\mathbf{x}}(\mathbf{x}_j, \mathbf{W}) \mathcal{N}_{\mathbf{x}}(\mathbf{u}, \boldsymbol{\Sigma}_x) d\mathbf{x} ,
$$

where, for notational convenience, we write the Gaussian covariance function as[9] $C_G(\mathbf{x}_i, \mathbf{x}_j) = c N_{\mathbf{x}_i}(\mathbf{x}_j, \mathbf{W})$, with $c = (2\pi)^{D/2} |\mathbf{W}|^{1/2} v$. Using the product of Gaussians formula,[10] we find

$$
l_i^{exG} = c N_{\mathbf{u}}(\mathbf{x}_i, \mathbf{W} + \boldsymbol{\Sigma}_x) .
\tag{26}
$$

And for l_{ij}^{exG}, using this product twice,

$$
l_{ij}^{exG} = c^2 N_{\mathbf{x}_i}(\mathbf{x}_j, 2\mathbf{W}) N_{\mathbf{u}} \left(\frac{\mathbf{x}_i + \mathbf{x}_j}{2}, \boldsymbol{\Sigma}_x + \frac{\mathbf{W}}{2} \right) .
\tag{27}
$$

---

[9] Note that $N(.,.)$ is used to denote the parametric form of the function, it does not correspond to a normal probability distribution $\mathcal{N}(.,.)$.

[10] Recall that $\mathcal{N}_x(a, A) \mathcal{N}_x(b, B) = z \mathcal{N}_x(d, D)$ with $D = (A^{-1} + B^{-1})^{-1}$, $d = D(A^{-1}a + B^{-1}b)$ and $z = \mathcal{N}_a(b, A + B) = \mathcal{N}_b(a, A + B)$.

Exact Predictive Mean Replacing l_i^{exG} in $m^{exG}(\mathbf{u}, \boldsymbol{\Sigma}_x)$, we have

$$m^{exG}(\mathbf{u}, \boldsymbol{\Sigma}_x) = \sum_{i=1}^{N} \beta_i c N_{\mathbf{u}}(\mathbf{x}_i, \mathbf{W} + \boldsymbol{\Sigma}_x) \tag{28}$$

and we can directly check that, as we would expect, $m(\mathbf{u}, \boldsymbol{\Sigma}_x = \mathbf{0}) = \mu_G(\mathbf{u})$.

It is useful to write $m^{exG}(\mathbf{u}, \boldsymbol{\Sigma}_x)$ as a *corrected* version of $\mu_G(\mathbf{u})$. Using the matrix inversion lemma, we have $(\mathbf{W} + \boldsymbol{\Sigma}_x)^{-1} = \mathbf{W}^{-1} - \mathbf{W}^{-1}(\mathbf{W}^{-1} + \boldsymbol{\Sigma}_x^{-1})^{-1}\mathbf{W}^{-1}$, leading to

$$l_i^{exG} = C_G(\mathbf{u}, \mathbf{x}_i) C_{corr}(\mathbf{u}, \mathbf{x}_i) \tag{29}$$

with

$$C_{corr}(\mathbf{u}, \mathbf{x}_i) = |\mathbf{I} + \mathbf{W}^{-1}\boldsymbol{\Sigma}_x|^{-1/2} \exp\left[\frac{1}{2}(\mathbf{u} - \mathbf{x}_i)^T \Delta^{-1}(\mathbf{u} - \mathbf{x}_i)\right] \tag{30}$$

where $\Delta^{-1} = \mathbf{W}^{-1}(\mathbf{W}^{-1} + \boldsymbol{\Sigma}_x^{-1})^{-1}\mathbf{W}^{-1}$. The predictive mean is then given by

$$m^{exG}(\mathbf{u}, \boldsymbol{\Sigma}_x) = \sum_{i=1}^{N} \beta_i C_G(\mathbf{u}, \mathbf{x}_i) C_{corr}(\mathbf{u}, \mathbf{x}_i) . \tag{31}$$

Compared to the noise-free $\mu_G(\mathbf{u})$, the covariances between the new noisy input and the training inputs, formerly given by $C_G(\mathbf{u}, \mathbf{x}_i)$, are now weighted by $C_{corr}(\mathbf{u}, \mathbf{x}_i)$, thus accounting for the uncertainty associated to $\mathbf{u}$.

Exact Predictive Variance Replacing l_{ij}^{exG} by its expression, we have

$$v^{exG}(\mathbf{u}, \boldsymbol{\Sigma}_x) = C_G(\mathbf{u}, \mathbf{u}) - c^2 \sum_{i,j=1}^{N} (K_{ij}^{-1} - \beta_i\beta_j) N_{\mathbf{x}_i}(\mathbf{x}_j, 2\mathbf{W}) N_{\mathbf{u}}\left(\frac{\mathbf{x}_i + \mathbf{x}_j}{2}, \boldsymbol{\Sigma}_x + \frac{\mathbf{W}}{2}\right)$$
$$- m^{exG}(\mathbf{u}, \boldsymbol{\Sigma}_x)^2$$

and again, we can show that for $\boldsymbol{\Sigma}_x = \mathbf{0}$, we have $v^{exG}(\mathbf{u}, \boldsymbol{\Sigma}_x = \mathbf{0}) = \sigma_G^2(\mathbf{u})$.[11]

---

[11] We have

$$v^{exG}(\mathbf{u}, \boldsymbol{\Sigma}_x = \mathbf{0}) = C_G(\mathbf{u}, \mathbf{u}) - c^2 \sum_{i,j=1}^{N} (K_{ij}^{-1} - \beta_i\beta_j) N_{\mathbf{x}_i}(\mathbf{x}_j, 2\mathbf{W}) N_{\mathbf{u}}\left(\frac{\mathbf{x}_i + \mathbf{x}_j}{2}, \frac{\mathbf{W}}{2}\right)$$
$$- m^{exG}(\mathbf{u}, \boldsymbol{\Sigma}_x = \mathbf{0})^2$$

with $m^{exG}(\mathbf{u}, \boldsymbol{\Sigma}_x = \mathbf{0})^2 = c^2 \sum_{i,j=1}^{N} \beta_i\beta_j N_{\mathbf{x}_i}(\mathbf{x}_j, 2\mathbf{W}) N_{\mathbf{u}}\left(\frac{\mathbf{x}_i + \mathbf{x}_j}{2}, \frac{\mathbf{W}}{2}\right)$, to be compared to the noise-free predictive variance that we can write $\sigma_G^2(\mathbf{u}) = C_G(\mathbf{u}, \mathbf{u}) - c^2 \sum_{i,j=1}^{N} K_{ij}^{-1} N_{\mathbf{x}_i}(\mathbf{x}_j, 2\mathbf{W}) N_{\mathbf{u}}\left(\frac{\mathbf{x}_i + \mathbf{x}_j}{2}, \frac{\mathbf{W}}{2}\right)$, using $N_{\mathbf{u}}(\mathbf{x}_i, \mathbf{W}) N_{\mathbf{u}}(\mathbf{x}_j, \mathbf{W}) = N_{\mathbf{x}_i}(\mathbf{x}_j, 2\mathbf{W}) N_{\mathbf{u}}\left(\frac{\mathbf{x}_i + \mathbf{x}_j}{2}, \frac{\mathbf{W}}{2}\right)$.

As done for the predictive mean, we can find another form for $v^{exG}(\mathbf{u}, \boldsymbol{\Sigma}_x)$ where the Gaussian covariance function appears weighted by a correction term. It can be shown that we can write l_{ij}^{exG} as

$$l_{ij}^{exG} = C_G(\mathbf{u}, \mathbf{x}_i) C_G(\mathbf{u}, \mathbf{x}_j) C_{corr_2}(\mathbf{u}, \bar{\mathbf{x}})$$

where $\bar{\mathbf{x}} = \frac{\mathbf{x}_i + \mathbf{x}_j}{2}$ and

$$C_{corr_2}(\mathbf{u}, \bar{\mathbf{x}}) = \left| \left(\frac{\mathbf{W}}{2} \right)^{-1} \boldsymbol{\Sigma}_x + \mathbf{I} \right|^{-1/2} \exp \left[\frac{1}{2} (\mathbf{u} - \bar{\mathbf{x}})^T \Lambda^{-1} (\mathbf{u} - \bar{\mathbf{x}}) \right] \qquad (32)$$

with $\Lambda^{-1} = \left(\frac{\mathbf{W}}{2} \right)^{-1} \left(\left(\frac{\mathbf{W}}{2} \right)^{-1} + \boldsymbol{\Sigma}_x^{-1} \right)^{-1} \left(\frac{\mathbf{W}}{2} \right)^{-1}$.

In terms of $\sigma_G^2(\mathbf{u})$, we can then write

$$v^{exG}(\mathbf{u}, \boldsymbol{\Sigma}_x) = \sigma_G^2(\mathbf{u}) + \sum_{i,j=1}^{N} K_{ij}^{-1} C_G(\mathbf{u}, \mathbf{x}_i) C_G(\mathbf{u}, \mathbf{x}_j)(1 - C_{corr_2}(\mathbf{u}, \bar{\mathbf{x}}))$$

$$+ \sum_{i,j=1}^{N} \beta_i \beta_j C_G(\mathbf{u}, \mathbf{x}_i) C_G(\mathbf{u}, \mathbf{x}_j)(C_{corr_2}(\mathbf{u}, \bar{\mathbf{x}}) - C_{corr}(\mathbf{u}, \mathbf{x}_i) C_{corr}(\mathbf{u}, \mathbf{x}_j)) ,$$

$$(33)$$

where we have used
$m^{exG}(\mathbf{u}, \boldsymbol{\Sigma}_x)^2 = \sum_{i,j=1}^{N} \beta_i \beta_j C_G(\mathbf{u}, \mathbf{x}_i) C_{corr}(\mathbf{u}, \mathbf{x}_i) C_G(\mathbf{u}, \mathbf{x}_j) C_{corr}(\mathbf{u}, \mathbf{x}_j)$.

Although we will not give the details of the calculations here, it can be shown that these predictive mean and variance tend to the approximate mean and variance presented in section 3 when $\boldsymbol{\Sigma}_x$ tends to zero (so that we can approximate e^x by $1 + x$). As Figure 5 for the approximate moments, Figure 6 shows the exact predictive mean and error-bars (triangles) obtained when predicting at noisy inputs (asterisks).

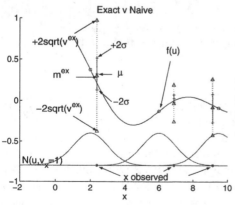

Fig. 6. As in Fig. 5, the triangles now indicate the exact predictive means with their error-bars, accounting for the uncertainty on the noisy inputs (asterisks).

5   Iterative k-step Ahead Prediction

Using the results derived in the previous sections, we now derive an algorithm for propagating the uncertainty as we predict ahead in time the output of a nonlinear dynamic system, represented by on one-step ahead non-linear auto-regressive (NAR) model.

At this point, it might be useful to recall the different notations used, depending on the situation, as done in Table 1. It is important not to forget that the predictive distribution corresponding to a noise-free $\mathbf{u}$ is Gaussian but it is not when predicting at $\mathbf{x} \sim \mathcal{N}_{\mathbf{x}}(\mathbf{u}, \boldsymbol{\Sigma}_x)$. We only compute its mean and variance, which is done exactly when the covariance function is e.g. Gaussian or linear, or approximately, in the *general* case.

Table 1. Notation used, depending on the type of covariance function (left column) and whether the prediction is at a noise-free or a noisy input. ('Where' in the document the corresponding equations can be found is indicated in small fonts.)

Covariance function	Prediction at $\mathbf{u}$	Prediction at $\mathbf{x} \sim \mathcal{N}_{\mathbf{x}}(\mathbf{u}, \boldsymbol{\Sigma}_x)$
General	$\mu(\mathbf{u}),\ \sigma^2(\mathbf{u})$	$m^{ap}(\mathbf{u}, \boldsymbol{\Sigma}_x)$ Eq. (13)
	Eqs. (1), at $\mathbf{u}$	$v^{ap}(\mathbf{u}, \boldsymbol{\Sigma}_x)$ Eq. (14)
Linear	$\mu_L(\mathbf{u}),\ \sigma_L^2(\mathbf{u})$	$m^{exL}(\mathbf{u}, \boldsymbol{\Sigma}_x)$ Eq. (18)
	Eqs. (15)	$v^{exL}(\mathbf{u}, \boldsymbol{\Sigma}_x)$ Eq. (21)
Gaussian	$\mu_G(\mathbf{u}),\ \sigma_G^2(\mathbf{u})$	$m^{exG}(\mathbf{u}, \boldsymbol{\Sigma}_x)$ Eq. (31)
	Eqs. (23)	$v^{exG}(\mathbf{u}, \boldsymbol{\Sigma}_x)$ Eq. (33)

5.1   Background

Given a discrete one-dimensional time-series $y_1, \ldots, y_t$, we wish to predict its value at, say, time $t + k$. Viewing the observed time-series as a projection of the dynamics of the underlying system, which lie in a higher dimensional space [13], we consider the following non-linear auto-regressive (NAR) model

$$y_{t+1} = f(\mathbf{x}_t) \quad \text{with} \quad \mathbf{x}_t = [y_t, y_{t-1}, \ldots, y_{t-L}]^T , \tag{34}$$

whose order, L, corresponds to the dimension of the reconstructed space (number of delayed outputs, called *lag* or *embedding dimension*). The state (or input) at time t is $\mathbf{x}_t$ and y_{t+1} is the corresponding output. Note that in practise, y_{t+1} is alone considered as noisy ($y_{t+1} = f(\mathbf{x}_t) + \epsilon_{t+1}$). Here, we simply assume that ϵ_{t+1} is a white noise but colored noise models can also be considered, as in [14].

Using this one-step ahead model, the iterative k-step ahead prediction task can be thought of as a missing or noisy data modelling problem[12] since what

---

[12] The missing variables can be seen as noisy variables for complete noise.

we want is to predict y_{t+k}, when y_{t+k-1} down to y_t are missing, provided the time-series is known up to time t. This problem has been the scope of much research (see e.g. [15, 16]) but has not yet been addressed for the GP model. A naive way of solving the iterative multiple-step ahead prediction task is simply to substitute a single value to the missing value (say the value of the time-series at another time-step, or a maximum likelihood estimate) but this approach has been shown not to be optimal and to lead to biased predictions [17, 15]. In [18], long-term predictions are improved by eliminating the systematic errors induced by each successive short term prediction, by considering a function of the estimates.

Using our approximation for the prediction at a noisy input, we suggest to incorporate the uncertainty about intermediate regressor values as we predict ahead in time. This results in an update of the uncertainty on the current prediction and therefore an improvement of each successive predictions.

5.2   Propagation of Uncertainty Algorithm

We assume that a zero-mean GP model was trained to minimize the one-step ahead predictions of a time-series known up to time t. By propagating the uncertainty as we predict ahead in time, we mean that for $y_{t+k} = f(y_{t+k-1}, \dots, y_{t+k-L})$, we consider the delayed $y_{t+k-1}, \dots, y_{t+k-L}$ as Gaussian random variables, with mean $m(.,.)$ and variance $v(.,.)$, computed either approximately or exactly, depending on the covariance function of the process.

Here is a sketch of how we proceed:

— Time $t+1$, $\mathbf{x}_{t+1} = [y_t, \dots, y_{t-L}]^T$: Since the state is formed on known values of the time-series, we simply have $y_{t+1} \sim \mathcal{N}(\mu(\mathbf{x}_{t+1}), \sigma^2(\mathbf{x}_{t+1}))$.
— Time $t+2$, $\mathbf{x}_{t+2} = [y_{t+1}, y_t, \dots, y_{t+1-L}]^T \sim \mathcal{N}(\mathbf{u}_{t+2}, \boldsymbol{\Sigma}_{t+2})$ with

$$\mathbf{u}_{t+2} = \begin{bmatrix} \mu(\mathbf{x}_{t+1}) \\ y_t \\ \vdots \\ y_{t+1-L} \end{bmatrix} \quad \text{and} \quad \boldsymbol{\Sigma}_{t+2} = \begin{bmatrix} \sigma^2(\mathbf{x}_{t+1}) & 0 & \dots & 0 \\ 0 & 0 & \dots & 0 \\ \vdots & \vdots & \vdots & \vdots \\ 0 & \dots & \dots & 0 \end{bmatrix}.$$

Within our analytical approximation, we only compute the mean and variance of y_{t+2} and consider $y_{t+2} \sim \mathcal{N}(m(\mathbf{u}_{t+2}, \boldsymbol{\Sigma}_{t+2}), v(\mathbf{u}_{t+2}, \boldsymbol{\Sigma}_{t+2}))$.
— Time $t+3$, $\mathbf{x}_{t+3} = [y_{t+2}, y_{t+1}, \dots, y_{t+2-L}]^T \sim \mathcal{N}(\mathbf{u}_{t+3}, \boldsymbol{\Sigma}_{t+3})$ with

$$\mathbf{u}_{t+3} = \begin{bmatrix} m(x_{t+2}) \\ \mu(x_{t+1}) \\ y_t \\ \vdots \\ y_{t+2-L} \end{bmatrix} \quad \text{and} \quad \boldsymbol{\Sigma}_{t+3} = \begin{bmatrix} v(x_{t+2}) & \text{Cov}[y_{t+2}, y_{t+1}] & 0 & \dots & 0 \\ \text{Cov}[y_{t+1}, y_{t+2}] & \sigma^2(x_{t+1}) & 0 & \dots & 0 \\ 0 & 0 & 0 & \dots & 0 \\ \vdots & \vdots & \vdots & \vdots & \vdots \\ 0 & \dots & \dots & \dots & 0 \end{bmatrix}.$$

Compute $y_{t+3} \sim \mathcal{N}(m(\mathbf{u}_{t+3}, \boldsymbol{\Sigma}_{t+3}), v(\mathbf{u}_{t+3}, \boldsymbol{\Sigma}_{t+3}))$.

Repeating this procedure up to the desired horizon k, and assuming $k > L$, at $t + k$, we have $\mathbf{x}_{t+k} = [y_{t+k-1}, y_{t+k-2}, \ldots, y_{t+k-L}]^T \sim \mathcal{N}(\mathbf{u}_{t+k}, \boldsymbol{\Sigma}_{t+k})$ and compute $y_{t+k} \sim \mathcal{N}(m(\mathbf{u}_{t+k}, \boldsymbol{\Sigma}_{t+k}), v(\mathbf{u}_{t+k}, \boldsymbol{\Sigma}_{t+k}))$. The input mean is then given by

$$\mathbf{u}_{t+k} = [m(x_{t+k-1}), m(x_{t+k-2}), \ldots, m(x_{t+k-L})]^T$$

and the input covariance matrix is

$$\boldsymbol{\Sigma}_{t+k} = \begin{bmatrix} v(x_{t+k-1}) & \mathrm{Cov}[y_{t+k-1}, y_{t+k-2}] & \ldots & \mathrm{Cov}[y_{t+k-1}, y_{t+k-L}] \\ \mathrm{Cov}[y_{t+k-2}, y_{t+k-1}] & v(x_{t+k-2}) & \ldots & \mathrm{Cov}[y_{t+k-2}, y_{t+k-L}] \\ \ldots & \ldots & \ldots & \ldots \\ \mathrm{Cov}[y_{t+k-L}, y_{t+k-1}] & \mathrm{Cov}[y_{t+k-L}, y_{t+k-2}] & \ldots & v(x_{t+k-L}) \end{bmatrix} .$$

We now need to compute the cross-covariance terms: In general, at time $t+l$, we have the random input vector $\mathbf{x}_{t+l} = [y_{t+l-1}, \ldots, y_{t+l-L}]^T \sim \mathcal{N}(\mathbf{u}_{t+l}, \boldsymbol{\Sigma}_{t+l})$. The $L \times L$ covariance matrix $\boldsymbol{\Sigma}_{t+l}$ has the delayed predictive variances on its diagonal and the cross-covariance terms correspond to the covariances between y_{t+l-i} and y_{t+l-j}, for $i, j = 1 \ldots L$ with $i \neq j$. Discarding the last (*oldest*) element of $\mathbf{x}_{t+l}$, we need to compute $\mathrm{Cov}[y_{t+l-i}, y_{t+l-j}] = \mathrm{Cov}[y_{t+l}, \mathbf{x}_{t+l}]$, that is

$$\mathrm{Cov}[y_{t+l}, \mathbf{x}_{t+l}] = E[y_{t+l}\mathbf{x}_{t+l}] - E[y_{t+l}]E[\mathbf{x}_{t+l}] \tag{35}$$

where $E[y_{t+l}] = m(\mathbf{u}_{t+l}, \boldsymbol{\Sigma}_{t+l})$ and $E[\mathbf{x}_{t+l}] = \mathbf{u}_{t+l}$. For the expectation of the product, we have

$$E[y_{t+l}\mathbf{x}_{t+l}] = \int\int y_{t+l}\mathbf{x}_{t+l}p(y_{t+l}, \mathbf{x}_{t+l})dy_{t+l}d\mathbf{x}_{t+l}$$

$$= \int\int y_{t+l}\mathbf{x}_{t+l}p(y_{t+l}|\mathbf{x}_{t+l})p(\mathbf{x}_{t+l})dy_{t+l}d\mathbf{x}_{t+l}$$

and since $\int y_{t+l}p(y_{t+l}|\mathbf{x}_{t+l})dy_{t+l} = \mu(\mathbf{x}_{t+l})$, we can write

$$E[y_{t+l}\mathbf{x}_{t+l}] = \int \mathbf{x}_{t+l}\mu(\mathbf{x}_{t+l})p(\mathbf{x}_{t+l})d\mathbf{x}_{t+l} .$$

Replacing $\mu(\mathbf{x}_{t+l})$ by its expression, we have

$$E[y_{t+l}\mathbf{x}_{t+l}] = \sum_i \beta_i \int \mathbf{x}_{t+l}C(\mathbf{x}_{t+l}, \mathbf{x}_i)p(\mathbf{x}_{t+l})d\mathbf{x}_{t+l} . \tag{36}$$

Depending on the form of $C(., .)$, we evaluate this integral exactly or approximately. Denoting $\mathbf{x}_{t+l}$ by $\mathbf{x}$ for notational convenience, let $I_i = \int \mathbf{x}C(\mathbf{x}, \mathbf{x}_i)p(\mathbf{x})d\mathbf{x}$ be the integral we wish to solve.

Gaussian Case In the case of the Gaussian covariance function, we have $m(., .) = m^{exG}(., .)$ and $v(., .) = v^{exG}(., .)$, as given by Eqs. (31) and (33).

Using a similar notation as in section 4.2, we need to solve

$$I_i^{exG} = c\int \mathbf{x}N_{\mathbf{x}}(\mathbf{x}_i, \mathbf{W})p(\mathbf{x})d\mathbf{x}$$

where $C(\mathbf{x}, \mathbf{x}_i) = cN_{\mathbf{x}}(\mathbf{x}_i, \mathbf{W})$, with $c = (2\pi)^{D/2}|\mathbf{W}|^{1/2}v$. As before, using the product of Gaussians, we find

$$I_i^{exG} = cN_{\mathbf{u}}(\mathbf{x}_i, \mathbf{W} + \boldsymbol{\Sigma}_x)[(\mathbf{I} + \mathbf{W}\boldsymbol{\Sigma}_x^{-1})^{-1}\mathbf{x}_i + (\mathbf{I} + \boldsymbol{\Sigma}_x\mathbf{W}^{-1})^{-1}\mathbf{u}]$$

where $cN_{\mathbf{u}}(\mathbf{x}_i, \mathbf{W} + \boldsymbol{\Sigma}_x) = C(\mathbf{u}, \mathbf{x}_i)C_{corr}(\mathbf{u}, \mathbf{x}_i)$, with $C_{corr}(\mathbf{u}, \mathbf{x}_i)$ given by (30). We can then write

$$E[y_{t+l}\mathbf{x}_{t+l}] = \sum_i \beta_i C(\mathbf{u}_{t+l}, \mathbf{x}_i)C_{corr}(\mathbf{u}_{t+l}, \mathbf{x}_i)[(\mathbf{I} + \mathbf{W}\boldsymbol{\Sigma}_{t+l}^{-1})^{-1}\mathbf{x}_i + (\mathbf{I} + \boldsymbol{\Sigma}_{t+l}\mathbf{W}^{-1})^{-1}\mathbf{u}_{t+l}].$$

After simplifications, the cross-covariance terms are given by

$$\mathrm{Cov}[y_{t+l}, \mathbf{x}_{t+l}] = \sum_i \beta_i C(\mathbf{u}_{t+l}, \mathbf{x}_i)C_{corr}(\mathbf{u}_{t+l}, \mathbf{x}_i)(\mathbf{I} + \mathbf{W}\boldsymbol{\Sigma}_{t+l}^{-1})^{-1}\mathbf{x}_i . \quad (37)$$

General Case When the covariance function is such that approximations are needed, the predictive mean and variance corresponding to a noisy input are given by $m(.,.) = m^{ap}(.,.)$, using Eq. (13) and $v(.,.) = v^{ap}(.,.)$, using (14).

For the computation of the cross-covariances, we resort to a second order Taylor approximation of the covariance function, as in section 3. We then have[13]

$$I_i^{ap} \approx \mathbf{u}^T C(\mathbf{u}, \mathbf{x}_i) + \mathbf{C}'(\mathbf{u}, \mathbf{x}_i)^T \boldsymbol{\Sigma}_x + \frac{1}{2}\mathbf{u}^T \mathrm{Tr}[\boldsymbol{\Sigma}_x \mathbf{C}''(\mathbf{u}, \mathbf{x}_i)] .$$

After simplifications, we obtain the following expression for the cross-covariance terms

$$\mathrm{Cov}[y_{t+l}, \mathbf{x}_{t+l}] = \sum_i \beta_i \mathbf{C}'(\mathbf{u}_{t+l}, \mathbf{x}_i)^T \boldsymbol{\Sigma}_{t+l}. \quad (38)$$

6   Numerical Examples

For clarity, we will denote the different approaches as follows:

- MC, for the Monte-Carlo approximation to the true predictive distribution corresponding to a noisy input;
- A, for the *Gaussian approximation* that computes only the mean and variance of this distribution, and specifically A_{ap} when these moments are computed using the Taylor approximation, and A_{ex} when they are computed exactly;

---

[13] This result was obtained by extending the one-dimensional case to L-dimensions. In 1D, we have

$$I_i^{ap} \approx \int x \left(C(u, x_i) + (x - u)C'(u, x_i) + \frac{1}{2}(x - u)^2 C''(u, x_i) \right) p(x)dx$$

$$\approx uC(u, x_i) + v_x C'(u, x_i) + \frac{1}{2}uv_x C''(u, x_i)$$

where we have used $\int x^2 p(x)dx = v_x + u^2$ and $\int x^3 p(x)dx = 3uv_x + u^3$.

– N, for the *naive* predictive mean and variances that do not account for the noise on the input.

We assess the performance of the different methods by computing the average squared error (E_1), over the test set, and average minus log predictive density (E_2), which measures the density of the actual true test output under the Gaussian predictive distribution and use its negative log as a measure of loss. To assess the performance of the Monte-Carlo approximation, we compute the squared error and minus log-likelihood loss for the predictions given by each sample and average over the number of samples. We also compute the average predictive mean (sample mean) and average predictive variance (sample variance) and compute the associated losses.

6.1 A Simple Comparison on a Static Example

On the static example previously used, we compare the different approaches for the prediction at a noisy input, when the true noise-free input is 2 (left) and 6 (right) and the input noise variance is 1. Figure 7 shows the predictive distribution given by MC (continuous), N (dashed), A_{ap} (dots) and A_{ex} (asterisks). Note how the naive approach leads to a narrow distribution (N), peaked around its mean value, since it does not account for the uncertainty on the input. The Monte-Carlo approximation to the true distribution highlights how the true distribution is non-Gaussian.

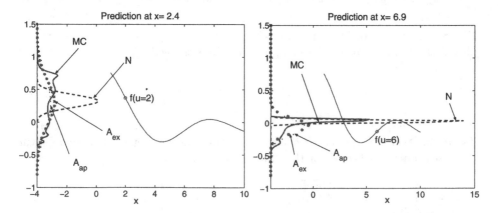

Fig. 7. Predictive distributions (on the y-axis) obtained when predicting at a noisy input: MC is the numerical approximation by simple Monte-Carlo, A_{ex} and A_{ap} correspond to the *Gaussian approximation* with moments computed exactly and approximately. N is the *naive* predictive distribution that does not account for the noise on the input.

For both the prediction at $x = 2.4$ (left) and $x = 6.9$ (right), Figure 8 shows the histogram of the losses (squared error E_1 on the left and minus log predic-

tive density E_2 on the right) computed for each of the 100 samples given by
the Monte-Carlo approximation. The minus-log predictive density loss is a very
useful quantitative measure to assess the 'goodness' or quality of an approach
as, unlike the squared error loss, it also accounts for the variance (or uncertainty) attached to the mean predictions. Table 2 summarizes the average losses
obtained for each method (average over three test points). In this table, the
losses reported for MC correspond to those obtained using the average sample
mean and variance (average over 100 samples). We can also compute the losses
associated to each sample and average those. We then obtain $E_1 = 0.42$ and
$E_2 = 25.09$.

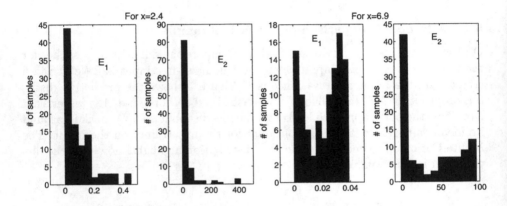

Fig. 8. Squared error (E_1) and minus log-likelihood (E_2) computed for 100 samples of the Monte-Carlo approximation (for the observed noisy $x = 2.4$, left and
$x = 6.9$, right).

Table 2. Average squared error E_1 and minus log-predictive density E_2 over
three test points obtained for the different approaches.

Loss	N	A_{ap}	A_{ex}	MC
E_1	0.009	0.004	0.005	0.004
E_2	7.685	−0.53	−0.635	−0.58

From this simple static example, for which the input noise variance is assumed
to be known, we can conclude that our *Gaussian approximation* leads to results
comparable to those obtained by simple Monte-Carlo, which approximates the
true distribution.

6.2   Dynamic Case

The Mackey-Glass chaotic time-series constitutes a well-known challenging benchmark for the multiple-step ahead prediction task, due to its strong non-linearity
[19]. We consider $\frac{dy(t)}{dt} = -by(t) + a\frac{y(t-\tau)}{1+y(t-\tau)^{10}}$, with $a = 0.2$, $b = 0.1$ and $\tau = 17$.
The series is re-sampled with period 1 and normalized. We then assume the
following NAR model $y_{t+1} = f(y_t, y_{t-1}, \ldots, y_{t-L})$, where $L = 16$ and we cor-
rupt the output y_{t+1} by a white noise with variance 0.001. Having formed the
input/output pairs, we train a zero-mean Gaussian Process with a Gaussian covariance function[14] on 100 points (taken at random). We first validate the model
on one-step ahead predictions: We obtain $E_1 = 4.41\ 10^{-4}$, $E_2 = -2.16$ where
the average is taken over $N_t = 1000$ test points. After performing a simulation
of the test set (i.e. N_t-steps ahead prediction, where N_t is the length of the test
set), we decide to make $k = 100$ steps ahead predictions (which corresponds to
the horizon up to which predictions are 'reasonably good').

This example is intended to illustrate the propagation of uncertainty algorithm, described in section 5.2. We assess the quality of the predictions obtained
using the approximate moments, given by the *Gaussian approximation*, by comparing them to the exact ones. We also compare the 'exact predictions' to those
given by the naive approach, that feeds back only the predictive means as we
predict ahead in time. Let t be the time up to which the time-series is known.
Fig. 9 (top plots) shows the mean predictions (left) with their associated uncertainties (right) from $t + 1$ to $t + 100$. The crosses indicate the exact moments
given by the *Gaussian approximation* (A_{ex}), the circles indicate the approximate
moments (A_{ap}) and the dots the naive moments (N) that ignore the uncertainty
induced by each successive prediction. We can see that up to around 60 steps
ahead, the predictive means given by the different approaches are very similar.
The uncertainty bars given by naive approach are very tight and the model is
overly confident about its mean predictions. On the other hand, both the exact
and approximate error-bars reflect well the fact that, as we predict ahead in
time, less information is available and the estimates (predictive means) become
more and more uncertain. On Fig. 9, the bottom left figure shows the 100-step
ahead predictive means with their uncertainty. The upper plot shows the predictive means given by the naive approach, with their 2σ error-bars which are so
tight that one cannot distinguish them from the means. The middle and bottom
plots show respectively the approximate and exact 100-step ahead means where
the shaded area corresponds to the uncertainty interval. On the right, we can see
the evolution of the average squared error (left) and minus log-predictive density
(right, on a log-scale), as the number of steps increases from one to 100. In this
case, both losses clearly indicate that as the number of steps increases, the naive
approach leads to poor predictions. These plots also show that, although not as
good as A_{ex}, the predictions given by the approximate moments A_{ap} are quite
encouraging.

---

[14] The covariance function is that given by Eq. (22) with $v = 1$.

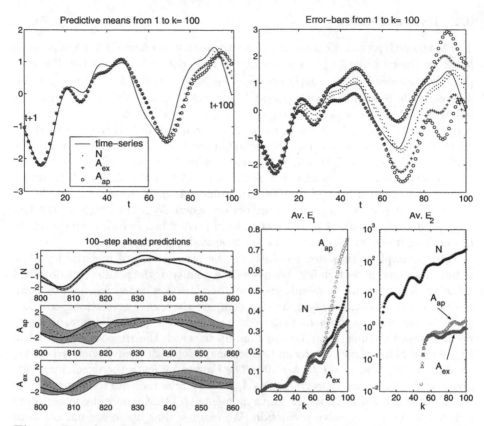

Fig. 9. Top plots: Iterative method in action on the Mackey-Glass time-series. Mean predictions (left) and uncertainty error bars (right) from 1 to 100 steps ahead, given by the exact moments A_{ap} (crosses), the approximate ones A_{ap} (circles) and the naive ones (dots). Bottom plots: 100-step ahead prediction of a portion of the time-series (left). From top to bottom: naive, approximate and exact means with the uncertainty region shaded. Right: Evolution of the average losses as the number of steps ahead increases from one to 100 (E_1 is the average squared error and E_2 the minus log-predictive density, on a log-scale)

We now turn to comparing the *Gaussian approximation* (exact moments) to the approximation of the true distribution by Monte-Carlo (MC). The Monte-Carlo approximation for the 100-step ahead prediction is done as follows: At $t+1$, compute $p(y_{t+1}|\mathcal{D}, \mathbf{x}_{t+1}) = \mathcal{N}(\mu(\mathbf{x}_{t+1}), \sigma^2(\mathbf{x}_{t+1}))$ where $\mathbf{x}_{t+1} = [y_t, y_{t-1}, \ldots, y_{t-16}]$. At $t+2$, draw a sample y_{t+1}^s from $p(y_{t+1}|\mathcal{D}, \mathbf{x}_{t+1})$, form the state $\mathbf{x}_{t+2} = [y_{t+1}^s, y_t, \ldots, y_{t-15}]$ and compute $p(y_{t+2}|\mathcal{D}, \mathbf{x}_{t+2}) = \mathcal{N}(\mu(\mathbf{x}_{t+2}), \sigma^2(\mathbf{x}_{t+2}))$. So on, up to $t+100$. Then, go back to $t+1$ and repeat the whole process. We repeat this $S = 1000$ times ($s = 1 \ldots S$), so that we finally obtain 1000 samples for each time-step. Finally, we do so for 100 different 'starting times t' (i.e. 100 test inputs), resulting in a $100 \times S \times k$ matrix of predictive means and variances, where S is the number of samples and k is the prediction horizon ($k = 100$). Fig. 10 shows the predictive uncertainties from $t + 1$ to $t + 100$.

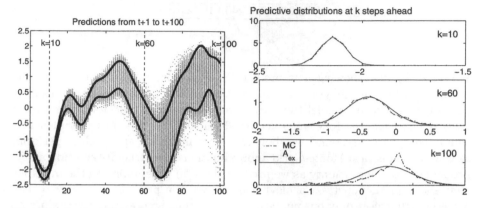

Fig. 10. Left: 1000 predictive error-bars from the Monte-Carlo approximation, from $t + 1$ to $t + k$, where $k = 100$ steps ahead. Also plotted, the predictive uncertainties given by the exact method A_{ex} (continuous lines). At $t + k$, for $k = 10, 60, 100$, we plot the corresponding predictive distribution(right plot), as it is approximated numerically by Monte-Carlo (dotted line) and analytically, by the Gaussian with exact moments (continuous).

This experiment clearly validates our analytical approximation of the true predictive distribution as we can see that the error bars given by the exact moments encompass those of the samples from the Monte-Carlo approximation. It is interesting noting how the approximation to the true distribution is long-tailed at 100 step-ahead.

Table 3, reports the average losses computed for the different approaches. (Note that since the Monte-Carlo approach uses only 100 test points, all losses are averaged over 100 points only.) The reported losses for MC correspond to the those computed using the average sample mean and variance. We can also compute the losses given by each single prediction and average them. We then obtain $E_1 = 0.72$ and $E_2 = 340.27$. These results for the Monte-Carlo approximation might look surprising but one should keep in mind that estimating the quality of this approximation with these losses is not really representative (since the distribution is not normal).

7 Conclusions

We have presented an original solution to the problem of iterative multiple-step ahead prediction of nonlinear dynamic systems within a NAR representation.

Table 3. Average (over 100 test points) squared error (E_1) and minus log predictive density (E_2) for the $k = 100$ step ahead predictions.

	N	A_{ap}	A_{ex}	MC
E_1	0.52	0.75	0.35	0.38
E_2	243.46	1.55	0.94	172.51

We do so by first showing how predicting at an uncertain or noisy input can be done within an analytical approximation of the predictive distribution of the Gaussian Process model (note that this approach is valid for other kernel-based models like the Relevance Vector Machines, see [20]). In experiments on simulated dynamic systems, we show that this analytical approach 1, performs as well as a numerical Monte-Carlo approximation of the true distribution and 2, propagating the uncertainty as we predict ahead in time improves the multiple-step ahead prediction task, achieving more realistic prediction variances than a method that uses only output estimates and thus ignores the uncertainty on current state.

In the derivation of the mean and variance of the predictive distribution, we show how exact or approximate moments are obtained, depending on the form of the covariance function. In the case of the Gaussian covariance function, for which exact moments are available, a numerical example proves that the approximate moments, computed using the Gaussian covariance function, lead to almost similar results as those obtained using the exact moments, which is encouraging for using the approximation.

Explicitly using the predictive variance has been recently successfully used in a control context [21] and also the propagation of uncertainty methodology, in a model predictive control framework where knowledge of the accuracy of the model predictions over the whole prediction horizon is required (see [22]).

In this chapter, we do not address the problem of learning in the presence of noisy inputs (we have assumed that the training inputs were noise-free). This is the subject of ongoing research. We suggest an approximation similar to that presented here: Assuming the input noise is white, the new non-Gaussian process can be approximated by a GP. We then derive its covariance function that accounts for the input noise variance, which is then learnt as an extra parameter.

Acknowledgements Thanks to Carl Edward Rasmussen who initiated this work. Many thanks to Joaquin Quiñonero-Candela for his feedback on this chapter and to Professor Mike Titterington for useful discussions on the subject. The authors gratefully acknowledge the support of the *Multi-Agent Control* Research Training Network - EC TMR grant HPRN-CT-1999-00107 and RMS is grateful for EPSRC grant *Modern statistical approaches to off-equilibrium modelling for nonlinear system control* GR/M76379/01.

References

[1] Williams, C.K.I., Rasmussen, C.E.: Gaussian Processes for Regression. In Touretzky, D.S., Mozer, M.C., Hasselmo, M.E., eds.: Advances in Neural Information Processing Systems. Volume 8., MIT Press (1996) 514–520

[2] Williams, C.K.I.: Prediction with Gaussian Processes: From linear regression to linear prediction and beyond. Technical Report NCRG-97-012, Dept of Computer Science and Applied Mathematics. Aston University. (1997)

[3] Williams, C.K.I.: Gaussian Processes. The handbook of Brain Theory and Neural Networks, Second edition, MIT Press (2002)

[4] Mackay, D.J.C.: Information theory, Inference and Learning Algorithms. Cambridge University Press (2003)

[5] Girard, A., Rasmussen, C., Quinonero-Candela, J., Murray-Smith, R.: Gaussian Process Priors With Uncertain Inputs – Application to Multiple-Step Ahead Time Series Forecasting. In Becker, S., Thrun, S., Obermayer, K., eds.: Advances in Neural Information Processing Systems. Volume 15., MIT Press (2003) 545–552

[6] Quinonero-Candela, J., Girard, A., Larsen, J., Rasmussen, C.E.: Propagation of Uncertainty in Bayesian Kernels Models – Application to Multiple-Step Ahead Forecasting. In: IEEE International Conference on Acoustics, Speech and Signal Processing. Volume 2. (2003) 701–4

[7] Girard, A., Rasmussen, C., Murray-Smith, R.: Gaussian Process Priors with Uncertain Inputs: Multiple-Step Ahead Prediction. Technical Report TR-2002-119, Computing Science Department, University of Glasgow (2002)

[8] Lindley, D.V.: Introduction to Probability and Statistics from a Bayesian viewpoint. Cambridge University Press (1969)

[9] Papoulis, A.: Probability, random variables, and stochastic processes. McGraw-Hill (1991)

[10] Rasmussen, C.E.: Evaluation of Gaussian Processes and other methods for nonlinear regresion. PhD thesis, University of Toronto (1996)

[11] Neal, R.M.: Bayesian learning for neural networks. PhD thesis, University of Toronto (1995)

[12] MacKay, D.J.C.: Bayesian methods for backpropagation networks. In Domany, E., van Hemmen, J.L., Schulten, K., eds.: Models of Neural Networks III. Springer-Verlag, New York (1994) 211–254

[13] Takens, F.: Detecting strange attractors in turbulence. In Rand, D., Young, L., eds.: Dynamical Systems and Turbulence. Volume 898., Springer-Verlag (1981) 366–381

[14] Murray-Smith, R., Girard, A.: Gaussian Process priors with ARMA noise models. In: Irish Signals and Systems Conference, Maynooth. (2001) 147–152

[15] Tresp, V., Hofmann, R.: Missing and Noisy Data in Nonlinear Time-Series Prediction. In S. F. Girosi, J. Mahoul, E.M., Wilson, E., eds.: Neural Networks for Signal Processing. Volume 24 of IEEE Signal Processing Society, New York. (1995) 1–10

[16] Tresp, V., Hofmann, R.: Nonlinear Time-Series Prediction with Missing and Noisy Data. Neural Computation 10 (1998) 731–747

[17] Ahmad, S., Tresp, V.: Some Solutions to the Missing Feature Problem in Vision. In Hanson, S.J., Cowan, J.D., Giles, C.L., eds.: Advances in Neural Information Processing Systems. Volume 5., Morgan Kaufmann, San Mateo, CA (1993) 393–400

[18] Judd, K., Small, M.: Towards long-term prediction. Physica D 136 (2000) 31–44

[19] Mackey, M.C., Glass, L.: Oscillation and chaos in physiological control systems. Science **197** (1977) 287–289
[20] Quinonero-Candela, J., Girard, A.: Prediction at an Uncertain Input for Gaussian Processes and Relevance Vector Machines – Application to Multiple-Step Ahead Time-Series Forecasting. Technical report, Informatics and Mathematical Modelling, Technical Univesity of Denmark (2002)
[21] Murray-Smith, R., Sbarbaro, D.: Nonlinear adaptive control using non-parametric Gaussian process prior models. In: 15th IFAC Triennial World Congress. International Federation of Automatic Control. (2002)
[22] Murray-Smith, R., Sbarbaro, D., Rasmussen, C.E., Girard, A.: Adaptive, Cautious, Predictive control with Gaussian Process priors. In: IFAC International Symposium on System Identification, Rotterdam. (2003)

Nonlinear Predictive Control with a Gaussian Process Model

Juš Kocijan[1,2] and Roderick Murray-Smith[3,4]

[1] Jozef Stefan Institute, Jamova 39, SI-1000 Ljubljana, Slovenia
jus.kocijan@ijs.si
[2] Nova Gorica Polytechnic, Nova Gorica
[3] Dept. Computing Science, University of Glasgow, Glasgow
[4] Hamilton Institute, National University of Ireland, Maynooth

Abstract. Gaussian process models provide a probabilistic non-parametric modelling approach for black-box identification of nonlinear dynamic systems. The Gaussian processes can highlight areas of the input space where prediction quality is poor, due to the lack of data or its complexity, by indicating the higher variance around the predicted mean. Gaussian process models contain noticeably less coefficients to be optimized. This chapter illustrates possible application of Gaussian process models within model predictive control. The extra information provided by the Gaussian process model is used in predictive control, where optimization of the control signal takes the variance information into account. The predictive control principle is demonstrated via the control of a pH process benchmark.

1 Introduction

Model Predictive Control (MPC) is a common name for computer control algorithms that use an explicit process model to predict the future plant response. According to this prediction in the chosen period, also known as the prediction horizon, the MPC algorithm optimizes the manipulated variable to obtain an optimal future plant response. The input of chosen length, also known as control horizon, is sent to the plant and then the entire sequence is repeated again in the next time period. The popularity of MPC is to a great extent owed to the ability of MPC algorithms to deal with constraints that are frequently met in control practice and are often not well addressed by other approaches. MPC algorithms can handle hard state and rate constraints on inputs and states that are usually, but not always incorporated in the algorithms via an optimization method. Linear model predictive control approaches [13] started appearing in the early eighties and are well-established in control practice (e.g. [18] for an overview). Nonlinear model predictive control (NMPC) approaches [1] started to appear about ten years later and have also found their way into control practice (e.g. [19,23]) though their popularity can not be compared to linear model predictive control. This fact is connected with the difficulty in nonlinear model construction and with the lack of the necessary confidence in the model. There

R. Murray-Smith, R. Shorten (Eds.): Switching and Learning, LNCS 3355, pp. 185–200, 2005.

were a number of contributions in the field of nonlinear model predictive control dealing with issues like stability, efficient computation, optimization, constraints and others. Some recent work in this field can be found in [2,12]. NMPC algorithms are based on various nonlinear models. Often these models are developed as first principles models, but other approaches, like black-box identification approaches are also popular. Various predictive control algorithms are based on neural network models e.g. [17], fuzzy models e.g. [8] or local model networks e.g. [6]. The quality of control depends on the quality of the model. New developments in NMPC approaches are coming from resolving various issues: from faster optimization methods to different process model. The contribution of this chapter is to describe a NMPC principle with a Gaussian process model. The Gaussian process model is an example of a probabilistic non-parametric model that also provides information about prediction uncertainties which are difficult to evaluate appropriately in nonlinear parametric models. The majority of work on Gaussian processes shown up to now considers modelling of static nonlinearities. The use of Gaussian processes in modelling dynamic systems is a recent development e.g. [15,14,3,21,10,11] and some control algorithms based on such are described in [16,5].

The chapter is organized as follows. Dynamic Gaussian process models are briefly introduced in the next section. The control algorithm principle is described in Section 3 and illustrated with the benchmark pH process control in Section 4. Conclusions are stated at the end of the chapter.

2 Modelling of Dynamic Systems with Gaussian Processes

A Gaussian process is an example of the use of a flexible, probabilistic, non-parametric model which directly provides us with uncertainty predictions. Its use and properties for modelling are reviewed in [22].

A Gaussian process is a collection of random variables which have a joint multivariate Gaussian distribution. Assuming a relationship of the form $y = f(\mathbf{x})$ between an input $\mathbf{x}$ and output y, we have $y^1, \ldots, y^n \sim \mathcal{N}(0, \Sigma)$, where $\Sigma_{pq} = \mathrm{Cov}(y_p, y_q) = C(\mathbf{x}_p, \mathbf{x}_q)$ gives the covariance between output points corresponding to input points $\mathbf{x}_p$ and $\mathbf{x}_q$. Thus, the mean $\mu(\mathbf{x})$ (usually assumed to be zero) and the covariance function $C(\mathbf{x}_p, \mathbf{x}_q)$ fully specify the Gaussian process. Note that the covariance function $C(.,.)$ can be any function with the property that it generates a positive definite covariance matrix.

A common choice is

$$C(\mathbf{x}_p, \mathbf{x}_q) \;=\; v_1 \exp\left[-\frac{1}{2}\sum_{d=1}^{D} w_d (x_p^d - x_q^d)^2\right] + v_0 \tag{1}$$

where $\boldsymbol{\Theta} = [w_1 \ldots w_D \; v_0 \; v_1]^T$ are the 'hyperparameters' of the covariance functions and D the input dimension. Other forms of covariance functions suitable for different applications can be found in [20]. For a given problem, the parameters are learned (identified) using the data at hand. After the learning, one can use the w parameters as indicators of 'how important' the corresponding input components (dimensions) are: if w_d is zero or near zero it means that the inputs in dimension d contain little information and could possibly be removed.

Consider a set of N D-dimensional input vectors $\mathbf{X} = [\mathbf{x}_1, \mathbf{x}_2, \ldots, \mathbf{x}_N]$ and a vector of output data $\mathbf{y} = [y^1, y^2, \ldots, y^N]^T$. Based on the data $(\mathbf{X}, \mathbf{y})$, and given a new input vector $\mathbf{x}^*$, we wish to find the predictive distribution of the corresponding output y^*. Unlike other models, there is no model parameter determination as such, within a fixed model structure. With this model, most of the effort consists in *tuning* the parameters of the covariance function. This is done by maximizing the log-likelihood of the parameters, which is computationally relatively demanding since the inverse of the data covariance matrix $(N \times N)$ has to be calculated at every iteration.

The described approach can be easily utilized for regression calculation. Based on training set $\mathbf{X}$ a covariance matrix $\mathbf{K}$ of size $N \times N$ is determined. As already mentioned before the aim is to find the distribution of the corresponding output y^* at some new input vector $\mathbf{x}^* = [x_1(N+1), x_2(N+1), \ldots, x_D(N+1)]^T$.

For a new test input $\mathbf{x}^*$, the predictive distribution of the corresponding output is $y^*|\mathbf{x}^*, (\mathbf{X}, \mathbf{y})$ and is Gaussian, with mean and variance

$$\mu(\mathbf{x}^*) = \mathbf{k}(\mathbf{x}^*)^T K^{-1} \mathbf{y} \tag{2}$$
$$\sigma^2(\mathbf{x}^*) = k(\mathbf{x}^*) - \mathbf{k}(\mathbf{x}^*)^T \mathbf{K}^{-1} \mathbf{k}(\mathbf{x}^*) + v_0 \tag{3}$$

where $\mathbf{k}(\mathbf{x}^*) = [C(\mathbf{x}^1, \mathbf{x}^*), \ldots, C(\mathbf{x}^N, \mathbf{x}^*)]^T$ is the $N \times 1$ vector of covariances between the test and training cases and $k(\mathbf{x}^*) = C(\mathbf{x}^*, \mathbf{x}^*)$ is the covariance between the test input and itself.

For multi-step ahead prediction we have to take account of the uncertainty of future predictions which provide the 'inputs' for estimating further means and uncertainties.

If we now consider a new random input, $\mathbf{x}^* \sim \mathcal{N}(\mu_{\mathbf{x}^*}, \Sigma_{\mathbf{x}^*})$, Girard *et. al.* [3], have shown that, within a Gaussian approximation the predictive distribution is again Gaussian with mean and variance

$$m(\mu_{\mathbf{x}^*}, \Sigma_{\mathbf{x}^*}) = E_{\mathbf{x}^*}[\mu(\mathbf{x}^*)] \tag{4}$$
$$v(\mu_{\mathbf{x}^*}, \Sigma_{\mathbf{x}^*}) = E_{\mathbf{x}^*}[\sigma^2(\mathbf{x}^*)] + E_{\mathbf{x}^*}[\mu(\mathbf{x}^*)^2] - (E_{\mathbf{x}^*}[\mu(\mathbf{x}^*)])^2 \tag{5}$$

The more detailed derivation can be found in the previous chapter of this book [4]. Equations (4) and (5) can be applied to calculation of multi-step ahead prediction with propagation of uncertainty.

Gaussian processes can, like neural networks, be used to model static nonlinearities and can therefore be used for modelling of dynamic systems if delayed input and output signals are used as regressors. In such cases an autoregressive model is considered, such that the current output depends on previous estimated outputs, as well as on previous control inputs.

$$\mathbf{x}(k) = [\hat{y}(k-1), \hat{y}(k-2), \ldots, \hat{y}(k-L), u(k-1),$$
$$u(k-2), \ldots, u(k-L)]^T$$
$$\hat{y}(k) = f(\mathbf{x}(k)) + \epsilon \tag{6}$$

Where k denotes consecutive number of data sample. Let $\mathbf{x}$ denote the state vector composed of the previous estimated outputs $\hat{y}$ and inputs u up to a given lag L and ϵ is white noise.

Iterative multi-step ahead prediction is done, as described in the previous chapter [4], by feeding back the predictive mean, as well as the predictive variance at each time-step, thus taking the uncertainty attached to each intermediate prediction into account. The Gaussian process model now not only describes the dynamic characteristics of the non-linear system, but at the same time provides information about the confidence in the predictions. The Gaussian process can highlight areas of the input space where prediction quality is poor, due to the lack of data, by indicating the higher variance around the predicted mean.

It is worthwhile noting that the derivatives of means and variances can be calculated in straightforward manner. For more details see [21].

3   Nonlinear Model Predictive Control

Nonlinear model predictive control as it was applied with the Gaussian process model can be in general described with a block diagram, as depicted in Fig. 1. The model used is fixed, identified off-line, which means that used control

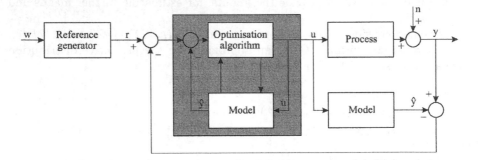

Fig. 1. Block diagram of model predictive control system

algorithm is not an adaptive one. The structure of the entire control loop is therefore less complex than in the case where the model changes with time. The following items describe the basic idea of predictive control:

- Prediction of the system output signal $y(k+j)$ is calculated for each discrete sample k for a large horizon in future $(j = N_1, \ldots, N_2)$. Predictions are denoted as $\hat{y}(k+j|k)$ and represent j-step ahead prediction, while N_1 and N_2 determine lower and upper bound of prediction horizon. Lower and upper bound of output signal prediction horizon determine coincidence horizon, within which a match between output and reference signal is expected. Output signal prediction is calculated from process model. Predictions are dependent also on the control scenario in the future $u(k+j|k), j = 0, \ldots, N_u-1$, which is intended to be applied from a moment k onwards.
- The reference trajectory is determined $r(k + j|k), j = N_1, \ldots, N_2$, which determines reference process response from present value $y(k)$ to the setpoint trajectory $w(k)$.
- The vector of future control signals $\mathbf{U}(k)$ containing $u(k + j|k), j = 0, \ldots,$ $N_u - 1$ is calculated by minimization of cost function (also called objective function) such that predicted error between $r(k + j|k)$ and $\hat{y}(k + j|k), j = N_1, \ldots, N_2$ is minimal. Structuring of future control samples can be used in some approaches.
- Only the first element $u(k|k)$ of the optimal control signal vector $u(k + j|k), j = 0, \ldots, N_u - 1$ is applied.

In the next sample a new measured output sample is available and the entire procedure is repeated. This procedure is called a receding horizon strategy.

The control objective is to be achieved by minimization of the cost function. The cost function penalizes deviations of the predicted controlled outputs $\hat{y}(k + j|k)$ from a reference trajectory $r(k + j|k)$. This reference trajectory may depend on measurements made up to time k. Its initial point may be the output measurement $y(k)$, but also a fixed set-point, or some predetermined trajectory. The minimization of cost function, in which future control signal $(\mathbf{U}(k))$ is calculated, can be subject to various constraints (e.g. input, state, rates, etc).

The process model for calculation of predicted outputs is in our case a Gaussian process model, which provides not only the mean value $\hat{y}(k + j|k)$ but also the corresponding variance.

There are many alternative ways of how NMPC with Gaussian process models can be realized.

Cost function. The cost function used (11) is just one of many possible ones. It is well known that selection of the cost function has a major impact on the amount of computation.

Optimization problem for $\Delta \mathbf{U}(k)$ instead of $\mathbf{U}(k)$. This is not just a change of formalism, but also enables other forms of MPC. One possibility is a DMC controller with nonlinear model, e.g. [8] - a frequently used principle, that together with appropriate cost function enables problem representation as a least squares problem that can be solved in one iteration in which an explicit solution is found. This is, as in the case with other special case simplifications, not a general case solution.

Process model. The process model can be determined off-line and fixed for the time of operation or determined on-line during the operation of controller

[16]. However, the problem of increasing covariance matrix dimension with incoming data has to be dealt with. Safety related issues also need to be considered thoroughly in the case of adaptive version application.

Soft constraints. Using constraint optimization algorithms is very demanding for computation and soft constrains, namely weights on constrained variables in cost function, can be used to decrease the amount of computation. More on this topic can be found in [9,24].

Linear MPC. It is worth to remark that even though this is a constrained nonlinear MPC problem it can be used in its specialized form as a robust linear MPC.

There are several issues of interest for applied NMPC. Let us mention some of them.

Efficient numerical solution. Nonlinear programming optimization algorithm is very demanding for computation. Various approximations and other approaches (e.g. approximation of explicit solution) exist to decrease computational load, mainly for special cases, like linear process models or special cost functions.

One possibility to decrease the computational load necessary for optimization is with the incorporation of prediction derivation (and variance) into optimization algorithm. When using Gaussian process models the prediction and variance derivation can be calculated in a straightforward manner.

Stability. At present no stability conditions have been derived for Gaussian processes as a representative of probabilistic non-parametric models.

Robustness. This issue has a major impact on the applicability of the algorithm in practice. The fact that the process model contains the information about the model confidence enables controller to optimize the manipulative variable to "avoid" regions where the confidence in the model is not high enough. This possibility itself makes the controller robust if applied properly. MPC robustness in the case of other algorithms is usually not some specially built feature of the MPC algorithms, but was more an issue of assessment for particular MPC algorithms.

4   Example

4.1   pH Process

A simplified schematic diagram of the pH neutralization process taken from [7] is given in Fig. 2. The process consists of an acid stream (Q_1), buffer stream (Q_2) and base stream (Q_3) that are mixed in a tank T_1. Prior to mixing, the acid stream enters the tank T_2 which introduces additional flow dynamics. The acid and base flow rates are controlled with flow control valves, while the buffer flow rate is controlled manually with a rotameter. The effluent pH (pH) is the measured variable. Since the pH probe is located downstream from the tank T_1, a time delay (T_d) is introduced in the pH measurement. In this study, the pH

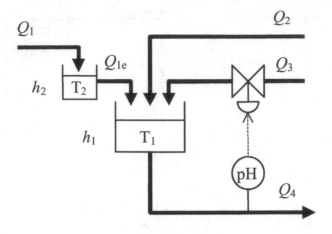

Fig. 2. The pH neutralization system scheme

is controlled by manipulating the base flow rate. A more detailed description of the process with mathematical model and necessary parameters is presented in [7].

The dynamic model of the pH neutralization system shown in Fig. 2 is derived using the conservation equations and equilibrium relations. The model also includes valve and transmitter dynamics as well as hydraulic relationships for the tank outlet flows. Modelling assumptions include perfect mixing, constant density, and complete solubility of the ions involved. The simulation model of pH process, which was used for necessary data generation contains therefore various nonlinear elements as well as implicitly calculated function which is value of highly nonlinear titration curve.

4.2   Model Identification

Based on responses and iterative cut-and-try procedure a sampling time of 25 seconds was selected. The sampling time was so large that the dead-time mentioned in the previous section disappeared.

The chosen identification signal of 400 samples was generated from a uniform random distribution and rate of 50 seconds.

Obtained hyperparameters of the third order Gaussian process model were:

$$\begin{aligned}
\boldsymbol{\Theta} &= [w_1, w_2, w_3, w_4, w_5, w_6, v_0, v_1] \\
&= [-6.0505, -2.0823, -0.4785, -5.3388, -3.4206, -8.7080, \\
&\quad 0.8754, -5.4164]
\end{aligned} \tag{7}$$

where hyperparameters from w_1 to w_3 denote a weight for each output regressor, from w_4 to w_6 denote a weight for each input regressor, v_0 is estimated noise variance and v_1 is the estimate of the vertical scale of variation.

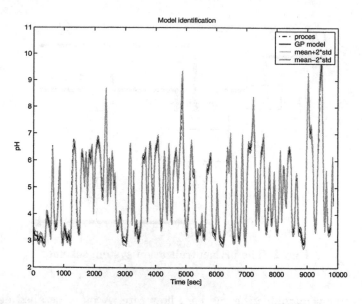

Fig. 3. Response of Gaussian process model on excitation signal used for identification

The region in which the model was obtained can be seen from Fig. 3. A very good fit can be observed for the identification input signal which was used for optimization. However, the obtained model contains information mainly in the region below pH=7 as can be concluded from the response in Fig. 3. The validation signal was from the same region as the identification signal. The identification and validation signal were obtained with generator of random noise with different initial values. Response of the model to validation signal and comparison with process response is depicted in Fig. 4. Fitting of the response for validation signal:

- average absolute test error

$$AE = 0.0691 \tag{8}$$

- average squared test error

$$SE = 0.0109 \tag{9}$$

- log density error

$$LD = -0.7130 \tag{10}$$

After model validation the model was utilized for control design. See [11] for more issues on pH process modelling with Gaussian process models.

4.3 Control

A moving-horizon minimization problem of the special form [13]

$$\min_{\mathbf{U}(k)} \ [r(k + P) - \hat{y}(k + P)]^2 \tag{11}$$

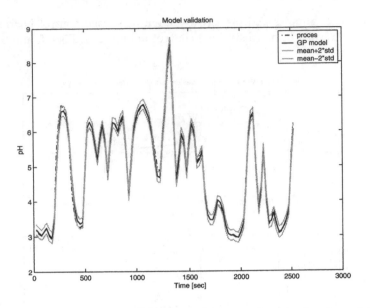

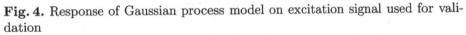

Fig. 4. Response of Gaussian process model on excitation signal used for validation

subject to:

$$\text{var } \hat{y}(k+P) \leq k_v \tag{12}$$

$$\mid \mathbf{U}(k) \mid \leq k_{ih} \tag{13}$$

$$\mid \dot{\mathbf{U}}(k) \mid \leq k_{ir} \tag{14}$$

$$\mid \mathbf{x}(k) \mid \leq k_{sh} \tag{15}$$

$$\mid \dot{\mathbf{x}}(k) \mid \leq k_{sr} \tag{16}$$

is used in our case, where $\mathbf{U}(k) = [u(k) \ldots u(k+P)]$ is input signal, P is the coincidence point (the point where a match between output and reference value is expected) and inequalities from (12) to (16) represent constraint on output variance k_v, input hard constraint k_{ih}, input rate constraint k_{ir}, state hard constraint k_{sh} and state rate constraint k_{sr} respectively. The process model is a Gaussian process.

The optimization algorithm, which is constrained nonlinear programming, is solved at each sample time over a prediction horizon of length P, for a series of moves which equals to control horizon. In our case control horizon was chosen to be one and to demonstrate constraint on variance the rest of constraints was not taken into the account. Nevertheless, all this modifications do not change the generality of solution, but they do affect the numerical solution itself.

The control algorithm described above was tested for the pH process with simulation. The reference trajectory r is defined so that it approaches the set-point exponentially from the current output value. The coincidence point was chosen to be 8 samples and, as already mentioned, the control horizon is one sample. The results of unconstrained control are given in Figs. 5 and 6.

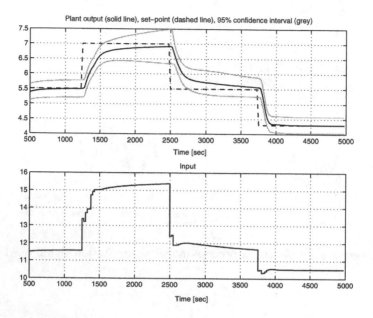

Fig. 5. Non-constrained case: response of Gaussian process model based control (upper figure) and control signal (bottom figure)

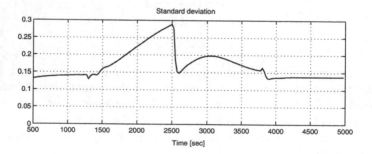

Fig. 6. Non-constrained case: standard deviation corresponding to the previous figure

It can be seen from different set-point responses that the model differs from the process in different regions. It can be clearly seen that the variance increases as output signal approaches regions which were not populated with enough identification data. It should be noted however that variances are sum of variances that correspond to information about regions where there are varying degrees of confidence in the model accuracy, depending upon local density of available identification data and of output response variances. When variances increase too much, one design option is that the response can be optimized with constrained control. Results can be seen in Figs. 7 and 8.

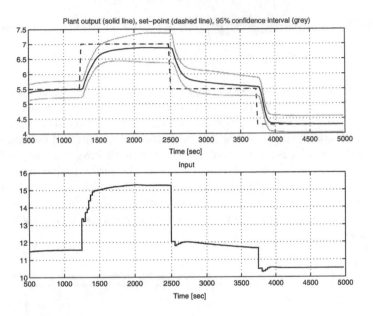

Fig. 7. Constrained case ($\sigma_{max} = 0.25$): response of Gaussian process model based control (upper figure) and control signal (bottom figure)

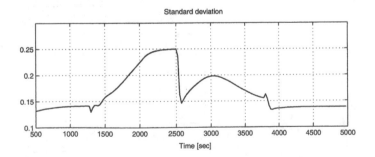

Fig. 8. Constrained case ($\sigma_{max} = 0.25$): standard deviation corresponding to the previous figure

It can be seen from Figs. 7 and 8 that the closed-loop system response now avoids the region with large variance, at the cost of an increase in steady-state error. This could be interpreted also as trade-off between designed performance and safety.

A possible alternative selection of cost function that avoids constrained optimization and is therefore computationally less demanding would be as follows.

$$\min_{\mathbf{U}(k)} \; E\{[r(k+P) - \hat{y}(k+P)]^2\} \tag{17}$$

Using the fact that $\text{var}\{\hat{y}\} = E\{\hat{y}^2\} - E\{\hat{y}\}^2$, the cost function can be written as:

$$\min_{\mathbf{U}(k)} \; [r(k+P) - E\{\hat{y}(k+P)\}]^2 + \text{var}\{\hat{y}(k+P)\} \tag{18}$$

Results with cost function (18) are given in Figs. 9 and 10. It can be again observed from Figs. 9 and 10 that the closed-loop system response avoids the region with large variance, at the cost of increased steady-state error, as was the case with constrained control, but with less computational burden than the constrained control case. The control strategy with cost function (18) is "to avoid" going into regions with higher variance. The term "higher variance" does not specify any specific value. In the case that controller does not seem to be cautious enough, a pragmatic calibration option is that the variance term can be weighted to enable shaping of the closed-loop response according to variance information:

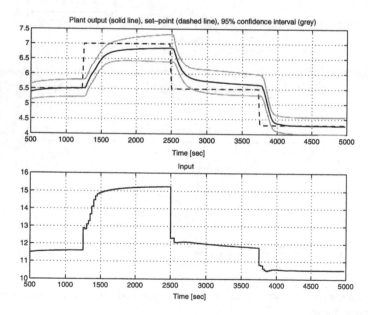

Fig. 9. Response of Gaussian process model based control with "soft constraints" (upper figure) and control signal (bottom figure)

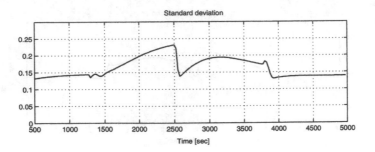

Fig. 10. Standard deviation corresponding to the previous figure

$$\min_{\mathbf{U}(k)} \; [r(k+P) - E\{\hat{y}(k+P)\}]^2 + \lambda \mathrm{var}\{\hat{y}(k+P)\} \qquad (19)$$

NMPC with the second cost function with weight on variance $\lambda = 2$, using unconstrained optimization, gives the results depicted in Figs. 11 and 12, showing a reduction in the standard deviation of the predictions of the closed-loop response, compared to Fig. 10, and minor changes in the mean behaviour.

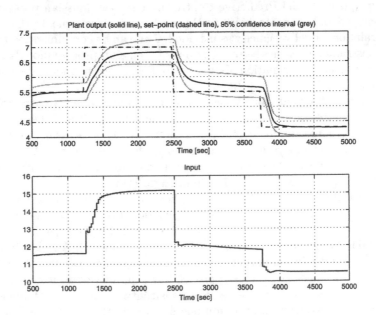

Fig. 11. Response of Gaussian process model based control with "soft constraints" (upper figure) and control signal (bottom figure)

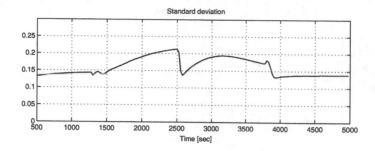

Fig. 12. Standard deviation corresponding to the previous figure

Beside the difference in the optimization algorithm, the presented options give also a design choice on how "safe" the control algorithm is. In the case when it is very undesirable to go into "unknown" regions the constrained version might be the better option.

5 Conclusions

The principle of Model Predictive Control based on a Gaussian process model was presented and illustrated with a pH process control example. In the example, a constraint on model variance was included. This can be complimented also with other constraints when necessary. The use of Gaussian process models makes it possible to include information about the confidence in the model depending on the region.

It was indicated that using Gaussian process models offers an attractive possibility for control design that results in a controller with a higher level of robustness due to information contained in the model. It is necessary to stress that the presented control strategy represents only a feasibility test for Gaussian process application for model predictive control and additional efforts are necessary before this approach will be applicable in engineering practice.

A practical challenge dealing with application of Gaussian process models in control applications is related to the computational burden associated with the number of training data (although recent work in [21] has shown how derivative observations can improve the situation in control contexts). Another interesting issue that is under investigation is disturbance rejection. Despite these current challenges, the Gaussian process approach has a number of exciting advantages. The simple model structure, the reduced sensitivity to the choice of model structure and the uncertainty information one obtains on the predictions are attractions of the Gaussian process approach. The principle shown is quite general and several modifications that accelerate computation can be used and are planned to be derived in the future.

Acknowledgments

This work was made possible by EC funded Multi-Agent Control Research Training Network HPRN-CT-1999-00107, and RM-S by EPSRC project GR/M76379/01, and Science Foundation Ireland grant 00/PI.1/C067. J. Kocijan acknowledges the support of Slovene Ministry of education, science and sport through project P2-0001. The authors are grateful for comments and useful suggestions from Agathe Girard, Daniel Sbarbaro and Gregor Gregorčič.

References

1. Allgöwer F., Badgwell T.A., Qin S.J., Rawlings J.B., Wright S.J., Nonlinear predictive control and moving horizon estimation - an introductory overview, In: Frank, P.M (edt.), Advances in control: highlights of ECC'99, Springer, (1999), 391-449.
2. Allgöwer F., Zheng A. (eds.), Nonlinear Model Predictive Control, Progress in system and control theory, Vol. 26, Birkhäuser Verlag, Basel, (2000).
3. Girard A., Rasmussen C.E., Quinonero Candela, J. and Murray-Smith R., Gaussian Process Priors With Uncertain Inputs & Application to Multiple-Step Ahead Time Series Forecasting, NIPS 15, Vancouver, Canada, MIT Press, (2003).
4. Girard A. and Murray-Smith R., Gaussian Process: Prediction at a noisy input and application to iterative multiple-step ahead forecasting of time-series, In: Switching and Learning in Feedback Systems, Eds R. Murray-Smith, R. Shorten, Springer, 2004.
5. Gregorčič G., Lightbody G., Internal model control based on a Gaussian process prior model, Proceedings of ACC'2003, Denver, CO, (2003), 4981-4986.
6. Johansen T.A., Foss B.A., Sorensen A.V., Non-linear predictive control using local models - applied to a batch fermentation process, Control Eng. Practice, 3(3), (1995), 389-396.
7. Henson M.A., Seborg D.E., Adaptive Nonlinear Control of a pH Neutralization Process, IEEE Trans. Control System Technology, Vol. 2, No. 3, (1994), 169-183.
8. Kavšek-Biasizzo K., Škrjanc I., Matko D., Fuzzy predictive control of highly nonlinear pH process, Computers & chemical engineering, Vol. 21, Supp. 1997, (1997), S613-S618.
9. Kerrigan E.C., Maciejowski J.M., Soft constraints and exact penalty functions in model predictive control, Control 2000 Conference, Cambridge, (2000).
10. Kocijan J., Girard A., Banko B., Murray-Smith R., Dynamic Systems Identification with Gaussian Processes, Proceedings of 4th Mathmod, Vienna, (2003), 776-784.
11. Kocijan J., Likar B., Banko B., Girard A., Murray-Smith R., Rasmussen C.E., A case based comparison of identification with neural network and Gaussian process models, Preprints of IFAC ICONS Conference, Faro, (2003), 137-142.
12. Kouvaritakis B., Cannon M. (eds.), Nonlinear predictive control, Theory and practice, IEE Control Engineering Series 61, IEE, (2001).
13. Maciejowski J.M., Predictive control with constraints, Pearson Education Limited, Harlow, (2002).
14. Murray-Smith, R., Girard, A., Gaussian Process priors with ARMA noise models, Irish Signals and Systems Conference, Maynooth, (2001), 147-152.
15. Murray-Smith, R and T. A. Johansen and R. Shorten, On transient dynamics, off-equilibrium behaviour and identification in blended multiple model structures, European Control Conference, Karlsruhe, (1999), BA-14.

16. Murray-Smith R. and Sbarbaro D., Nonlinear adaptive control using nonparametric Gaussian process prior models, In: Proc. IFAC Congress, Barcelona, (2002).
17. Nørgaard M., Ravn O., Poulsen N.K., Hansen L.K., Neural networks for modelling and control of dynamic systems, Springer, London, (2000).
18. Qin S.J., Badgwell T.A., An overview of industrial model predictive control technology, In: Kantor J.C., Garcia C.E., Carnahan B. (eds.) Fifthe international conference on Chemical process control, AChE and CACHE, (1997), 232-56.
19. Qin S.J., Badgwell T.A., An overview of nonlinear model predictive control applications, In: Allgöwer F., Zheng A. (eds.), Nonlinear model predictive control, Birkhauser Verlag, (2000), 369-392.
20. Rasmussen C.E., Evaluation of Gaussian Processes and other Methods for Non-Linear Regression, Ph.D. Disertation, Graduate department of Computer Science, University of Toronto, Toronto, (1996).
21. Solak, E., Murray-Smith R., Leithead, W.E., Leith, D.J., and Rasmussen, C.E., Derivative observations in Gaussian Process models of dynamic systems, NIPS 15, Vancouver, Canada, MIT Press, (2003).
22. Williams C.K.I., Prediction with Gaussian processes: From linear regression to linear prediction and beyond, In: Learning in Graphical Models, Jordan, M.I., (ed.), Kluwer Academic, Dordrecht, (1998), 599-621.
23. Young R.E., Bartusiak R.D., Fontaine, R.W., Evolution of an industrial nonlinear model predictive controller, Preprints on Chemical Process Control - CPC VI, CACHE, Tucson, AZ, (2001), 399-401.
24. Zheng A., Morari M., Stability of model predictive control with mixed constraints, IEEE Trans. Autom. Control, 40(19), (1995), 1818-1823.

Control of Yaw Rate and Sideslip in 4-Wheel Steering Cars with Actuator Constraints*

Miguel A. Vilaplana, Oliver Mason, Douglas J. Leith, and William E. Leithead

Hamilton Institute
National University of Ireland, Maynooth, Co. Kildare, Ireland

Abstract. In this paper we present a new steering controller for cars equipped with 4-wheel steer-by-wire. The controller commands the front and rear steering angles with the objective of tracking reference yaw rate and sideslip signals corresponding to the desired vehicle handling behaviour. The structure of the controller is based on a simplified model of the lateral dynamics of 4-wheel steering cars. We show that the proposed structure facilitates the design of a robust steering controller valid for varying vehicle speed. The controller, which has been designed using classical techniques according to the Individual Channel Design (ICD) methodology, incorporates an anti-windup scheme to mitigate the effects of the saturation of the rear steering actuators. We analyse the robust stability of the resulting non-linear control system and present simulation results illustrating the performance of the controller on a detailed non-linear vehicle model.

1 Introduction

The concept of generic prototype vehicles has emerged as a promising solution to an outstanding challenge in the development of ride and handling characteristics for advanced passenger cars: the bridging of the gap between numerical simulations based on a vehicle model –a virtual prototype– and experiments on a proof-of-concept prototype vehicle. A generic prototype vehicle would be equipped with advanced computer-controlled actuators enabling it to modify its ride and handling characteristics. Examples of such advanced actuators are four and rear steer-by-wire, brake-by-wire and active suspensions. An integrated chassis controller would command those actuators to track a set of reference signals corresponding to a desired ride and handling behaviour. Currently, moving-base driving simulators are used to emulate the ride and handling behaviour of virtual prototypes prior to building real ones. However, the achievable accelerations of such simulators severely constrain their ability to realistically recreate the full range of vehicle motion. Generic prototype vehicles could allow for the realistic recreation of the ride and handling characteristics of virtual prototypes, thereby enabling engineers to experience and evaluate their behaviour prior to making the decision of building expensive proof-of-concept prototypes.

---

* The authors thank Jens Kalkkuhl of DaimlerChrysler Research and Technology for his assistance with this research.

R. Murray-Smith, R. Shorten (Eds.): Switching and Learning, LNCS 3355, pp. 201–222, 2005.

In this paper, we present a steering controller that enables cars equipped with 4-wheel steer-by-wire to display predefined handling characteristics. This steering controller is intended as a first step towards an integrated chassis controller for a generic prototype vehicle. The proposed steering controller commands front and rear steering angles with the objective of tracking reference yaw rate and sideslip angle signals obtained online from the driver's inputs to steering wheel and pedals. These reference signals describe the lateral dynamic response to those inputs of a virtual prototype with the desired handling characteristics. In addition, the steering controller automatically rejects disturbances in sideslip and yaw rate, such as those caused by μ-split braking manoeuvres or lateral wind gusts. We assume that the controlled output variables, i.e. yaw rate and sideslip angle, are measured (in practice, the latter may typically be estimated using, for example, a Kalman filter).

A substantial body of research on the control of 4-wheel steering cars already exists and a variety of control structures have been proposed in the literature. Most of these structures rely on the use of gain-scheduled feedforward control to command the rear steering angle [1]. In such control structures, some of which have been implemented on production passenger cars, the rear steering angle is computed as a function of the front steering angle that results from the driver's input to the steering wheel. The different control laws depend on the performance objectives, which are usually related to the improvement of the manoeuvrability and cornering stability of the vehicle. The work described in [2] proposes to combine feedforward and feedback control to command the rear steering angle, while the front steering angle remains under direct control of the driver. The control objective is to follow a predefined model of the vehicle dynamics. In order to achieve a satisfactory degree of robustness the feedback controller is designed using μ synthesis. An example of an steering controller specifically designed for cars equipped with 4-wheel steer-by-wire is presented in [3]. The controller structure is based on the cross-feedback of the measured yaw rate to the front steering angle. This structure decouples the control of the lateral acceleration from the control of the yaw rate. Two outer feedback loops are used so that front wheel steering is used to track the desired lateral acceleration and rear wheel steering is used to regulate the damping of the resulting yaw dynamics.

The structure of the steering controller presented in this paper is based on a simplified linear model of the lateral dynamics of 4-wheel steering cars at constant speed. The main elements of the controller structure are a linear input transformation and a speed-dependent inner feedback loop. When applied to the simplified model mentioned above, this structure partially decouples the sideslip and yaw rate responses to the new controllable inputs, with the yaw rate response being speed-invariant. Thus, the proposed structure acts as an implicit gain scheduling on the vehicle speed. The control design is based on a more accurate model of the vehicle lateral dynamics. This model includes the steering actuator dynamics as well as the communication time delay between controller and actuators. When applied to this model, the proposed controller structure results in approximate partial decoupling of the sideslip and yaw rate responses, with a nearly speed-invariant yaw rate response. The resulting 2-by-2 multivariable control design problem is restated as two single-input, single-output (SISO) control design problems according to the ICD paradigm. Assuming certain bandwidth restrictions, individual linear controllers for the resulting sideslip and yaw rate channels are designed using classical techniques. The resulting steering controller satisfies robustness and disturbance rejection requirements. The controller is augmented with a feedforward element to improve the response to reference inputs and with an anti-windup scheme to mitigate the effects of the saturation of the rear steering actuators. The resulting non-linear steering controller is valid for varying vehicle speed and shows excellent performance robustness to model uncertainty.

Since the proposed steering controller is intended as the foundation for an integrated chassis controller, the main design criteria are transparency, simplicity and modularity. We have adopted the ICD methodology in an attempt to satisfy these criteria. ICD exploits the full potential of diagonal control within a classical Nyquist-Bode framework, opening the way for modular and transparent design based on individual SISO channels that arise naturally from the control specifications. In addition, we deal with the issue, often overlooked in the literature, of ensuring that the steering controller remains robustly stable and performs satisfactorily in the event of rear steering actuator saturation.

In this paper, we focus on describing the controller structure and analysing its robust stability considering the possible saturation of the rear steering actuators. A detailed description of the control design process can be found in [4].

The remainder of this paper is structured as follows. First, we describe the simplified model of the lateral dynamics of 4-wheel steering cars used to define the structure of the proposed steering controller. Based on this model, we derive the controller structure and state the resulting multivariable control design problem. Subsequently, we introduce the ICD methodology in the context of the problem at hand. Next, we apply the proposed controller structure to a more accurate model of the lateral dynamics of 4-wheel steering cars and restate our multivariable control design problem in terms of individual channels according to ICD. Then, we briefly explain the design process. Subsequently, we analyse the robust stability of the resulting non-linear control system using some new results from the theory of common quadratic Lyapunov functions. Finally, simulation results obtained with a detailed non-linear model of a Mercedes S-Class are given to illustrate the performance and robustness of the steering controller.

2   Simplified Linear Model of the Lateral Dynamics of 4-Wheel Steering Cars

Throughout this paper, it is assumed that the essential features of the lateral dynamics of the car can be described using the single-track model [3]. In the single-track model, the two wheels at each axle are lumped into a single imaginary wheel located at the centre of the respective axle. The resulting front and rear wheels are interconnected by a one-dimensional rigid element with the car's mass and moment of inertia around the vertical axis. The forces acting on each wheel of the single-track model correspond to the combined forces acting on the left and right wheel at the corresponding axle. Only the lateral motion of the car is considered when using the single-track model. It is assumed that the centre of gravity of the single-track model is at road level so that the roll, pitch and heave dynamics can be neglected. Additionally, it is assumed that the longitudinal speed is constant. Figure 1 depicts the single-track model indicating the main elements necessary for the analysis of its lateral dynamics.

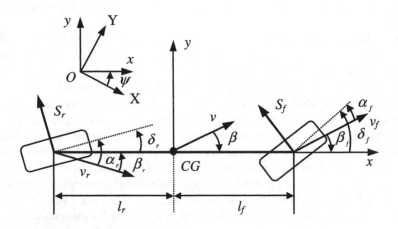

Fig. 1. Single-track model of a 4-wheel steering car

In Fig. 1, the set of reference axes CG-xy, with origin at the centre of gravity CG, is fixed to the vehicle and O-XY is an inertial reference frame; v is the velocity of the vehicle with respect to O-XY; v_f and v_r are the velocities at the front and rear axle, respectively, with respect to O-XY; ψ is the yaw angle and β is the sideslip angle. It is assumed that the front (respectively, rear) steering angle of the single-track model, δ_f (respectively, δ_r) in Fig. 1, corresponds to the steering angle at the two front (respectively, rear) wheels. Since we are not concerned with the longitudinal motion of the single-track model, we only consider tyre-road interaction forces perpendicular to the wheel plane, i.e. cornering forces. The force S_f (respectively, S_r) in Fig. 1 represents the combined cornering forces acting on the front (respectively, rear) axle.

To derive the equations governing the linearised lateral dynamics of the single-track model, we assume that the front and rear steering angles are small, which in turn results in the angles β, β_f, α_f, β_r and α_r in Fig. 1 also being small. Under this assumption, the application of the equations of motion of a rigid body to the single-track model results in

$$\dot{\beta} = \dot{\psi} - \frac{S_f + S_r}{mv_x} \tag{1}$$

$$\ddot{\psi} = \frac{S_f l_f - S_r l_r}{I_{zz}} \tag{2}$$

where m is the mass of the vehicle, I_{zz} is its moment of inertia with respect to the vertical axis, l_f (respectively, l_r) is the distance from the centre of gravity to the front (respectively, rear) axle and v_x is the projection of the velocity vector along the $CG - x$ axis, i.e. the vehicle longitudinal velocity, which we hereafter refer to as the vehicle speed.

For small α_f and α_r, S_f and S_r can be approximated by the following equations [5]:

$$S_f = K_f \alpha_f \tag{3}$$
$$S_r = K_r \alpha_r \tag{4}$$

Considering the kinematics of the single-track model as a rigid body, the angles α_f and α_r are calculated as follows:

$$\alpha_f = \delta_f + \beta - \frac{l_f \dot{\psi}}{v_x} \tag{5}$$

$$\alpha_r = \delta_r + \beta + \frac{l_r \dot{\psi}}{v_x} \tag{6}$$

The constant K_f in (3) is obtained by adequately reducing the combined cornering stiffness of the two front tyres to take into account the caster effect. Conventional steering systems are designed so that the tyre-road contact patch trails behind the steering axis, resulting in a self-aligning torque on the front axle known as the caster effect. We have to consider the caster effect as it is assumed that the front steer-by-wire function is integrated with a conventional steering system. This construction allows for the introduction of a safety management system that reverts to normal steering in case of failure of the steer-by-wire function. The constant K_r in (4) is simply the combined cornering stiffness of the rear tyres. No caster effect is generated at the rear axle since it is assumed that each rear wheel is steered individually by an electro-hydraulic actuator.

Equations (1), (2), (3), (4), (5) and (6) can be rearranged into the state-space representation of a linear time-invariant system with two inputs (δ_f and δ_r) and two outputs (β and $\dot{\psi}$). The resulting state-space representation is given below:

$$\dot{x} = Ax + Bu \tag{7}$$
$$y = Cx + Du \tag{8}$$

where

$$u = \begin{bmatrix} \delta_f \\ \delta_r \end{bmatrix}, \quad y = x = \begin{bmatrix} \beta \\ \dot{\psi} \end{bmatrix}, \tag{9}$$

$$A = \begin{bmatrix} -\dfrac{K_f + K_r}{m v_x} & \dfrac{K_f l_f - K_r l_r}{m v_x^2} + 1 \\[3mm] \dfrac{K_f l_f - K_r l_r}{I_{zz}} & -\dfrac{K_f l_f^2 + K_r l_r^2}{I_{zz} v_x} \end{bmatrix}, \quad B = \begin{bmatrix} -\dfrac{K_f}{m v_x} & -\dfrac{K_r}{m v_x} \\[3mm] \dfrac{K_f l_f}{I_{zz}} & -\dfrac{K_r l_r}{I_{zz}} \end{bmatrix}, \tag{10}$$

$$C = \begin{bmatrix} 1 & 0 \\ 0 & 1 \end{bmatrix} \quad \text{and} \quad D = \begin{bmatrix} 0 & 0 \\ 0 & 0 \end{bmatrix} \tag{11}$$

The matrix transfer function of the system with the state-representation (7)-(8) is given by

$$G(s) = C(sI - A)^{-1}B + D = \begin{bmatrix} g_{11}(s) & g_{12}(s) \\ g_{21}(s) & g_{22}(s) \end{bmatrix} \tag{12}$$

The linear time-invariant system introduced above describes the lateral dynamics of the single-track model around the trajectory given by zero sideslip, zero yaw rate, zero steering angles and constant vehicle speed.

3   Structure of the Steering Controller

The steering controller is to be designed to track reference signals corresponding to normal driving situations. In such situations, the tyres are far from their adhesion limit and the cornering forces behave approximately in a linear fashion according to 3 and 4. The transfer function $G(s)$ in 12 can then be used to model the car lateral dynamics and the design of a steering controller for constant vehicle speed can be tackled by solving the classical 2-by-2 multivariable control design problem depicted in Fig. 2.

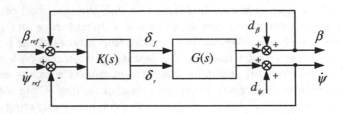

Fig. 2. Design of a linear multivariable steering controller for fixed vehicle speed

A linear controller $K(s)$ designed based on $G(s)$ would only be valid for the corresponding vehicle speed. In principle, a set of local controllers corresponding to different vehicle speeds could be combined using gain-scheduling techniques into a non-linear controller valid for varying vehicle speed. In order to simplify the design process, we take a different approach and state the control design problem in terms of the virtual plant that results from pre-compensating $G(s)$ with a constant matrix gain, i.e. linearly transforming the inputs, and then introducing a speed-dependent matrix gain in a feedback path around the pre-compensated plant. By modifying $G(s)$ in this manner and basing the design on the resulting virtual plant, we impose a structure that facilitates the design of a steering controller valid for varying vehicle speed. This is due to the fact that the virtual plant to be controlled, which we denote as $\tilde{G}(s)$, yields a speed-invariant yaw rate response. The derivation of the controller structure is explained in detail below.

3.1 Linear Input Transformation

Suppose that the inputs to the plant G(s) are the result of the following linear transformation:

$$\begin{bmatrix} \delta_f \\ \delta_r \end{bmatrix} = E \begin{bmatrix} \Delta_1 \\ \Delta_2 \end{bmatrix} \tag{13}$$

where $E \in \mathbb{R}^{2 \times 2}$. Considering 3.1, the resulting dynamical equation for the single-track model with respect to the new inputs is:

$$\dot{x} = Ax + BE\Delta = Ax + B_1\Delta, \quad \text{with} \quad \Delta = \begin{bmatrix} \Delta_1 \\ \Delta_2 \end{bmatrix} \tag{14}$$

If we choose

$$E = -\frac{1}{\frac{K_r}{K_f}\left(1 + \frac{l_r}{l_f}\right)} \begin{bmatrix} \dfrac{K_r l_r}{K_f l_f} & -\dfrac{K_r}{K_f} \\ -1 & -1 \end{bmatrix} \tag{15}$$

the resulting matrix B_1 is diagonal:

$$B_1 = \begin{bmatrix} -\dfrac{K_f}{mv_x} & 0 \\ 0 & \dfrac{K_f K_f}{I_{zz}} \end{bmatrix} \tag{16}$$

The chosen matrix E correspond to the inputs:

$$\Delta_1 = \delta_f + \frac{K_r}{K_f}\delta_r \tag{17}$$

$$\Delta_2 = \delta_f - \frac{K_r l_r}{K_f l_f}\delta_r \tag{18}$$

A physical interpretation of these new inputs is in terms of a mode given by Δ_1, which excites the sideslip by steering front and rear wheels in the same direction, and a mode given by Δ_2, which excites the yaw rate by steering front and rear wheels in opposite directions. It can be argued that by using Δ_1 and Δ_2 and as control actions the 4-wheel steering vehicle is controlled in a "natural" way, separating the dynamics into their linear and rotational components. The resulting dynamical equation of the yaw rate with respect to the new inputs is:

$$\frac{I_{zz}}{K_f l_f}\ddot{\psi} + \frac{K_f l_f^2 + K_r l_r^2}{K_f l_f v_x}\dot{\psi} = \Delta_2 + \left(1 - \frac{K_r l_r}{K_f l_f}\right)\beta \tag{19}$$

Taking Laplace transforms of both sides in 3.1 and rearranging results in:

$$\dot{\psi}(s) = \frac{K_1}{s + p(v_x)}\Delta_2(s) + \frac{K_1 K_2}{s + p(v_x)}\beta(s) \tag{20}$$

where

$$K_1 = \frac{K_f l_f}{I_{zz}}, \quad K_2 = 1 - \frac{K_r l_r}{K_f l_f}, \quad \text{and} \quad p(v_x) = \frac{K_f l_f^2 + K_r l_r^2}{I_{zz} v_x} \tag{21}$$

The yaw rate dynamics are characterised by a speed-varying first order pole at frequency $p(v_x)$ and are coupled with the sideslip dynamics.

3.2   Speed-Dependent Feedback Element

We introduce a feedback element of the form:

$$\Delta = \tilde{\Delta} - Fy \tag{22}$$

which results in the new vector of controllable inputs $\tilde{\Delta} = \begin{bmatrix} \tilde{\Delta}_1 \\ \tilde{\Delta}_2 \end{bmatrix}$. The matrix $F \in \mathbb{R}^{2\times 2}$ is given by

$$F = \begin{bmatrix} 0 & 0 \\ K_2 & K_v(v_x) \end{bmatrix} \tag{23}$$

with K_2 from (21) and $K_v(v_x)$ defined as

$$K_v(v_x) = K_0 - \frac{p(v_x)}{K_1} \tag{24}$$

with K_1 from (21) and K_0 an arbitrary constant. Since $y = x$, the state-space equation can be written as follows:

$$\dot{x} = Ax + B_1(\tilde{\Delta} - Fx) = (A - B_1 F)x + B_1 \tilde{\Delta} \tag{25}$$

where

$$A - B_1 F = \begin{bmatrix} -\dfrac{K_f + K_r}{mv_x} & \dfrac{K_f l_f - K_r l_r}{mv_x^2} + 1 \\ 0 & -\dfrac{K_0 K_f l_f}{I_{zz}} \end{bmatrix} \tag{26}$$

The corresponding matrix transfer function with respect to the new controllable inputs is upper-triangular:

$$\tilde{G}(s) = C(sI - \tilde{A})^{-1} B_1 + D = \begin{bmatrix} \tilde{g}_{11}(s) & \tilde{g}_{12}(s) \\ 0 & \tilde{g}_{22}(s) \end{bmatrix} \tag{27}$$

The resulting dynamical equation of the yaw rate with respect to the new controllable inputs $\tilde{\Delta}_1$ and $\tilde{\Delta}_2$ is speed-invariant, taking the form:

$$\ddot{\psi} = -K_0 K_1 \dot{\psi} + K_1 \tilde{\Delta}_2 \tag{28}$$

We choose K_0 to be:

$$K_0 = \frac{K_f l_f^2 + K_r l_r^2}{K_f l_f v_{x0}} \tag{29}$$

with v_{x0} an arbitrary fixed vehicle speed. Then, taking Laplace transforms of both sides of 3.2 results in:

$$\dot{\psi}(s) = \frac{K_1}{s + p(v_{x0})} \tilde{\Delta}_2 \qquad (30)$$

The introduction of the feedback element described above results in the yaw rate dynamics depending only on one of the two inputs to be controlled, $\tilde{\Delta}_2$. Besides, the yaw rate response to it is speed-invariant and characterised by a fixed first order pole at frequency $p(v_{x0})$.

3.3   Control Design Problem with Diagonal Controller

Considering the above, we base the control design on the virtual plant $\tilde{G}(s)$. Since we intend to apply the ICD design methodology , we assume that $\tilde{G}(s)$ is to be controlled by a diagonal controller. Consequently, the multivariable control problem in Fig. 2 can be restated as shown in Fig. 3, which depicts the proposed controller structure.

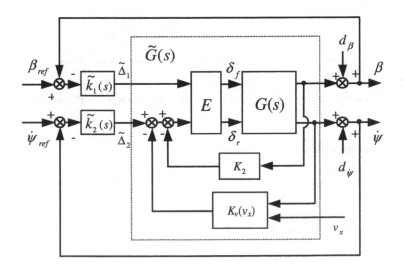

Fig. 3. Multivariable control design problem in terms of $\tilde{G}(s)$

4   Individual Channel Decomposition According to ICD

ICD [6] is a frequency-domain approach to the analysis and design of linear multivariable control systems that provides a solid framework for the application of concepts and techniques from classical linear control, such as Nyquist and

Bode plots and gain and phase margins, to multivariable control design problems. Within the ICD framework, an m-input, m-output feedback system with a diagonal controller is decomposed, without loss of information, into m equivalents SISO feedback control systems called channels. Each individual channel originates from the pairing of a reference input to its corresponding output. Consequently, a channel has its own performance specifications expressed in terms of its response to the corresponding reference input. ICD is very much an application-oriented approach capable of exploring the potential and limitations of diagonal control for a given system. According to the ICD methodology, the multivariable control problem in Fig. 3 can be decomposed into the two channels shown in Fig. 4.

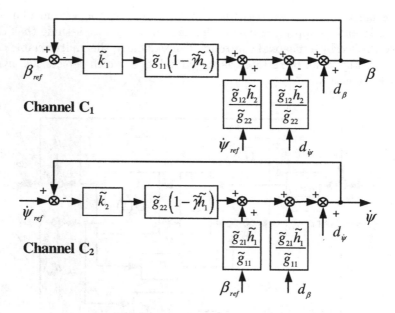

Fig. 4. Multivariable control design problem in terms of $\tilde{G}(s)$

The channel decomposition in Fig. 4 is based on the following functions:

$$\tilde{\gamma}(s) = \frac{\tilde{g}_{12}(s)\tilde{g}_{21}(s)}{\tilde{g}_{11}(s)\tilde{g}_{22}(s)} \tag{31}$$

$$\tilde{h}_1(s) = \frac{\tilde{k}_1(s)\tilde{g}_{11}(s)}{1 + \tilde{k}_1(s)\tilde{g}_{11}(s)}, \quad \tilde{h}_2(s) = \frac{\tilde{k}_2(s)\tilde{g}_{22}(s)}{1 + \tilde{k}_2(s)\tilde{g}_{22}(s)} \tag{32}$$

The closed-loop response of the channels to the reference inputs and $\beta_{ref}(s)$ and $\dot{\psi}_{ref}(s)$ are given by:

$$\beta(s) = \tilde{t}_{11}(s)\beta_{ref}(s) + \tilde{t}_{12}(s)\dot{\psi}_{ref}(s) \tag{33}$$

$$\dot{\psi}(s) = \tilde{t}_{21}(s)\beta_{ref}(s) + \tilde{t}_{22}(s)\dot{\psi}_{ref}(s) \tag{34}$$

where

$$\tilde{t}_{ii}(s) = \frac{\tilde{c}_i(s)}{1 + \tilde{c}_i(s)}, \quad i = 1, 2 \tag{35}$$

$$\tilde{t}_{ij}(s) = \frac{\tilde{g}_{ij}(s)\tilde{h}_j(s)}{\tilde{g}_{jj}(s)}(1 + \tilde{c}_i(s))^{-1}, \quad i = 1, 2; \ j = 1, 2; \ i \neq j \tag{36}$$

The term $\tilde{c}_i(s)$ in 36 and 36 is the open loop transmittance of channel i, which is defined as

$$\tilde{c}_i(s) = \tilde{k}_i(s)\tilde{g}_{ii}(s)(1 - \tilde{\gamma}(s)\tilde{h}_j(s)), \quad i = 1, 2; \ j = 1, 2; \ i \neq j \tag{37}$$

The closed-loop responses of the channels to the disturbance inputs $d_\beta(s)$ and $d_{\dot{\psi}}(s)$ are as follows:

$$\beta(s) = \tilde{s}_{11}(s)\beta_{ref}(s) + \tilde{s}_{12}(s)\dot{\psi}_{ref}(s) \tag{38}$$

$$\dot{\psi}(s) = \tilde{s}_{21}(s)\beta_{ref}(s) + \tilde{s}_{22}(s)\dot{\psi}_{ref}(s) \tag{39}$$

where

$$\tilde{s}_{ii}(s) = \frac{1}{1 + \tilde{c}_i(s)}, \quad i = 1, 2 \tag{40}$$

$$\tilde{s}_{ij}(s) = -\tilde{t}_{ij}(s) \tag{41}$$

Robust stability of the multivariable control system is equivalent to the robust stability of the channels providing that the Nyquist plots of the two multivariable structure functions $\tilde{\gamma}(s)\tilde{h}_j(s)$ for $j = 1, 2$ remain far from the (1,0) point.

5   Control Design for a More Accurate Model of the Lateral Dynamics of 4-Wheel Steering Cars

We now consider a more accurate linear model of the car lateral dynamics. We base the design of the linear controllers $\tilde{k}_1$ and $\tilde{k}_2$ on the virtual plant $\tilde{G}(s)$ that results from applying the proposed controller structure (see Fig. 3) to the matrix transfer function of this new model. While still relying on the single-track approximation, we augment the simple model described by (7)-(8) to include the following:

1. The tyre force dynamics and caster effect at the front axle modelled as:

$$\dot{S}_f = a\left(C\left(\delta_f - 2S_f\frac{n_s}{C_L} + \beta - \frac{l_f\dot{\psi}}{v_x}\right) - S_f\right) \tag{42}$$

where where S_f is the cornering force generated at the front axle, a is a constant that depends on the vehicle speed, C is the nominal tyre cornering stiffness, δ_f is the output of the front steering actuators, n_s is a caster parameter and C_L is an elasticity constant of the steering system.

2. Tyre force dynamics at the rear axle modelled as:

$$\dot{S}_r = a\,(C\alpha_r - S_r) \tag{43}$$

where S_r is the cornering force generated at the rear axle.

3. Front and rear steering actuators modelled as second order systems:

$$\begin{bmatrix} \dot{\delta}_f \\ \ddot{\delta}_f \end{bmatrix} = \begin{bmatrix} A^f_{11} & A^f_{12} \\ A^f_{21} & A^f_{21} \end{bmatrix} \begin{bmatrix} \delta_f \\ \dot{\delta}_f \end{bmatrix} + \begin{bmatrix} b^f_1 \\ b^f_2 \end{bmatrix} \delta^i_f \tag{44}$$

$$\begin{bmatrix} \dot{\delta}_r \\ \ddot{\delta}_r \end{bmatrix} = \begin{bmatrix} A^r_{11} & A^r_{12} \\ A^r_{21} & A^r_{21} \end{bmatrix} \begin{bmatrix} \delta_r \\ \dot{\delta}_r \end{bmatrix} + \begin{bmatrix} b^r_1 \\ b^r_2 \end{bmatrix} \delta^i_r \tag{45}$$

where δ^i_f and δ^i_r are the input to the front and rear steering actuators, respectively, and δ_f and δ_r are the output of the front and rear steering actuators, respectively.

4. Communication time delay of 20 ms between controller and actuators modelled using Padè approximation.

5.1   Control Specifications

The main requirements for the controlled 4-wheel steering car are:

1. Tracking sideslip and yaw rate reference signals with the highest possible closed-loop bandwidth. These reference signals are obtained in real-time from the driver's inputs to the steering wheel and pedals.
2. Rejecting any disturbances in sideslip and yaw rate with the highest possible bandwidth to avoid interference with the driver's reactions.
3. Maintaining tracking and disturbance rejection performance for vehicle speeds between 10 and 60 m/s and for driving situations involving speed changes, such as acceleration and braking.
4. Robustness with respect to changes in the car parameters, in particular with respect to changes in the tyre stiffness under different road conditions.
5. Satisfactory performance in the event of the saturation of the rear actuators.

5.2   Control Design

We have designed the steering controller considering the more accurate model of the car lateral dynamics introduced above. The model parameters are those corresponding to a Mercedes S Class. Details on the design of the controllers $\tilde{k}_1(s)$ and $\tilde{k}_2(s)$ can be found in [4]. Here we provide a summary of the design process. Before actually designing $\tilde{k}_1$ and $\tilde{k}_2$, two tasks have to be carried out:

1. In order to improve the cross-channel disturbance rejection in the sideslip channel (disturbances from the reference yaw rate to the sideslip response),

a low-pass filter is added to the cross-feedback term. This results in a K_2 in Fig' 3 taking the form:

$$K_2(s) = \left(1 - \frac{K_r l_r}{K_f l_f}\right) \frac{1}{\frac{s}{s_0} + 1} \tag{46}$$

where the value of the pole frequency, s_0, is to be selected.

2. The value of v_{x0} in $K_v(v_x)$ has to be chosen. The choice of v_{x0} is related to the robustness of the system.

Once s_0 and v_{x0} have been selected, we can write the transfer function matrix $\tilde{G}(s)$ for any given vehicle speed. With the more accurate model of the car lateral dynamics introduced above, the resulting $\tilde{G}(s)$ is approximately upper-triangular and the yaw rate response can be considered speed-invariant up to a certain frequency. By imposing a bandwidth separation of approximately 7 rad/s between the two channels, the controllers $\tilde{k}_1(s)$ and $\tilde{k}_2(s)$ can be designed based on $\tilde{g}_{11}(s)$ and $\tilde{g}_{22}(s)$, respectively, using classical Bode plot-based SISO techniques. Simple controllers of the form

$$\tilde{k}_1(s) = -\frac{K_{1I}}{s}, \quad \text{Integrator} \tag{47}$$

$$\tilde{k}_2(s) = K_{2p} + \frac{K_{2I}}{s}, \quad \text{PI controller} \tag{48}$$

$$\tag{49}$$

achieve an excellent degree of robustness and satisfactory performance regarding the rejection of cross-channel and external disturbances. These controllers result in a low bandwidth sideslip channel (approx 3 rad/s) and a high bandwidth yaw rate channel (approx 10 rad/s). The speed-dependent feedback term $K_v(v_x)$ acts as an implicit gain scheduling scheme that combines linear controllers parameterised by the vehicle speed into a non-linear controller valid for varying speed.

Having designed $\tilde{k}_1(s)$ and $\tilde{k}_2(s)$ to achieve robustness and disturbance rejection performance, we then add a linear feedforward element to the steering controller to improve its tracking performance. The feedforward element adequately speeds up and shapes the responses to reference signals.

6   Anti-windup Scheme

The steering controller has been designed without considering the possible saturation of the steering actuators. While the maximum allowable front rear steering angle ($\pm 40^0$) is not likely to be reached in the driving situations in which the controller will operate, possible rear actuator saturation has to be considered since the maximum allowable rear steering angle is restricted to only $\pm 5^0$ due to space constraints. When the rear actuators saturate, the feedback loops are broken and the system runs in open-loop because the output of the actuators remain constant independently of the output of the system. Since the steering controller performs integrating action, the error continues to be integrated and the integral terms may become very large (they "wind up"). This may lead to large transients, excessive overshoots or even instability. An anti-windup scheme has been incorporated into the steering controller to mitigate the effects of the saturation of the rear actuators. The proposed scheme is inspired by conventional anti-windup methods and works as follows. The rear steering angle signal commanded by the controller is subtracted from the average of the measured rear steering angles. The resulting signal is fed back to the input of the controller $\tilde{k}_1(s)$ through a gain K_{AW}. As it will be shown in the simulation results below, this scheme prevents the integrators in both $\tilde{k}_1(s)$ and $\tilde{k}_2(s)$ from winding up and allows the steering controller to retain full control of the yaw rate. Fig. 5 below shows the full steering controller, including feedforward and anti-windup, as it would be implemented in a real car.

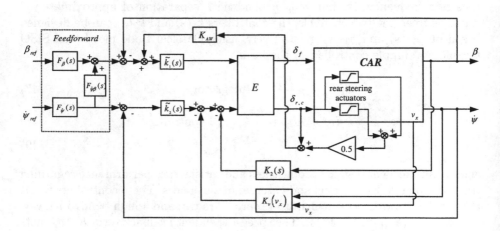

Fig. 5. Full steering controller

7   Robust Stability Analysis

In this section, we analyse the robust stability of the control system considering the possible saturation of the rear steering actuators. In the analysis, we do not consider the feedforward element of the steering controller, as it does not affect the stability of the overall control system, and we use the more accurate model of the car lateral dynamics introduced in Section 5 without the communication time delay. To study the stability of the resulting feedback control system, we transform it into an equivalent one whose forward path contains a SISO linear time-invariant subsystem and whose feedback path contains a saturation nonlinearity. This nonlinearity models the constraints on the rear steering actuators. The equivalent system, which is depicted in Fig. 6, is an example of a Lur'e system. Thus, asymptotic stability results developed for Lur'e systems, such as the Circle Criterion ([7],[8]), can be used in the analysis.

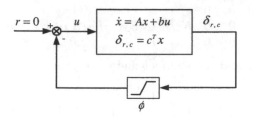

Fig. 6. Equivalent control system for stability analysis: a Lur'e problem

The state space representation of subsystem in the forward path of the feedback system in Fig. 6 is given by:

$$\dot{x} = Ax + bu \tag{50}$$
$$\delta_{r,c} = c^T x \tag{51}$$

where $\delta_{r,c}$ is the rear steering angle demanded by the controller and the matrix $A \in \mathbb{R}^{11 \times 11}$ has the following structure

$$A = \left[\begin{array}{c|c} A_{11} & A_{12} \\ \hline A_{21} & A_{22} \end{array} \right] \tag{52}$$

where the block matrices $A_{11} \in \mathbb{R}^{4 \times 4}$, $A_{12} \in \mathbb{R}^{4 \times 7}, A_{21} \in \mathbb{R}^{7 \times 4}$ and $A_{22} \in \mathbb{R}^{7 \times 7}$ are given by

$$A_{11} = \begin{bmatrix} 0 & 1 & -\frac{2}{mv_x} & -\frac{2}{mv_x} \\ 0 & 0 & \frac{2l_f}{I_{zz}} & -\frac{2l_r}{I_{zz}} \\ aC & -\frac{aCl_f}{v_x} & -a(1+\frac{2Cn_s}{C_L}) & 0 \\ aC & \frac{aCl_r}{v_x} & 0 & -a \end{bmatrix}, \quad A_{12} = \begin{bmatrix} 0 & 0 & 0 & 0 & 0 & 0 & 0 \\ 0 & 0 & 0 & 0 & 0 & 0 & 0 \\ aC & 0 & 0 & 0 & 0 & 0 & 0 \\ 0 & 0 & aC & 0 & 0 & 0 & 0 \end{bmatrix} \tag{53}$$

$$A_{21} = \begin{bmatrix} 0 & -b_1^f E_{12}(K_{2p} + K_v) & 0 & 0 \\ 0 & -b_2^f E_{12}(K_{2p} + K_v) & 0 & 0 \\ 0 & 0 & 0 & 0 \\ 0 & 0 & 0 & 0 \\ Ks_0 & 0 & 0 & 0 \\ -K_{1I} & K_{1I}K_{AW}E_{22}(K_{2p} + K_v) & 0 & 0 \\ 0 & -K_{2I} & 0 & 0 \end{bmatrix} \tag{54}$$

$$A_{22} = \begin{bmatrix} A_{11}^f & A_{12}^f & 0 & 0 & -b_1^f E_{12} & b_1^f E_{11} & b_1^f E_{12} \\ A_{21}^f & A_{22}^f & 0 & 0 & -b_2^f E_{12} & b_2^f E_{11} & b_2^f E_{12} \\ 0 & 0 & A_{11}^r & A_{12}^r & 0 & 0 & 0 \\ 0 & 0 & A_{21}^r & A_{22}^r & 0 & 0 & 0 \\ 0 & 0 & 0 & 0 & -s_0 & 0 & 0 \\ 0 & 0 & 0 & 0 & K_{1I}K_{AW}E_{22} & -K_{1I}K_{AW}E_{21} & -K_{1I}K_{AW}E_{22} \\ 0 & 0 & 0 & 0 & 0 & 0 & 0 \end{bmatrix} \quad (55)$$

The state vector x and the vectors b and c are given by

$$x = \begin{bmatrix} \beta \\ \dot\psi \\ S_f \\ S_r \\ \delta_f \\ \dot\delta_f \\ \delta_r \\ \dot\delta_r \\ K_2 \\ \tilde u_1 \\ \tilde u_{2I} \end{bmatrix}, \quad b = \begin{bmatrix} 0 \\ 0 \\ 0 \\ 0 \\ 0 \\ 0 \\ -b_1^r \\ -b_2^r \\ 0 \\ -K_{1I}K_{AW} \\ 0 \end{bmatrix}, \quad c = \begin{bmatrix} 0 \\ -E_{22}(K_{2p}+K_v) \\ 0 \\ 0 \\ 0 \\ 0 \\ 0 \\ 0 \\ -E_{22} \\ E_{21} \\ E_{22} \end{bmatrix} \quad (56)$$

The steering angles δ_f and δ_r in the state vector x are the output of the actuators. The state $\tilde u_1$ is the output of the controller $\tilde k_1$ and the state $\tilde u_{2I}$ is the output of the integrator in the controller $\tilde k_2$.

The saturation nonlinearity ϕ in the feedback path of the system in Fig. can be modelled as:

$$\phi(\delta_{r,c}) = k_{sat}(\delta_{r,c})\delta_{r,c} \quad (57)$$

with

$$k_{sat}(\delta_{r,c}) = \begin{cases} 1 & \text{if } |\delta_{r,c}| \le \delta_{sat} \\ \dfrac{\delta_{max}}{|\delta_{r,c}|} & \text{if } |\delta_{r,c}| > \delta_{sat} \end{cases} \quad (58)$$

Here δ_{sat} is the absolute value of the steering angles at which the rear actuators saturate. Note that the function $k_{sat}(\delta_{r,c})$ can be written as a function of the state vector x and $0 \le k_{sat}(x) \le 1$ for all x.

The closed-loop state-equation of the system in Fig. 6 can be written as

$$\dot x = (A - k_{sat}(x)bc^T)x \quad (59)$$

Now, if there is a positive definite matrix P such that

$$A^T P + PA <, \quad (A - bc^T)^T P + P(A - bc^T) < 0 \quad (60)$$

then $V(x) = x^T Px$ will define a Lyapunov function for the system (59), thus assuring its asymptotic stability. This follows because all of the matrices that arise in (59) are convex combinations of the two matrices $A, A - bc^T$. Thus if there is a solution P to (60), this guarantees the asymptotic stability of (59).

The Circle Criterion provides a frequency domain condition that can be used to test for the existence of a solution to (60). It has recently been shown that it is also possible to test for the existence of such a solution using a simple time-domain condition ([9],[10]). Specifically, there is a positive definite P satisfying (60) if and only if the matrices A and $(A - bc^T)$ are Hurwitz, i.e. their eigenvalues lie in the open left half of the complex plane, and their product $A(A - bc^T)$ has no negative real eigenvalues. We use this fact to analyse the robust stability of our control system with respect to parametric uncertainty. A major advantage of the time-domain condition is its simplicity, as it only requires the calculation of one set of eigenvalues as opposed to checking a frequency domain condition for infinitely many values of a variable.

Figure 7 summarises the results of the analysis. To generate Fig. 7 we proceeded as follows. First the steering controller was tuned for the nominal values of the car model parameters corresponding to a Mercedes S-Class. The real values of those parameters are uncertain, each of them lying within an interval around its nominal value. For a given fixed vehicle speed, we calculated the entries of A, b and c for a large number of values of the car model parameters randomly chosen from their respective uncertainty intervals. We checked that for all those values of the parameters, the matrices A and $(A - bc^T)$ remained Hurwitz. We then calculated the eigenvalues of $A(A - bc^T)$ for all the values of the parameters considered. We repeated the process outlined above for three different vehicle speeds: 20 m/s, 35 m/s and 50 m/s. In Figure , we plot the two eigenvalues closest to the real negative axis obtained with the different random values of the car model parameters for the three different speeds considered. As it can be seen in the figure, the eigenvalues of the matrix product $A(A - bc^T)$ remain well clear of the real negative axis. In light of the above, we conclude that the control system in Fig. 6 is robustly asymptotically stable for the speeds considered. We can then affirm that our original control system is robustly BIBO stable for those speeds.

8   Simulation Results

The full steering controller has been discretised and implemented on a detailed non-linear simulation model of a Mercedes S-Class equipped with 4-wheel steer-by-wire. The simulation results shown below illustrate the controller's performance and robustness.

8.1   Tracking of Reference Signals at Different Vehicle Speeds

The references to be tracked are as follows:

1. Yaw rate reference: Ramp of slope 0.05 rad/s^2 during 1 s and then maintain constant.

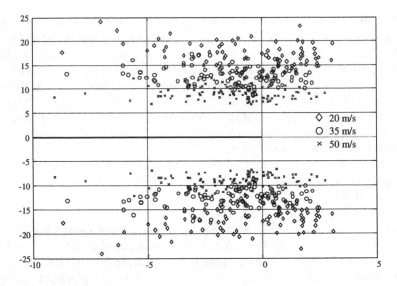

Fig. 7. Illustration of the robust asymptotic stability of the control system

2. Sideslip reference: Ramp of slope -0.005 rad/s during 1 s and then maintain constant.

Figure 8 shows the responses of the control system to these references for different values of the vehicle speed.

8.2   Disturbance Rejection in μ-split Braking

In a μ-split braking situation the car brakes with the tyres at opposite sides of the vehicle on different local road conditions. This results in the tyres at one side of the car see an adhesion coefficient (μ) different from the one seen by the tyres at the other side. An example of μ-split braking is a car braking with the two wheels at one side over a patch of ice and the other two on dry asphalt. In μ-split braking the torque created by the difference between the braking forces at either side of the vehicle introduces disturbances in both yaw rate and sideslip. These disturbances may induce the car to spin and cause the driver to lose control of the vehicle. The proposed steering controller automatically rejects any disturbances in sideslip and yaw rate generated in a μ-split braking situation. To illustrate this capability, consider the following example. A car travels along a straight level road at a speed of 60 m/s. At some point, the driver starts braking without turning the steering wheel. Suppose that the two wheels at the left hand side of the car are on dry asphalt ($\mu \approx 1$) and the two on the right hand side are on ice ($\mu \approx 0.2$). Since the driver keeps the steering wheel straight, the reference signals to be tracked by the steering controller are zero rad/s yaw rate and zero rad sideslip. Figure 9 illustrates the result of

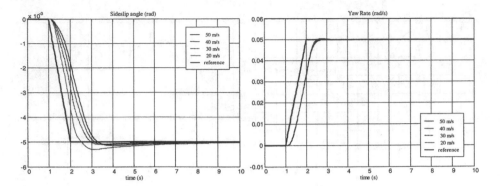

Fig. 8. Tracking performance of the steering controller

simulating this manoeuvre with and without the steering controller switched on. It can be seen that without the steering controller the car spins. On the other hand, with the controller in place the disturbances are quickly rejected and the car barely deviates from its intended straight path. The performance of the steering controller in this manoeuvre demonstrates the robustness of the control system–the cornering stiffness of a tyre during braking decreases as a result of the longitudinal slip [5]–as well as its ability to operate with varying vehicle speed.

8.3   Rear Actuator Saturation

The following manoeuvre is considered. Suppose again that the car travels on a road with a μ-split surface so that the two wheels at the left hand side of the car are on dry asphalt ($\mu \approx 1$) and the two on the right hand side are on ice ($\mu \approx 0.2$). While turning at 50 m/s, the driver applies the brakes moderately for 1 s without moving the steering wheel. The simulation results shown in Fig. 10 below illustrate the response of the controlled car with and without anti-windup. As it can be seen in the figure, the driver applies the brakes between time = 8 s and time = 9 s. The steering controller attempts to automatically reject the disturbances introduced during braking while tracking the reference sideslip and yaw rate signals. This results in the saturation of the rear actuators. Without anti-windup, the controller is not able to recover from the disturbances and spin out of control. On the other hand, the full steering controller (with anti-windup) is able to retain control of the car.

9   Conclusions

In this paper, we have presented a new steering controller for cars equipped with 4-wheel steer-by-wire. The controller allows the car to track given reference

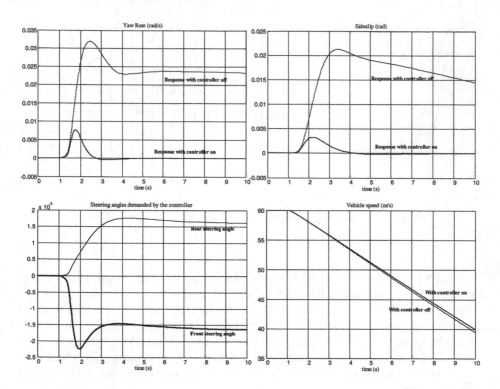

Fig. 9. Disturbance rejection performance of the steering controller in a μ-split braking manoeuvre

sideslip and yaw rate signals while rejecting external disturbances. Schematically, the controller comprises five distinct functional elements: a linear input transformation and a feedback element scheduled with the vehicle speed, which together render the yaw rate dynamics nearly speed-invariant with respect to the new controllable inputs; a linear diagonal controller valid for all operating vehicle speeds, which provides robustness and disturbance rejection performance; a feedforward element, which improves tracking performance; an anti-windup scheme, which allows the controller to perform satisfactorily when the rear actuators saturate. We have analysed the robust stability of the control system using recent results from the theory of common quadratic Lyapunov functions. The performance and robustness of the control system have been demonstrated through simulation. Future work will include a detailed robustness and integrity analysis together with validation experiments with the controller implemented on a real car equipped with 4-wheel steer-by-wire.

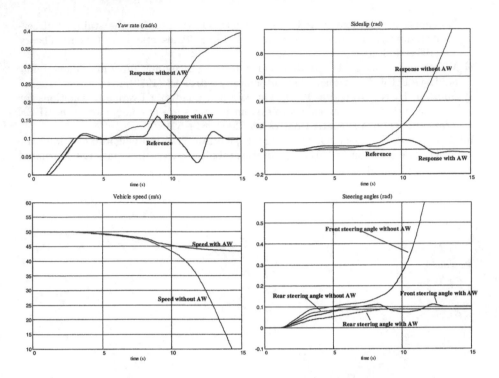

Fig. 10. Performance of the controller when the rear actuators saturate

References

[1] Furukawa, Y., Yuhara, N., Sano, S., Takeda, H., Matsushita, Y.: A review of four-wheel steering studies from the viewpoint of vehicle dynamics and control. Vehicle System Dynamics **18** (1989) 151–186

[2] Hirano, Y., Fukatani, K.: Development of robust active rear steering control. Proceedings of the International Symposium on Advanced Vehicle Control AVEC'96 (1996) 359–376

[3] Ackermann, J.: Robust decoupling, ideal steering dynamics and yaw stabilization of 4WS cars. Automatica **30** (1994) 1761–1768

[4] Vilaplana, M. A., Leith, D. J., Leithead, W. E., Kalkkuhl, J.: Control of sideslip and yaw rate in cars equipped with 4-wheel steer-by-wire. SAE Technical paper 2004-01-2076. Proceeding of the SAE 2004 Automotive Dynamics, Stability and Controls Conference and Exhibition (2004)

[5] Guillespie, T.D.: Fundamentals of vehicle dynamics. Society of Automotive Engineers (1992)

[6] Leithead, W. E., O'Reilly, J.: m-Input m-output feedback control by Individual Channel Design - Part 1: Structural issues. International Journal of Control **56** (1992) 1347-1397

[7] Narendra, K. S., Goldwyn, R. : A geometrical criterion for the stability of certain nin-linear, non-autonomous systems. IEEE Transactions on Circuit Theory **11** (1964) 406–407

[8] Willems, J.: The circle criterion and quadratic Lyapunov functions for stability analysis. IEEE Transactions on Automatic Control **18** (1973) 184

[9] Shorten, R. N., Narendra, K. S.: On common quadratic Lyapunov functions for pairs of stable LTI systems whose system matrices are in companion form. IEEE Transactions on Automatic Control **48** (2003) 618–621

[10] Shorten, R. N., Mason, O., O'Cairbre, F., Curran, P.: A unifying framework for the SISO circle criterion and other quadratic stability criteria. International Journal of Control **77** (2004) 1–8

A Second-Order Cone Bounding Algorithm for Robust Minimum Variance Beamforming*

Ngai Wong[1], Venkataramanan Balakrishnan[2], and Tung-Sang Ng[1]

[1] Department of Electrical and Electronic Engineering
The University of Hong Kong
Pokfulam Road, Hong Kong
{nwong,tsng}@eee.hku.hk
[2] School of Electrical and Computer Engineering
Purdue University
West Lafayette, IN 47907-1285, USA
ragu@ecn.purdue.edu

Abstract. We present a geometrical approach for designing robust minimum variance (RMV) beamformers against steering vector uncertainties. Conventional techniques enclose the uncertainties with a convex set; the antenna weights are then designed to minimize the maximum array output variance over this set. In contrast, we propose to cover the uncertainty by a second-order cone (SOC). The optimization problem, with optional robust interference rejection constraints, then reduces to the minimization of the array output variance over the intersection of the SOC and a hyperplane. This is cast into a standard second-order cone programming (SOCP) problem and solved efficiently. We study the computationally efficient case wherein the uncertainties are embedded in complex-plane trapezoids. The idea is then extended to arbitrary uncertainty geometries. Effectiveness of the proposed approach over other schemes and its fast convergence in signal power estimation are demonstrated with numerical examples.

1 Introduction

Antenna arrays constitute an important part in modern communication systems, serving to introduce extra degrees of freedom in beampattern synthesis, spatial filtering and/or detection of incoming signals. The design of antenna arrays when precise system parameters are available is a well-studied problem; for instance, the celebrated minimum variance (MV) beamformer, designed using Capon's method [1], has the property that the variance of the combined (i.e., weighted and summed) array output is minimized, while a unity gain is maintained in the look direction. However, in practical situations, exact models of the antenna array are unavailable. Uncertainties in the steering vector of the desired signal arise due

---

* This work was supported in part by the Hong Kong Research Grants Council, the University Research Committee of The University of Hong Kong, and the National Science Foundation of the United States of America under Grant No. ECS-0200320.

R. Murray-Smith, R. Shorten (Eds.): Switching and Learning, LNCS 3355, pp. 223–247, 2005.

to a multitude of reasons including array calibration errors, uncertain angle-of-arrival (AOA), amplifier imperfections and environmental inhomogeneities [2, 3, 4, 5, 6, 7, 8, 9, 10, 11, 12, 13, 14]. These uncertainties, when not accounted for in the design process, can lead to severely degraded performance. For example, the performance of the MV beamformer is known to be sensitive and susceptible to mismatches in the presumed and actual steering vectors [14]. Hence we have the "robust antenna weight design problem," i.e., the design of antenna weights such that the performance can be guaranteed in spite of the presence of uncertainties.

Some approaches towards robust antenna weight design can be found in [15, 16, 17, 18, 19, 20, 21] and the references therein. For example, point and derivative constraints [15, 16, 17] imposed on the mainbeam can be used to design antenna arrays that offer tolerance against AOA mismatch, but their performance subject to other kinds of mismatches is hard to predict. The eigenspace-based beamformer in [18] is effective, although only when the signal-to-noise ratio (SNR) is high. Other methods in [17, 19, 20, 21] share a similar framework wherein a certain form of weighted diagonal loading or quadratic penalty is added to the objective or cost function. The weight determination of that penalty, however, is not clear in practice. Further, these techniques assume, either explicitly or implicitly, that the uncertainty is isotropic (i.e., equally probable around the nominal steering vector) which is generally not the case. In other words, these methods or algorithms may result in overly conservative designs at the expense of other considerations such as power, complexity, and feasibility.

Recently, a number of techniques based on mathematical programming have been proposed for the robust antenna weight design of *narrowband* systems called robust MV (RMV) beamforming [2, 3, 4, 5, 6, 7, 8, 9, 10, 11, 12]. The basic idea underlying these techniques is to model the steering vector uncertainties as a convex set or part of a convex set. The antenna weights are then determined so as to minimize the maximum array output variance (or an upper bound thereof) over the steering vector uncertainty set. In [3, 4, 5, 6, 7], the uncertainty set is covered by a hypersphere[3] or an ellipsoid around the nominal steering vector. It can be shown that this class of beamforming techniques belongs to the diagonal loading approach, of which the amount of loading can be directly determined from the uncertainty set. The resulting optimization problem is a second-order cone programming (SOCP) problem [22, 23], which can be solved efficiently via interior point algorithms, e.g., [24, 25, 26, 27], or by the *Lagrange multiplier* method, e.g., [4, 6]. Simulations have shown the superiority of this SOCP beamforming approach over other popular robust beamformers in adaptive arrays [3]. Nonetheless, uncertainty modeling using a worst-case hypersphere [3] is still isotropic and does not exploit the structure of the uncertainty, and may sometimes lead to impractical designs of high power requirement or even programming infeasibility.

---

[3] Here hypersphere and ellipsoid (flat ellipsoid) respectively refer to the n-dimensional counterparts of a Euclidean ball and the injective (non-injective) affine mapping of a Euclidean ball. A polyhedral cone is the set $C = \{x | Ax \leq 0\}$, i.e., C is the intersection of finitely many linear half-spaces. Specific details can be found in [2, 3, 4, 5, 6, 7, 8].

Ellipsoidal uncertainty modeling [4, 5, 6, 7] provides tighter uncertainty modeling and generally produces more accurate results in applications such as signal power estimation [6, 7]. A different design approach is to encompass the uncertainty set by a polyhedral cone [8]. A drawback is that the use of a polyhedral cone with limited extreme rays (the basis rays of a cone) can again result in overly conservative constraints, while increasing the number of extreme rays will cause an exponential growth in the programming complexity and prohibit its use in larger arrays. Moreover, determination of the polyhedral cone angle in relation to the uncertainty set was not pursued further in [8].

The main contribution of this paper is that it extends the idea of a polyhedral cone to a second-order cone (SOC), and develops a constructive way, employing either a simple heuristic or a theoretically optimal SOCP approach, to obtain a tight SOC bounding the uncertainty set (also see [2]). The convexity of the optimization constraint is exploited such that the optimization process can be largely reduced from the whole uncertainty set to the intersection of the bounding SOC and a hyperplane *outside* the set. A special case of modeling steering vector uncertainties using complex-plane trapezoids is studied in detail. For practical reasons, extension of the proposed scheme to robust interference rejection is also considered. The corresponding narrowband beamforming task is formulated and efficiently solved as an SOCP problem. Numerical examples show that this *SOC RMV beamformer* exhibits tight uncertainty modeling, very low power requirement, and fast convergence in signal power estimation.

The paper is organized as follows. In Sect. 2, preliminaries in MV beamforming are reviewed. Sect. 3 proposes a generic algorithm for RMV beamforming utilizing a geometrical SOC bounding idea. Reduction of the optimization process from a convex set to the circumference of a hyperellipse is described. An application of the proposed SOC RMV beamforming algorithm is demonstrated, wherein steering vector uncertainties are embedded in complex-plane trapezoids. Simplification of the techniques and their extension to arbitrary uncertainty geometries are also discussed. Sect. 4 presents numerical examples and verifies the effectiveness and power efficiency of the proposed approach over other schemes. Finally, Sect. 5 presents the conclusion.

We close this section with a description of the notations used. The set of real numbers is denoted by $\mathbb{R}$ and the set of complex numbers by $\mathbb{C}$. $\mathbb{R}^N$ and $\mathbb{C}^N$ denote the set of real and complex vectors, respectively, with N components. The set Ω is *convex* if $\mathbf{v}_1, \mathbf{v}_2 \in \Omega$ implies $\rho_1 \mathbf{v}_1 + \rho_2 \mathbf{v}_2 \in \Omega$ for every real $\rho_1, \rho_2 \geq 0$ that satisfy $\rho_1 + \rho_2 = 1$. A general convex optimization problem is the minimization of a linear function over a convex set Ω, namely,

$$\min(\mathbf{c}^* \mathbf{x}) \text{ subject to } \mathbf{x} \in \Omega \tag{1}$$

where $\mathbf{c}$ and $\mathbf{x}$ are complex or real vectors, and $(\circ)^*$ denotes conjugate transpose which is equivalent to transpose, $(\circ)^T$, for real vectors. A second-order cone or SOC is a special convex set whose definition, for an "upright" SOC of dimension $2N$ and a cone angle parameter λ, is

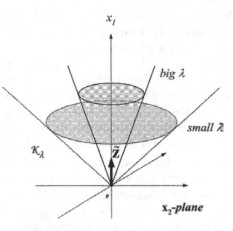

Fig. 1. An upright SOC with a variable cone angle. $\tilde{\mathbf{z}}$ is the unit vector along the symmetry axis

$$\mathcal{K}_\lambda = \left\{ \begin{bmatrix} x_1 \\ \mathbf{x_2} \end{bmatrix} \middle| x_1, \lambda \in \mathbb{R}, \mathbf{x_2} \in \mathbb{R}^{2N-1}, \lambda \geq 0, x_1 \geq \lambda \|\mathbf{x_2}\| \right\} , \qquad (2)$$

where $\|\circ\|$ denotes the Euclidean norm. The conceptual visualization of an SOC is shown in Fig. 1. Clearly, λ is a parameter that controls the cone angle, namely, a large λ corresponds to a "narrow" cone and vice versa. And $\tilde{\mathbf{z}} = [1\ 0\ \cdots\ 0]^T \in \mathbb{R}^{2N}$ is the unit vector in the direction of the symmetry axis of $\mathcal{K}_\lambda$. A second-order cone programming or SOCP problem [22, 23] with real-valued variables (indicated by tildes on top, as will be followed throughout this paper) takes the form of

$$\min(\tilde{\mathbf{c}}^T \tilde{\mathbf{x}}) \text{ subject to } \tilde{\mathbf{x}} \in \mathcal{K}_\lambda . \qquad (3)$$

It should be noted that although SOCP represents a subclass of the more general semidefinite programming (SDP) [28] (namely, the optimization of a linear function over linear matrix inequalities [29]), dedicated SOCP solvers, e.g., [25, 26], should be used [22] because of their much better worst-case complexity than general SDP solvers such as [27].

2   Background in Minimum Variance Beamforming

The output $\mathbf{x}(t) \in \mathbb{C}^N$ of an N-element antenna array is

$$\mathbf{x}(t) = \mathbf{a}(\theta)s(t) + \mathbf{A_i}\mathbf{S_i}(t) + \mathbf{n}(t) , \qquad (4)$$

where $\mathbf{a}(\theta) \in \mathbb{C}^N$ is the steering vector of the desired *narrowband* signal $s(t)$ arriving from an angle θ, $\mathbf{A_i}$ is an $N \times L$ matrix whose lth column, $\mathbf{a}(\theta_l)$, is the steering vector of the lth interfering signal in $\mathbf{S_i}(t) = [\, s_1(t) \, \cdots \, s_L(t) \,]^T$, and $\mathbf{n}(t) \in \mathbb{C}^N$ is the additive noise component. The combined output of the array subject to a complex weight $\mathbf{w}$ is

$$y(t) = \mathbf{w}^*\mathbf{x}(t) \ . \tag{5}$$

The interference-plus-noise covariance matrix $\mathbf{R_{in}}$ is defined as

$$\mathbf{R_{in}} = \mathbf{E}\left((\mathbf{A_i}\mathbf{S_i}(t) + \mathbf{n}(t))(\mathbf{A_i}\mathbf{S_i}(t) + \mathbf{n}(t))^*\right) \ , \tag{6}$$

whereas the sample covariance matrix $\mathbf{R_x}$ is defined, and approximated by M recently received samples (called *snapshots*), as

$$\mathbf{R_x} = \mathbf{E}(\mathbf{xx}^*) \approx \frac{1}{M}\sum_{p=1}^{M}\mathbf{x}(p)\mathbf{x}(p)^* \ . \tag{7}$$

One of the metrics for the performance of a beamformer is the signal-to-interference-plus-noise ratio (SINR) designated as

$$SINR = \frac{|\mathbf{w}^*\mathbf{a}(\theta)|^2\,\sigma_{\mathrm{s}}^2}{\mathbf{w}^*\mathbf{R_{in}}\mathbf{w}} \ , \tag{8}$$

with σ_{s}^2 being the signal power.

2.1   Capon Beamformer

The MV beamformer is obtained by solving

$$\min(\mathbf{w}^*\mathbf{R_x}\mathbf{w}) \text{ subject to } \mathbf{w}^*\mathbf{a}(\theta_{\mathrm{p}}) = 1 \ , \tag{9}$$

where θ_{p} and $\mathbf{a}(\theta_{\mathrm{p}})$ are the presumed (or nominal) AOA and steering vector respectively. If this presumed steering vector matches the physical steering vector, we have the optimal solution of (9) given by the Capon's method [1]

$$\mathbf{w_{mv}} = \frac{\mathbf{R_x}^{-1}\mathbf{a}(\theta_{\mathrm{p}})}{\mathbf{a}(\theta_{\mathrm{p}})^*\mathbf{R_x}^{-1}\mathbf{a}(\theta_{\mathrm{p}})} \ . \tag{10}$$

In beampattern synthesis, it may be desirable to allow for AOA uncertainty by maintaining unity gain in a small spread of angles [4]. This is done in the MV beamforming by adding extra equality constraints. For example, defining the matrix $\mathbf{C} = [\,\mathbf{a}(\theta_{\mathrm{p}})\,\mathbf{a}(\theta_{\mathrm{p}1})\,\mathbf{a}(\theta_{\mathrm{p}2})\,\cdots\,]$ where $\theta_{\mathrm{p}i}$'s are angles around θ_{p}, and replacing the optimization constraint in (9) by

$$\mathbf{C}^*\mathbf{w} = \mathbf{d} \tag{11}$$

where $\mathbf{d}$ is a column vector of ones, the optimal weight vector is now [4, 12]

$$\mathbf{w_{mv}} = \mathbf{R_x}^{-1}\mathbf{C}(\mathbf{C}^*\mathbf{R_x}^{-1}\mathbf{C})^{-1}\mathbf{d} \ . \tag{12}$$

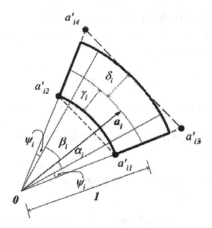

Fig. 2. Uncertainty region of a steering vector element (annulus sector in bold line) bounded by a trapezoid of vertices a'_{i1}, a'_{i2}, a'_{i3} and a'_{i4}. Here α_i, β_i, γ_i, δ_i, $\psi_i \geq 0$

This formulation can also be used to introduce nulling at the interference angles if we define, with respect to (4), $\mathbf{C} = [\mathbf{a}(\theta)\ \mathbf{A_i}]$ and $\mathbf{d} = [1\ \xi_1\ \cdots\ \xi_L]^T$ where $\xi_l \geq 0$, $l = 1, 2, \cdots, L$, are the desired interference gains (some small real values or zero) for signals coming from θ_l. Roughly speaking, the introduction of each equality constraint at a certain angle reduces one degree of freedom in the choice of the weight vector. Therefore, smaller arrays are more likely to yield infeasible designs when the constraints are stringent.

2.2   Robustness Against Signal Steering Vector Uncertainties

Let the signal steering vector be $\mathbf{a} = [a_1\ \cdots\ a_N]^T \in \mathbb{C}^N$. Referring to Fig. 2, an element a_i of $\mathbf{a}$ may be subject to phase uncertainties, α_i, β_i, due to uncertain AOA, and phase and gain uncertainties, ψ_i, γ_i and δ_i, due to amplifier tolerance. Thus in practice a_i may assume any value inside the bolded annulus sector in Fig. 2. Let $\mathbf{\Omega} \subset \mathbb{C}^N$ be the set that contains all the possible $\mathbf{a}$s (corresponding to all possible combinations of a_i, $i = 1, 2, \cdots, N$). An RMV beamformer [2,3, 4,5,6,7,8] is then designed by maintaining at least a unity gain for all members in $\mathbf{\Omega}$:

$$\min(\mathbf{w}^*\mathbf{R_x}\mathbf{w}) \text{ subject to } \mathrm{Re}(\mathbf{w}^*\mathbf{a}) \geq 1,\ \forall \mathbf{a} \in \mathbf{\Omega} \qquad (13)$$

where $\mathrm{Re}(\circ)$ and $\mathrm{Im}(\circ)$ (to appear later) give the real and imaginary parts of the argument.

2.3   Robustness Against Interference Uncertainties

In theory, the programming solution to (13) (namely, minimizing the output power subject to signal protection) automatically achieves interference rejection. But in practice the tolerance in the amplifier implementation may render different gains and phases from the designed values. In mobile or imperfect channel scenario, drifting of the interference angle(s) may also occur between updates of weights in an adaptive array. To maintain a high SINR, as will be seen in the numerical examples, it is of value to explore robust interference rejection. Similar to the case of the nominal steering vector, uncertainties in interference rejection can be lumped as uncertainties in the interference steering vectors. Suppose $\mathbf{a}_l$ $(l = 1, 2, \cdots, L)$ is contained in the uncertainty sets $\Omega_l \subset \mathbb{C}^N$, then it is desirable that the array look direction constraint and the interference rejection constraints hold simultaneously, namely,

$$\min(\mathbf{w}^*\mathbf{R_x}\mathbf{w}) \text{ subject to}$$
$$\begin{cases} \text{Re}(\mathbf{w}^*\mathbf{a}) \geq 1 \ , \ \forall \mathbf{a} \in \Omega \\ \|\mathbf{w}^*\mathbf{a}_l\| \leq \xi_l \ , \ \forall \mathbf{a}_l \in \Omega_l \ , \ l = 1, 2, \cdots, L \ . \end{cases} \tag{14}$$

The inequality settings, instead of equalities, in the interference rejection constraints lend themselves to compatibility in programming formulation as will become evident later.

2.4   Solution Via Convex Optimization

One numerical approach towards the solution of (14) is based on convex optimization. The first step is to embed the uncertainty sets Ω and Ω_l, $l = 1, 2, \cdots, L$, in convex sets (if they are not already convex); see for example, [2,3,4,5,6,7,8,9]. An example is to use the convex hulls, in the form of convex polytopes[4], of Ω and Ω_l. Then, from convexity, it suffices to check that the look direction constraint and interference rejection constraints are satisfied on the vertices of these uncertainty convex hulls. Therefore, by choosing the enclosing convex sets appropriately, an infinite set of optimization constraints can be reduced to those on the vertices of a hull, or on the curved boundary of an arbitrary convex geometry. Nonetheless, the complexity of the hull, in terms of its number of vertices, still increases exponentially with the number of antenna element N and prohibits practical computation. In [3,4,5,6,7,8], hyperspheres, nondegenerate and degenerate (or flat) ellipsoids, and polyhedral cones are respectively used to enclose the uncertainties, and the programming problem is cast as an SOCP or a quadratic programming problem of order linearly dependent on N. It should, however, be noted that the approach based on hyperspheres does not exploit the uncertainty structure and may result in overly conservative designs with high power requirement or even render the design problem infeasible. Also, robust interference rejection is not addressed in these works. In contrast, we propose an approach that exploits the uncertainty structure and provides robustness against steering vector uncertainties regarding both the desired and interfering

---

[4] A polytope is a finite region of n-dimensional space enclosed by a finite number of hyperplanes. And the convex hull of a set of points is the smallest convex set that includes the points.

signals. The final beamforming problem is also an SOCP problem of size linearly dependent on N.

For convenience of computation and coding, complex quantities are often transformed into real quantities. Indicating real-valued matrices and vectors by tildes, we define

$$
\tilde{\mathbf{w}} = \begin{bmatrix} \mathrm{Re}(\mathbf{w}) \\ \mathrm{Im}(\mathbf{w}) \end{bmatrix}, \ \tilde{\mathbf{a}} = \begin{bmatrix} \mathrm{Re}(\mathbf{a}) \\ \mathrm{Im}(\mathbf{a}) \end{bmatrix}, \ \tilde{\mathbf{a}}_l = \begin{bmatrix} \mathrm{Re}(\mathbf{a}_l) \\ \mathrm{Im}(\mathbf{a}_l) \end{bmatrix},
$$
$$
\tilde{\mathbf{R}}_\mathbf{x} = \begin{bmatrix} \mathrm{Re}(\mathbf{R}_\mathbf{x}) & -\mathrm{Im}(\mathbf{R}_\mathbf{x}) \\ \mathrm{Im}(\mathbf{R}_\mathbf{x}) & \mathrm{Re}(\mathbf{R}_\mathbf{x}) \end{bmatrix} .
\tag{15}
$$

Starting with look direction constraint only, (13) can be equivalently written as

$$
\min(\tilde{\mathbf{w}}^T \tilde{\mathbf{R}}_\mathbf{x} \tilde{\mathbf{w}}) \text{ subject to } \tilde{\mathbf{w}}^T \tilde{\mathbf{a}} \geq 1, \ \forall \tilde{\mathbf{a}} \in \tilde{\mathbf{\Omega}}
\tag{16}
$$

where $\tilde{\mathbf{\Omega}}$ is a set derived from $\mathbf{\Omega}$ by similarly stacking the real and imaginary parts of each element in $\mathbf{\Omega}$. To incorporate the interference rejection constraints in (14), we note that the magnitudes of the combined gains $\|\mathbf{w}^* \mathbf{a}_l\|$, $l = 1, 2, \cdots, L$, involve a quadratic relationship of the real and imaginary parts of $\mathbf{w}^* \mathbf{a}_l$ that describes a circle of radius ξ_l. To reduce the constraints into linear ones for SOCP formulation, two sufficient (stronger) conditions governing the real and imaginary parts are imposed, namely,

$$
\begin{cases} -\frac{\xi_l}{\sqrt{2}} \leq \tilde{\mathbf{w}}^T \tilde{\mathbf{a}}_l \leq \frac{\xi_l}{\sqrt{2}} \\ -\frac{\xi_l}{\sqrt{2}} \leq \tilde{\mathbf{w}}^T \begin{bmatrix} \mathbf{0} & \mathbf{I} \\ -\mathbf{I} & \mathbf{0} \end{bmatrix} \tilde{\mathbf{a}}_l \leq \frac{\xi_l}{\sqrt{2}} \end{cases}, \ \forall \tilde{\mathbf{a}}_l \in \tilde{\mathbf{\Omega}}_l
\tag{17}
$$

for $l = 1, 2, \cdots, L$, where $\mathbf{0}$ and $\mathbf{I}$ are zero and identity matrices of compatible dimensions, and $\tilde{\mathbf{\Omega}}_l$ being analogous to $\tilde{\mathbf{\Omega}}$. It can be seen that (17) confines the real and imaginary parts of $\mathbf{w}^* \mathbf{a}_l$ to be within a square inscribed in the circle of radius ξ_l. Accordingly, (17) can be appended to the constraint list in (16) to achieve robust interference rejection. Since increasing the number of constraints may also lead to infeasibility in the design problem, robust interference rejection is more likely to be realized in larger arrays where more freedom is available.

2.5 Signal Power Estimation and Array Output Power

A main goal in many antenna array applications is to estimate the signal power σ_s^2 [6, 7]. In traditional beamforming, this is simply given by

$$
\sigma_\mathrm{s}^2 \approx \mathbf{w}_\mathrm{mv}^* \mathbf{R}_\mathbf{x} \mathbf{w}_\mathrm{mv} .
\tag{18}
$$

A much more accurate estimate proposed in [6, 7], with the elimination of a "scaling ambiguity" by taking into account $\|\mathbf{a}(\theta)\| = \sqrt{N}$, can be shown to be

$$
\sigma_\mathrm{s}^2 \approx \frac{1}{N} \frac{\|\mathbf{R}_\mathbf{x} \mathbf{w}_\mathrm{rmv}\|^2}{\mathbf{w}_\mathrm{rmv}^* \mathbf{R}_\mathbf{x} \mathbf{w}_\mathrm{rmv}}
\tag{19}
$$

where $\mathbf{w}_{\mathrm{rmv}}$ is the solution to the RMV beamformers in [3,4,5,6,7] or the present work.

Another power related issue is the array output power. A set of appropriately designed antenna weights will significantly suppress interference, therefore from (4) and (5) we have

$$y(t) = \mathbf{w}^*\mathbf{x}(t) \approx \mathbf{w}^*\mathbf{a}(\theta)s(t) + \mathbf{w}^*\mathbf{n}(t) \tag{20}$$

where $\mathbf{a}(\theta)s(t)$ is a column vector with time (phase) shifted versions of the desired signal $s(t)$ (e.g., see [7]). If we further assume that the signal is random over time and uncorrelated with the noise (assumed to be white Gaussian), then the mean array output power is

$$E\left(y(t)y(t)^*\right) = \sigma_s^2 \left\|\mathbf{w}\right\|^2 + \sigma_n^2 \left\|\mathbf{w}\right\|^2 \tag{21}$$

where σ_n^2 is the noise power. The constraint of the Capon beamformer (namely, $\mathbf{w}^*\mathbf{a}(\theta_p) = 1$) will put the first term on the right of (21) to σ_s^2 only, but in the robust formulation the more general form in (21) holds. A major implication is that if an analog beamformer is built, the power of the array output is then proportional to $\left\|\mathbf{w}\right\|^2$ (see also [9,22]). While in digital implementation, the input needs to be normalized by $\left\|\mathbf{w}\right\|^2$ to prevent overflow due to finite wordlengths. In both cases, $\left\|\mathbf{w}\right\|^2$ ($= \left\|\tilde{\mathbf{w}}\right\|^2$ in (15)) serves as a metric that should be kept as low as possible. Since $\left\|\mathbf{w}\right\|^2$ can also be interpreted as the power output of an array subject to unit-power signal and zero noise and interference, it is given a unit of *watt*.

3   An SOC Bounding Approach

As we have observed, the main drawback in directly solving (16), either standalone or with additional constraints in (17), is the exponential growth in the problem size when the number of antenna elements, N, grows. Our main contribution is an algorithm, called the SOC RMV beamforming algorithm, that reformulates the original problem so that the order of constraints grows linearly with N. In addition, the uncertainty structure and convexity in the optimization constraints are exploited, thereby leading to accurate and power-efficient beamformers. To simplify notations, we assume in the rest of this paper that Ω and Ω_l (and consequently $\tilde{\Omega}$ and $\tilde{\Omega}_l$) are some convex sets that encompass the signal and interference steering vector uncertainties $\mathbf{a}$ and $\mathbf{a}_l$ ($\tilde{\mathbf{a}}$ and $\tilde{\mathbf{a}}_l$), $l = 1, 2, \cdots, L$.

Two theorems central to our proposed algorithm, which are related to the convexity in the optimization constraints, are given here:

Theorem 1. *(Robust look direction constraint) Let $\tilde{\Omega} = Co\{\tilde{\mathbf{a}}_1, \tilde{\mathbf{a}}_2, \cdots, \tilde{\mathbf{a}}_n\}$, where Co denotes the convex hull of a set, i.e., all convex combinations of its elements. Moreover, let $\mathbf{0} \notin \tilde{\Omega}$. Let $\mathcal{K}_\lambda$ be any SOC with $\tilde{\Omega} \subset \mathcal{K}_\lambda$. Suppose that $\mathcal{H}$ is a hyperplane separating $\mathbf{0}$ and $\tilde{\Omega}$. Define the hyperellipse $\tilde{\varepsilon}$ by $\tilde{\varepsilon} = \mathcal{K}_\lambda \cap \mathcal{H}$ and let $\partial\tilde{\varepsilon}$ denote its boundary. Under these conditions, consider some $\tilde{\mathbf{w}} \in \mathbb{R}^{2N}$. If $\tilde{\mathbf{w}}^T\tilde{\mathbf{a}} \geq 1$ for all $\tilde{\mathbf{a}} \in \partial\tilde{\varepsilon}$, then $\tilde{\mathbf{w}}^T\tilde{\mathbf{a}} \geq 1$ for all $\tilde{\mathbf{a}} \in \tilde{\Omega}$.*

Remark 1. The implication of Theorem 1 is that the condition $\tilde{\mathbf{w}}^T \tilde{\mathbf{a}} \geq 1$ for all $\tilde{\mathbf{a}} \in \partial \tilde{\varepsilon}$ is sufficient for robust look direction constraint in (16) to hold.

Remark 2. The condition $\mathbf{0} \notin \tilde{\boldsymbol{\Omega}}$ automatically holds for any physically meaningful set of steering vectors.

Proof. The hyperplane $\mathcal{K}$ can be parameterized as

$$\mathcal{H} = \left\{ \tilde{\mathbf{a}} \mid \tilde{\mathbf{b}}^T \tilde{\mathbf{a}} = 1 \right\} \tag{22}$$

for some $\tilde{\mathbf{b}} \in \mathbb{R}^{2N}$. Since $\mathcal{H}$ separates $\mathbf{0}$ and $\tilde{\boldsymbol{\Omega}}$, we must have $\tilde{\mathbf{b}}^T \tilde{\mathbf{a}}_i \geq 1$, $i = 1, 2, \cdots, n$. Define $\tau_i = (\tilde{\mathbf{b}}^T \tilde{\mathbf{a}}_i)^{-1}$. Then, by the definition of $\mathcal{K}_\lambda$, we must have $\tau_i \tilde{\mathbf{a}}_i \in \mathcal{K}_\lambda$. Moreover, $\tau_i \tilde{\mathbf{b}}^T \tilde{\mathbf{a}}_i = 1$, or $\tau_i \tilde{\mathbf{a}}_i \in \mathcal{H}$. Thus $\tau_i \tilde{\mathbf{a}}_i \in \tilde{\varepsilon}$.

Now suppose that for some $\tilde{\mathbf{w}} \in \mathbb{R}^{2N}$, $\tilde{\mathbf{w}}^T \tilde{\mathbf{a}} \geq 1$ for all $\tilde{\mathbf{a}} \in \partial \tilde{\varepsilon}$. Then, as $\tilde{\varepsilon}$ is a convex set, we must have $\tilde{\mathbf{w}}^T \tilde{\mathbf{a}} \geq 1$ for all $\tilde{\mathbf{a}} \in \tilde{\varepsilon}$, and in particular, we must have

$$\tau_i \tilde{\mathbf{w}}^T \tilde{\mathbf{a}}_i \geq 1 \ , \ i = 1, 2, \cdots, n \ .$$

Consequently

$$\tilde{\mathbf{w}}^T \tilde{\mathbf{a}}_i \geq \tau_i^{-1} \geq 1 \ , \ i = 1, 2, \cdots, n \ , \tag{23}$$

concluding the proof.   □

Theorem 2. *(Robust interference rejection) Let* $\tilde{\boldsymbol{\Omega}}_l = Co\{\tilde{\mathbf{a}}_{l1}, \tilde{\mathbf{a}}_{l2}, \cdots, \tilde{\mathbf{a}}_{ln}\}$ *and* $\mathbf{0} \notin \tilde{\boldsymbol{\Omega}}_l$. *Let* $\mathcal{K}_\lambda$ *be any SOC with* $\tilde{\boldsymbol{\Omega}}_l \subset \mathcal{K}_\lambda$. *Suppose* $\mathcal{H}$ *is a hyperplane such that* $\mathbf{0}$ *and* $\tilde{\boldsymbol{\Omega}}_l$ *lie on the same side of it, with the Euclidean distance of* $\mathbf{0}$ *from* $\mathcal{H}$ *exceeding that from any point in* $\tilde{\boldsymbol{\Omega}}_l$ *(i.e.,* $\tilde{\boldsymbol{\Omega}}_l$ *is "between"* $\mathbf{0}$ *and* $\mathcal{H}$). *Define the hyperellipse* $\tilde{\varepsilon}_l$ *by* $\tilde{\varepsilon}_l = \mathcal{K}_\lambda \cap \mathcal{H}$ *and let* $\partial \tilde{\varepsilon}_l$ *denote its boundary. Under these conditions, consider some* $\tilde{\mathbf{w}} \in \mathbb{R}^{2N}$. *If* $|\tilde{\mathbf{w}}^T \tilde{\mathbf{a}}_l| \leq \mu_l$ *for all* $\tilde{\mathbf{a}}_l \in \partial \tilde{\varepsilon}_l$, *then* $|\tilde{\mathbf{w}}^T \tilde{\mathbf{a}}_l| \leq \mu_l$ *for all* $\tilde{\mathbf{a}}_l \in \tilde{\boldsymbol{\Omega}}_l$.

Remark 3. Let $\tilde{\mathbf{w}}' = \begin{bmatrix} \mathbf{0} & -\mathbf{I} \\ \mathbf{I} & \mathbf{0} \end{bmatrix} \tilde{\mathbf{w}}$ and $\mu_l = \frac{\xi_l}{\sqrt{2}}$, Theorem 2 implies that the conditions $|\tilde{\mathbf{w}}^T \tilde{\mathbf{a}}_l| \leq \frac{\xi_l}{\sqrt{2}}$ and $|\tilde{\mathbf{w}}'^T \tilde{\mathbf{a}}_l| \leq \frac{\xi_l}{\sqrt{2}}$ for all $\tilde{\mathbf{a}}_l \in \partial \tilde{\varepsilon}_l$, $l = 1, 2, \cdots, L$, are sufficient for the robust interference rejection constraints in (17) to hold.

Proof. Follows similarly to that of Theorem 1, and is therefore omitted.   □

With reference to Fig. 3, the proposed SOC RMV beamforming algorithm is summarized in the following three steps:

1. Fit an SOC around the hull of $\tilde{\boldsymbol{\Omega}}$. If robust interference rejection is needed, also find SOCs around the hulls of $\tilde{\boldsymbol{\Omega}}_l$, $l = 1, 2, \cdots, L$.

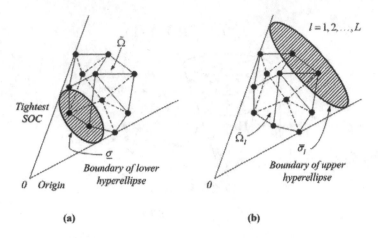

Fig. 3. A second-order cone encompassing: (a) $\tilde{\Omega}$ with a lower hyperellipse; (b) $\tilde{\Omega}_l$ with an upper hyperellipse

2. Intersect the SOC with a hyperplane tangent to the bottom of the hull of $\tilde{\Omega}$, thus forming a hyperellipse with boundary $\underline{\sigma}$ (Fig. 3(a)). In the case of robust interference rejection, hyperplanes tangent to the top of the hulls of $\tilde{\Omega}_l$ are found, forming hyperellipses of boundaries $\bar{\sigma}_l$, $l = 1, 2, \cdots, L$ (Fig. 3(b)).

3. Transform (16) into an SOCP problem and optimize with respect to the stronger conditions $\tilde{\mathbf{a}} \in \underline{\sigma}$, and $\tilde{\mathbf{a}}_l \in \bar{\sigma}_l$, $l = 1, 2, \cdots, L$, in (17) for robust interference rejection.

It can be seen that the constraints in step 3 represent two sets of stronger conditions. Specifically, by Theorem 1, if $\tilde{\mathbf{w}}^T \tilde{\mathbf{a}} \geq 1$ in (16) is satisfied for all $\tilde{\mathbf{a}} \in \underline{\sigma}$, it is automatically satisfied for all $\tilde{\mathbf{a}}$ on the hyperellipse, as well as all $\tilde{\mathbf{a}} \in \tilde{\Omega}$ above the hyperellipse. Similarly, if the conditions in (17) are satisfied for all $\tilde{\mathbf{a}}_l \in \bar{\sigma}_l$, $l = 1, 2, \cdots, L$, then by Theorem 2 they are automatically satisfied for all $\tilde{\mathbf{a}}_l$ on the hyperellipse, as well as all $\tilde{\mathbf{a}}_l \in \tilde{\Omega}_l$ below the hyperellipse. The following demonstrates an application of the proposed algorithm wherein the steering vector uncertainties are modeled by complex-plane trapezoids.

3.1 Parametrizing the SOCs Bounding $\tilde{\Omega}$ and $\tilde{\Omega}_l$

In this step, Ω and Ω_l, $l = 1, 2, \cdots, L$, are obtained by modeling steering vector uncertainties using complex-plane trapezoids. Subsequently $\tilde{\Omega}$ and $\tilde{\Omega}_l$ are derived by stacking the real and imaginary parts of each element in Ω and Ω_l. Illustration is provided only for the location of the SOC bounding $\tilde{\Omega}$, while that for the case of $\tilde{\Omega}_l$ proceeds in exactly the same way. Recalling from Sect. 2 and revisiting Fig. 2, an element a_i of $\mathbf{a} = [a_1 \cdots a_N]^T$ may assume any value inside the annulus sector due to phase uncertainties, α_i, β_i, resulting from uncertain

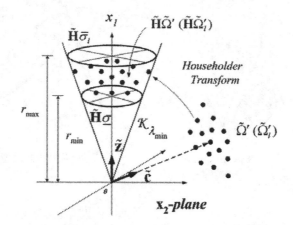

Fig. 4. Rotation of $\tilde{\Omega}'$ or $\tilde{\Omega}'_l$ into the bounding SOC using Householder transform

AOA, and phase and gain uncertainties, ψ_i, γ_i and δ_i, resulting from amplifier imperfections. A sensible way, which also serves as a stronger condition, is to encompass the annulus sector using a trapezoid with vertices a'_{i1}, a'_{i2}, a'_{i3}, a'_{i4} as in Fig. 2. The actual a_i may then be regarded as a convex combination of a'_{i1}, a'_{i2}, a'_{i3} and a'_{i4}, $i = 1, 2, \cdots, N$. By defining the set $\Omega' \subset \mathbb{C}^N$ as the union of these vertices,

$$\Omega' = \left\{ \mathbf{v}_j = [a'_{1k_1} \; a'_{2k_2} \; \cdots \; a'_{Nk_N}]^T \in \mathbb{C}^N \; \middle| \; \begin{matrix} k_i = 1, 2, 3 \text{ or } 4 \\ i = 1, 2, \cdots, N \\ j = 1, 2, \cdots, 4^N \end{matrix} \right\}, \qquad (24)$$

it is clear that every point in Ω' constitutes a vertex of the minimum convex hull of Ω. As discussed before, optimization over Ω can be replaced by optimizing over every point in Ω'. Apparently, the real-valued counterpart of Ω', $\tilde{\Omega}' \subset \mathbb{R}^{2N}$, is formed by stacking the real and imaginary parts of every point in Ω', i.e., if $\mathbf{v} \in \Omega'$, then $\tilde{\mathbf{v}} \in \tilde{\Omega}'$ is defined as $\tilde{\mathbf{v}} = [\text{Re}(\mathbf{v}^T) \; \text{Im}(\mathbf{v}^T)]^T$. Obviously, $\tilde{\Omega}'$ also constitutes the vertices of the minimum convex hull of $\tilde{\Omega}$. Instead of directly finding the SOC that encloses $\tilde{\Omega}'$, $\tilde{\Omega}'$ is first rotated, using the angle and distance preserving *Householder transform* [8], into the orientation of the upright SOC to find the SOC that just contains the rotated $\tilde{\Omega}'$, denoted as $\tilde{\mathbf{H}}\tilde{\Omega}'$ (or $\tilde{\mathbf{H}}\tilde{\mathbf{v}}$ for all $\tilde{\mathbf{v}} \in \tilde{\Omega}'$) in Fig. 4. Defining a unit vector $\tilde{\mathbf{c}} = [\text{Re}(\mathbf{c}^T) \; \text{Im}(\mathbf{c}^T)]^T \in \mathbb{R}^{2N}$ in the direction of $\tilde{\Omega}'$, we have

$$\tilde{\mathbf{H}} = \begin{cases} \mathbf{I} - 2\frac{(\tilde{\mathbf{c}}-\tilde{\mathbf{z}})(\tilde{\mathbf{c}}-\tilde{\mathbf{z}})^T}{(\tilde{\mathbf{c}}-\tilde{\mathbf{z}})^T(\tilde{\mathbf{c}}-\tilde{\mathbf{z}})}, & \tilde{\mathbf{c}} \neq \tilde{\mathbf{z}} \\ \mathbf{I}, & \tilde{\mathbf{c}} = \tilde{\mathbf{z}} \end{cases} \qquad (25)$$

where $\|\tilde{\mathbf{c}}\| = \|\mathbf{c}\| = 1$. $\tilde{\mathbf{H}}$ is a symmetric orthogonal matrix such that $\tilde{\mathbf{H}} = \tilde{\mathbf{H}}^T = \tilde{\mathbf{H}}^{-1}$. Now the question remains as to how to choose the unit vector $\tilde{\mathbf{c}}$ for the tightest SOC that encompasses $\tilde{\mathbf{H}}\tilde{\Omega}'$, denoted as $\mathcal{K}_{\lambda_{\min}}$. Two approaches are in place that offer a tradeoff between accuracy and computational load.

Optimal Method. This method finds the $\tilde{\mathbf{c}}$ that gives the tightest possible SOC. To do this we notice from (2) that an SOC, $\mathcal{K}_\lambda$, that contains a particular $\tilde{\mathbf{v}}$ must satisfy

$$
\begin{aligned}
\lambda &\le \left([1\, 0\, \cdots\, 0]\, \tilde{\mathbf{H}}\tilde{\mathbf{v}} \right) / \left\| \mathrm{diag}\, [0\, 1\, \cdots\, 1]\, \tilde{\mathbf{H}}\tilde{\mathbf{v}} \right\| \\
&= \left((\tilde{\mathbf{H}}\tilde{\mathbf{c}})^T \tilde{\mathbf{H}}\tilde{\mathbf{v}} \right) / \left\| \left(\mathbf{I} - \tilde{\mathbf{H}}\tilde{\mathbf{c}}(\tilde{\mathbf{H}}\tilde{\mathbf{c}})^T \right) \tilde{\mathbf{H}}\tilde{\mathbf{v}} \right\| \\
&= (\tilde{\mathbf{v}}^T \tilde{\mathbf{c}}) / \left\| \tilde{\mathbf{v}} - (\tilde{\mathbf{v}}^T \tilde{\mathbf{c}})\tilde{\mathbf{c}} \right\| \\
&= 1 / \sqrt{ \left(\|\tilde{\mathbf{v}}\| / \tilde{\mathbf{v}}^T \tilde{\mathbf{c}} \right)^2 - 1 } \ .
\end{aligned}
\tag{26}
$$

The second line in (26) stems from the fact that $\tilde{\mathbf{H}}\tilde{\mathbf{c}} = \tilde{\mathbf{z}} = [1\, 0\, \cdots\, 0]^T$, and the third line has the geometrical interpretation that it is the projection of $\tilde{\mathbf{v}}$ onto $\tilde{\mathbf{c}}$ divided by the norm of components in $\tilde{\mathbf{v}}$ that are orthogonal to $\tilde{\mathbf{c}}$. Apparently, λ is independent of $\tilde{\mathbf{H}}$ because the cone angle is preserved by the transform. Finding the tightest SOC is equivalent to choosing a $\tilde{\mathbf{c}}$ that maximizes the minimum λ, denoted by $\lambda_{\min}$, such that $\mathcal{K}_{\lambda_{\min}}$ contains $\tilde{\mathbf{H}}\tilde{\mathbf{v}}$ for all $\tilde{\mathbf{v}} \in \tilde{\Omega}'$. This statement can be restated as a programming problem in variable $\tilde{\mathbf{c}}$:

$$
\max_{\|\tilde{\mathbf{c}}\|=1} \left(\min_{\tilde{\mathbf{v}} \in \tilde{\Omega}'} \left(\frac{\tilde{\mathbf{v}}^T \tilde{\mathbf{c}}}{\|\tilde{\mathbf{v}}\|} \right) \right) \text{ where } \tilde{\mathbf{v}}^T \tilde{\mathbf{c}} > 0 \ ,
\tag{27}
$$

or qualitatively, to choose a unit vector $\tilde{\mathbf{c}}$ such that the minimum projection of those unit vectors $\tilde{\mathbf{v}}/\|\tilde{\mathbf{v}}\|$ ($\tilde{\mathbf{v}} \in \tilde{\Omega}'$) onto $\tilde{\mathbf{c}}$ is maximized. Clearly, $\tilde{\mathbf{c}}$ and $\tilde{\mathbf{v}}/\|\tilde{\mathbf{v}}\|$ are all on the surface of the unit sphere. Also, the projection of $\tilde{\mathbf{v}}/\|\tilde{\mathbf{v}}\|$ ($\tilde{\mathbf{v}} \in \tilde{\Omega}'$) onto $\tilde{\mathbf{c}}$ can equivalently be regarded as the projection, which is a real quantity, of $\mathbf{v}/\|\mathbf{v}\|$ ($\mathbf{v} \in \Omega'$) onto $\mathbf{c}$. With reference to Fig. 2 and by symmetry argument, the vector $\mathbf{c}$ that gives the tightest SOC must have each of its components lying along the symmetry axis of the corresponding trapezoid. Using $(\circ)_i$ to denote the ith component of a vector and $\arg(\circ)$ to denote the angle of a complex quantity, we have

$$
\arg((\mathbf{c})_i) = \arg(a'_{i1} + a'_{i2}) \text{ where } i = 1, 2, \cdots, N \ .
\tag{28}
$$

Since only the projection is of interest, the number of $\mathbf{v}_j$ in (24) to be considered is largely reduced. This is because due to symmetry about $(\mathbf{c})_i$, a vector in (24) with a particular $k_i = 1$ produces the same projection as another vector with that particular $k_i = 2$, and the same holds for the case of $k_i = 3$ and $k_i = 4$. So there are effectively 2^N $\mathbf{v}_j$s ($j = 1, 2, \cdots, 2^N$) of interest whose $k_i = 1$ or 3 ($i = 1, 2, \cdots, N$). Next, to find the magnitudes of those components in $\mathbf{c}$, we define a real vector $\hat{\mathbf{c}}$ that holds the element-wise magnitudes of $\mathbf{c}$, i.e.,

$$
(\hat{\mathbf{c}})_i = |(\mathbf{c})_i| \text{ where } i = 1, 2, \cdots, N \ ,
\tag{29}
$$

and another set of real vectors $\hat{\mathbf{v}}_j$ such that

$$(\hat{\mathbf{v}}_j)_i = \left| \left(\frac{\mathbf{v}_j}{\|\mathbf{v}_j\|} \right)_i \right| \cos \left(\frac{\alpha_i + \beta_i + 2\psi_i}{2} \right) \text{ where } i = 1, 2, \cdots, N , \quad (30)$$

for $j = 1, 2, \cdots, 2^N$ as described earlier. The problem of maximizing the minimum projection in (27) is then equivalent to the SOCP [22, 23] problem:

$$\max(\tau) \text{ (or } \min(-\tau)) \text{ subject to}$$
$$\begin{cases} 0 < \tau \leq \hat{\mathbf{v}}_j^T \hat{\mathbf{c}} \text{ where } j = 1, 2, \cdots, 2^N \\ \|\hat{\mathbf{c}}\| \leq 1 \end{cases} \quad (31)$$

for which the optimization variables are τ and $\hat{\mathbf{c}}$. The condition $\|\hat{\mathbf{c}}\| = 1$ will automatically be satisfied because the last constraint in (31) is tight for any optimal solution. It can be seen that (31) is essentially a linear programming problem except for the last quadratic constraint. Such SOCP structure can be solved using, say, the SOCP solver in [26]. The optimal τ thus obtained constitutes an optimal solution of (27) and can be substituted back into (26) to get the maximum $\lambda_{\min}$, namely,

$$\lambda_{\min} = \frac{1}{\sqrt{\tau^{-2} - 1}} . \quad (32)$$

The direction vector $\mathbf{c}$ (and thus $\tilde{\mathbf{c}}$) corresponding to this tightest SOC, $\mathcal{K}_{\lambda_{\min}}$, is then obtained through combining $\hat{\mathbf{c}}$ and (28). Referring to Fig. 4, a simple and obvious choice for the hyperplane intersecting the SOC (to be used in step 2 of the algorithm) is the one that is normal to the symmetry axis of the SOC. The parameter $r_{\min}$, which specifies the height of the supporting hypercircle resulting from the intersection, is calculated from the minimum projection of $\mathbf{v}$ ($\tilde{\mathbf{v}}$) onto $\mathbf{c}$ ($\tilde{\mathbf{c}}$). This is obtained in a straightforward manner by

$$r_{\min} = \left[\text{Re}([a_{11}' \, a_{21}' \cdots a_{N1}']) \, \text{Im}([a_{11}' \, a_{21}' \cdots a_{N1}']) \right] \begin{bmatrix} \text{Re}(\mathbf{c}) \\ \text{Im}(\mathbf{c}) \end{bmatrix} . \quad (33)$$

The tightest SOC that contains a particular Ω_l' (the interference counterpart of Ω'), $l = 1, 2, \cdots, L$, is found in the same way except now the height of the upper hypercircle is of interest, which is

$$r_{\max} = \left[\text{Re}([a_{13}' \, a_{23}' \cdots a_{N3}']) \, \text{Im}([a_{13}' \, a_{23}' \cdots a_{N3}']) \right] \begin{bmatrix} \text{Re}(\mathbf{c}) \\ \text{Im}(\mathbf{c}) \end{bmatrix} \quad (34)$$

where the a_{i3}'s and $\mathbf{c}$ stand for the uncertainty vectors and orientation for $\mathbf{a}_l$ and $\tilde{\Omega}_l'$, respectively.

Centroid Method. A simple heuristic that largely reduces the computation of $\mathbf{c}$ from its exponential dependence (see (31)) to linear dependence on N, at a small expense of accuracy, is to approximate the optimal $\mathbf{c}$ by the normalized (unit-length) centroid of all points in Ω'. Defining

$$\mathbf{a'}_k = [\, a'_{1k}\, a'_{2k}\, \cdots\, a'_{Nk}\,]^T,\, k = 1, 2, 3, 4 \;,\tag{35}$$

$\mathbf{c}$ is given by

$$\mathbf{c} = \left(\sum_{k=1}^{4} \mathbf{a'}_k\right) \Bigg/ \left\|\sum_{k=1}^{4} \mathbf{a'}_k\right\| \;.\tag{36}$$

Using the same notational convention as in the optimal method, $\lambda_{\min}$ is obtained by finding the particular $\hat{\mathbf{v}}_j$ that decorrelates with $\hat{\mathbf{c}}$ as much as possible. This can be achieved in just $N - 1$ comparison steps: First, the components of $\mathbf{c}$ are along the symmetry axes of the uncertainty trapezoids, so the $\hat{\mathbf{v}}_j$ definition in (30) still applies. Besides, $\hat{\mathbf{c}}$ is now a predetermined quantity as given by (36). The next step is to arrange the magnitude components of $\hat{\mathbf{c}}$ in a particular order so that they form a descending sequence; $\lambda_{\min}$ is then given by a $\hat{\mathbf{v}}_j$ whose N components are chosen to form an ascending sequence in that particular order. Since there are N such choices of $\hat{\mathbf{v}}_j$, $\lambda_{\min}$ can be determined within $N - 1$ comparisons. In contrast, it generally requires $4N$ comparisons to find the radius of the smallest hypersphere (centered at the presumed steering vector) [3] bounding the annulus sector in Fig. 2. Likewise, $r_{\min}$ and $r_{\max}$ are obtained by (33) and (34) respectively. Experiments have shown that with trapezoidal uncertainty modeling, the RMV beamformers designed using this centroid method perform almost identically as those obtained by the optimal way. Therefore, in practical situations the centroid approximation of $\mathbf{c}$ should always be used when computation is of concern, especially when N is large.

3.2   Transformation of Constraints into SOC Formulation

As discussed, the robust look direction constraint in (16) can be realized under a stronger condition, namely, on the boundary $\underline{\sigma}$ of the lower hypercircle. In Fig. 4, the boundary $\underline{\sigma}$, in a rotated manner, is

$$\tilde{\mathbf{H}}\underline{\sigma} = \begin{bmatrix} r_{\min} \\ \underline{\mathbf{u}} \end{bmatrix} \subset K_{\lambda_{\min}} \text{ where } \underline{\mathbf{u}} \subset \mathbb{R}^{2N-1},\; \|\underline{\mathbf{u}}\| = \frac{r_{\min}}{\lambda_{\min}} \;.\tag{37}$$

Noting $\underline{\sigma} = \tilde{\mathbf{H}}(\tilde{\mathbf{H}}\underline{\sigma})$, the gain constraint in (16) becomes

$$\tilde{\mathbf{w}}^T (\tilde{\mathbf{H}} \begin{bmatrix} r_{\min} \\ \underline{\mathbf{u}} \end{bmatrix}) \geq 1 \;.\tag{38}$$

Let $\tilde{\mathbf{H}}_1 \in \mathbb{R}^{1 \times 2N}$ be the first row of $\tilde{\mathbf{H}}$, and $\tilde{\mathbf{H}}_2 \in \mathbb{R}^{(2N-1) \times 2N}$ be $\tilde{\mathbf{H}}$ without the first row, (38) can be rewritten as

$$-\underline{\mathbf{u}}^T \tilde{\mathbf{H}}_2 \tilde{\mathbf{w}} \leq r_{\min} \tilde{\mathbf{H}}_1 \tilde{\mathbf{w}} - 1 \;.\tag{39}$$

When $\underline{u} = -(r_{\min}/\lambda_{\min})\tilde{\mathbf{H}}_2\tilde{\mathbf{w}}/\left\|\tilde{\mathbf{H}}_2\tilde{\mathbf{w}}\right\|$, the maximum of the left hand side of (39) is achieved. And the robust look direction constraint takes the form of an SOCP constraint:

$$\left\|\frac{r_{\min}}{\lambda_{\min}}\tilde{\mathbf{H}}_2\tilde{\mathbf{w}}\right\| \leq r_{\min}\tilde{\mathbf{H}}_1\tilde{\mathbf{w}} - 1 \ . \tag{40}$$

By the same token, the robust interference rejection constraints in (17) can be realized under a stronger condition, namely, on the boundary $\bar{\sigma}_l$ of the upper hypercircle. In Fig. 4, the boundary $\bar{\sigma}_l$, in a rotated manner, is given by

$$\tilde{\mathbf{H}}\bar{\sigma}_l = \begin{bmatrix} r_{\max} \\ \bar{\mathbf{u}} \end{bmatrix} \subset K_{\lambda_{\min}} \text{ where } \bar{\mathbf{u}} \subset \mathbb{R}^{2N-1}, \ \|\bar{\mathbf{u}}\| = \frac{r_{\max}}{\lambda_{\min}} \ . \tag{41}$$

Let $\tilde{\mathbf{H}}_{l1} \in \mathbb{R}^{1 \times 2N}$ be the first row of $\tilde{\mathbf{H}}$, and $\tilde{\mathbf{H}}_{l2} \in \mathbb{R}^{(2N-1) \times 2N}$ be $\tilde{\mathbf{H}}$ without the first row, the first equation in (17) can verified to be equivalent to

$$\left\|\frac{r_{\max}}{\lambda_{\min}}\tilde{\mathbf{H}}_{l2}\tilde{\mathbf{w}}\right\| \leq \min\left(-r_{\max}\tilde{\mathbf{H}}_{l1}\tilde{\mathbf{w}}, \ r_{\max}\tilde{\mathbf{H}}_{l1}\tilde{\mathbf{w}}\right) + \frac{\xi_l}{\sqrt{2}} \ . \tag{42}$$

Also, define

$$\tilde{\mathbf{J}} = \tilde{\mathbf{H}} \begin{bmatrix} \mathbf{0} & -\mathbf{I} \\ \mathbf{I} & \mathbf{0} \end{bmatrix}, \tag{43}$$

and let $\tilde{\mathbf{J}}_{l1} \in \mathbb{R}^{1 \times 2N}$ be the first row of $\tilde{\mathbf{J}}$, and $\tilde{\mathbf{J}}_{l2} \in \mathbb{R}^{(2N-1) \times 2N}$ be $\tilde{\mathbf{J}}$ without the first row, the second equation in (17) is equivalent to

$$\left\|\frac{r_{\max}}{\lambda_{\min}}\tilde{\mathbf{J}}_{l2}\tilde{\mathbf{w}}\right\| \leq \min\left(-r_{\max}\tilde{\mathbf{J}}_{l1}\tilde{\mathbf{w}}, \ r_{\max}\tilde{\mathbf{J}}_{l1}\tilde{\mathbf{w}}\right) + \frac{\xi_l}{\sqrt{2}} \ . \tag{44}$$

3.3   Beamformer Optimization Problem in SOCP Format

Finally, let $\tilde{\mathbf{R}}_{\mathbf{x}} = \tilde{\mathbf{U}}^T\tilde{\mathbf{U}}$ be the Cholesky factorization of $\tilde{\mathbf{R}}_{\mathbf{x}}$, the objective function of the SOC RMV beamforming problem in (16) can be rewritten as $\left\|\tilde{\mathbf{U}}\tilde{\mathbf{w}}\right\|^2$. As minimizing $\left\|\tilde{\mathbf{U}}\tilde{\mathbf{w}}\right\|^2$ is the same as minimizing $\left\|\tilde{\mathbf{U}}\tilde{\mathbf{w}}\right\|$, by introducing an auxiliary variable ε, (16) is cast into a standard SOCP problem of order linearly dependent on N:

$$\min(\varepsilon) \text{ subject to}$$

$$\left\|\tilde{\mathbf{U}}\tilde{\mathbf{w}}\right\| \leq \varepsilon, \ \left\|\frac{r_{\min}}{\lambda_{\min}}\tilde{\mathbf{H}}_2\tilde{\mathbf{w}}\right\| \leq r_{\min}\tilde{\mathbf{H}}_1\tilde{\mathbf{w}} - 1 \ . \tag{45}$$

When robust interference rejection is needed, the constraints in (42) and (44) can be appended to the constraint list in (45) as:

$$
\begin{cases}
\left\| \frac{r_{\max}}{\lambda_{\min}} \tilde{\mathbf{H}}_{l2} \tilde{\mathbf{w}} \right\| \leq -r_{\max} \tilde{\mathbf{H}}_{l1} \tilde{\mathbf{w}} + \frac{\xi_l}{\sqrt{2}} \\
\left\| \frac{r_{\max}}{\lambda_{\min}} \tilde{\mathbf{H}}_{l2} \tilde{\mathbf{w}} \right\| \leq r_{\max} \tilde{\mathbf{H}}_{l1} \tilde{\mathbf{w}} + \frac{\xi_l}{\sqrt{2}} \\
\left\| \frac{r_{\max}}{\lambda_{\min}} \tilde{\mathbf{J}}_{l2} \tilde{\mathbf{w}} \right\| \leq -r_{\max} \tilde{\mathbf{J}}_{l1} \tilde{\mathbf{w}} + \frac{\xi_l}{\sqrt{2}} \\
\left\| \frac{r_{\max}}{\lambda_{\min}} \tilde{\mathbf{J}}_{l2} \tilde{\mathbf{w}} \right\| \leq r_{\max} \tilde{\mathbf{J}}_{l1} \tilde{\mathbf{w}} + \frac{\xi_l}{\sqrt{2}}
\end{cases}
\tag{46}
$$

for $l = 1, 2, \cdots, L$. SOCP solvers utilizing interior-point algorithms, e.g., [25,26], can then be used to solve for the weights of this SOC RMV beamformer. The complexity of each iteration step is $O(N^3)$, and because the number of iterations is typically around ten, the complexity of this SOCP solver approach is still $O(N^3)$. Another way of solving (45), possibly with (46), is by the Lagrange multiplier method [4, 6] whose complexity is also $O(N^3)$. However, for each low-rank update of $\mathbf{R_x}$, the latter approach allows update of the weight design problem with a complexity of only $O(N^2)$, while the former approach requires recomputation every time [6].

Three additional comments are in order:

1. The final SOCP problem of the proposed beamformer is of the same order as other robust schemes in [3, 4, 5, 6, 7] using other uncertainty bounding geometries. However, in the SOCP problem setup, finding the hypersphere, flat ellipsoid, or the SOC (centroid approach) that enclose the uncertainty set all require $O(N)$ work. In contrast, finding the tightest SOC (optimal approach) and the minimum volume ellipsoid [4], provided SOCP and SDP solvers are used respectively, would require $O(\rho N^3)$ and $O(\rho N^4)$ work in every iteration (e.g., [22]), where ρ is proportional to the number of vertices in the uncertainty set. Consequently, the first three schemes are more practical when computational speed is of concern or array size is large.

2. If irregular, arbitrary-shape (but convex) polytopes are used to model the uncertainty set, the maximum $\lambda_{\min}$ can be obtained in the following way: First, find the minimum enclosing sphere (MES) of all points $\tilde{\mathbf{v}} / \|\tilde{\mathbf{v}}\|$ ($\tilde{\mathbf{v}} \in \tilde{\Omega}'$ or $\tilde{\Omega}'_l$) on the unit sphere, which can again be cast as a standard SOCP problem:

$$
\begin{aligned}
&\min(\text{radius}) \text{ subject to} \\
&\|\text{point}_i - \text{center}\| \leq \text{radius}, \quad \forall \text{point}_i ,
\end{aligned}
\tag{47}
$$

or solved using other techniques such as the Welzl's algorithm [30] in linear time. Then $\tilde{\mathbf{c}}$ is simply the unit vector pointing towards the center of this MES, while $\lambda_{\min}$, $r_{\min}$, and $r_{\max}$ are immediately inferred from the intersection of this MES and the unit sphere. The major difficulty with this approach, however, is the poor scalability due to the exponential increase in the number of hull vertices.

3. The proposed SOC RMV beamforming approach does not require the antenna array to be linear as no special restrictions are placed on the steering vectors. RMV beamforming for general non-uniform arrays with different element patterns still proceed in the same way. The proposed SOC bounding scheme also provides a deterministic and systematic way to construct the optimization constraints given the tolerance in AOA and array amplifiers. Due to the SOCP formulation of the beamforming problem, additional requirements like power restriction on the antenna weights and beampattern tuning [9] are readily incorporated. Furthermore, simple and tight uncertainty modeling with the centroid approach enables real-time setup and computation in adaptive arrays.

4   Numerical Examples

The first example studies a 5-element uniform array separated by half wavelengths. We start with a simple case of no interfering signal. Suppose a far-field narrowband signal of unit-power is impinging on the array. The signal AOA is $+20°$ with an uncertainty of $±2.5°$. The SNR is $10\,dB$ and the noise is white Gaussian and uncorrelated with the signal. The array amplifiers are of unity gain with an uncertainty of $±0.05$ and a phase uncertainty of $±3°$. The traditional non-robust Capon MV [1], and the robust hypersphere [3], full (nondegenerate) ellipsoid [4], flat (degenerate) ellipsoid [6], as well as the proposed beamformers are designed accordingly. The hyperspherical and the full ellipsoidal uncertainty bounding schemes are designed such that the annulus sector (Fig. 2) of each steering vector component is bounded within the uncertainty set. The hypersphere radius thus calculated is 0.8287. The flat ellipsoidal bounding is designed in a way as in [6], in which a "rank-two" flat ellipsoid is formed such that the steering vectors at the two uncertain AOA extremes are within the ellipsoid. Note that a flat ellipsoid assumes certain linear combinations of uncertainty [4] and may not include all steering vector combinations as in other robust schemes. In the proposed SOC RMV scheme, the optimal and the centroid methods give almost the same SOC, parameterized by $\lambda_{min} = 2.3783$ and $r_{min} = 1.9922$ for the centroid approach, and $\lambda_{min} = 2.4345$ and $r_{min} = 1.9822$ for the optimal approach. Fig. 5 shows the performance of various beamformers against AOA mismatch. As expected, the proposed SOC RMV beamformers corresponding to the centroid and optimal methods perform virtually the same, and thus only the one from centroid method is shown. The results are based on the theoretical covariance matrix, i.e., $M \to \infty$ in (7), e.g., see [6]. Fig. 5(a) shows that the proposed and the hypersphere schemes give the best SINR robustness, with their peak SINRs being comparable to the peak value of the Capon MV beamformer. The full and flat ellipsoid schemes have lower SINR performance but it improves when the mismatch is near the extremes. Not surprisingly, the Capon MV beamformer suffers from an abrupt decrease in SINR when the actual AOA deviates from the nominal one. Fig. 5(b) shows that the flat ellipsoidal bounding method produces the tightest results (gain ≥ 1) with respect to the specified range of uncertainty, while the hypersphere bounding method results in an "over-design" due to its inherently conservative nature. The proposed and the full ellipsoid schemes are much tighter compared to the hypersphere scheme. Fig. 5(c) investigates the accuracy of the signal power estimation, with the estimate from (18) being used for the Capon MV beamformer, and (19) being employed for the other schemes. Consistent with the results in [6], the flat ellipsoid scheme produces the most accurate estimate (0 dB) over the uncertainty range, while the proposed beamformer performs similarly to the hypersphere scheme. A major drawback of the hypersphere method is the increased power metric, proportional to $\|\mathbf{w}\|^2$ (see (21)) as illustrated in Fig. 5(d), that may cause the design to be practically infeasible. In contrast, the proposed beamformer shows a value close to the optimal value of the Capon MV beamformer. The performance of other robust schemes are in between. Next, we consider the convergence rate of signal power estimation when the sample covariance matrix is used. The results are plotted against the number of snapshots (M in (7)) in Fig. 6. Under this case of no interference, the convergence rates of all robust schemes are basically the same. Fig. 6(a) shows the power estimation in the absence of AOA mismatch, while Fig. 6(b) demonstrates how AOA mismatch can deteriorate the estimation accuracy of the Capon MV beamformer.

The second example considers a 10-element uniform array separated by half wavelengths. The unit-power signal has an AOA of $+10°\pm2.5°$. Four interference signals of power 6 dB lower than the signal power are coming from $-70°$, $-30°$, $+50°$, and $+70°$. The noise and amplifier tolerance are the same as in the previous example. The proposed beamformer designed with the centroid and optimal approaches are again similar ($\lambda_{\min} = 0.6054$ and $r_{\min} = 2.2699$ for the centroid approach, and $\lambda_{\min} = 0.6271$ and $r_{\min} = 2.1519$ for the optimal approach) and only the results from the centroid approach are shown. The hypersphere radius in this case is 2.3847. The full ellipsoidal bounding scheme is not implemented due to its high computational complexity. Fig. 7 shows similar observations for various schemes as in Fig. 5. However, in Fig. 7(c), it can be seen that the power estimation accuracy of the flat ellipsoid scheme is strongly dependent on the actual AOA mismatch. As shown in Fig. 7(d), the variation in the power metric of different schemes is much larger due to the increased number of antennae. It can be seen that the proposed scheme maintains a near-optimal value over the whole uncertainty range. Fig. 8 further reveals that in terms of signal power estimation, the proposed scheme enjoys the fastest convergence among others. In fact, our numerical experiments show that the proposed scheme consistently gives the fastest convergence.

Next, consider a case with a unit-power signal of AOA $+10° \pm 1.2°$. There are two unit-power interference signals from $-30° \pm 0.1°$ and $+50° \pm 0.1°$. The noise assumption is as before and the amplifier gain and phase uncertainties are ±0.05 and $\pm0.1°$ respectively. To address the issues of amplifier and interference uncertainties, we carry out robust interference rejection as discussed in Sect. 2.3. It is required that the signal gain be at least 20 dB higher than that of the interfering signals (i.e., $\xi_1, \xi_2 < 0.1$). Fig. 9 shows 1000 random beampatterns for the

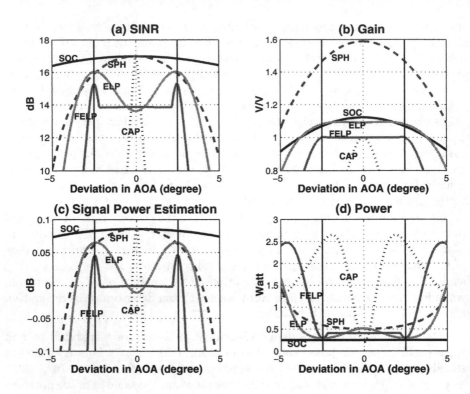

Fig. 5. (a)-(d). Performance of a 5-element array against AOA mismatch: Capon MV (CAP), hypersphere (SPH), full ellipsoid (ELP), flat ellipsoid (FELP), and the proposed (SOC) beamformers. The two vertical solid lines in each plot denote the AOA uncertainty range

Capon MV and the proposed SOC RMV beamformers (centroid approach) in which the signal AOA, interfering signal angles, and amplifier gains and phases vary randomly within their specified uncertainty ranges. The Capon MV beamformer is designed with the point nulling constraints embedded in (12), and the proposed beamformer is designed by incorporating (46) into (45). It can be seen that the worst-case performance of the Capon MV beamformer is severely degraded: the overall gain is significantly higher, admitting more noise and interference power to degrade the SINR and accuracy of signal power estimation. In contrast, the proposed scheme performs favorably against uncertainties and the beampatterns remain almost invariant. It should be noted that under these design criteria, modeling steering vector uncertainties of the desired and interfering signals using hyperspheres has rendered the SOCP problem infeasible. Table 1 gives the figures of merits for various schemes under the signal, interference, and implementation uncertainties. Specifically, the antenna weights are designed with a covariance matrix arising from a random set of data in the

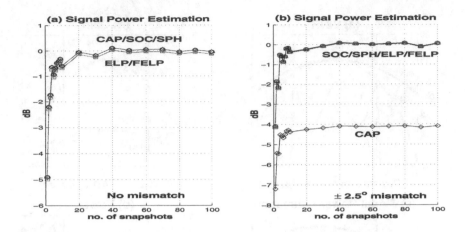

Fig. 6. Signal power estimation for a 5-element array using sample covariance matrix: (a) no AOA mismatch. (b) 2.5° AOA mismatch. Each data point is the average of 100 Monte Carlo simulations. Featured designs: Capon MV (CAP), hypersphere (SPH), full ellipsoid (ELP), flat ellipsoid (FELP), and the proposed (SOC) beamformers

uncertainty set, and then the performance of the array is measured subject to another random set of data from the uncertainty set. The results are averaged over 1000 Monte Carlo simulations. It is seen that the flat ellipsoid scheme, with its simplified uncertainty structure assumption, is most susceptible to amplifier parameter variations. While the proposed scheme with robust interference rejection delivers the highest SINR, most accurate power estimation, and low power metric. Finally, given the fact that SOC RMV beamformers resulting from the optimal method and the centroid method perform almost identically, the use of the centroid approximation in all practical cases is well justified.

Table 1. Figures of merits for different schemes under signal, interference, and implementation uncertainties. 1000 Monte Carlo simulations are averaged for each entry (RIR stands for robust interference rejection)

	Power(W)	SINR(dB)	Power Estimate(dB)
Capon	1.9840	6.0621	-2.9282
Hypersphere	0.2597	18.9755	0.0768
Flat Ellipsoid	0.6754	11.7388	-0.2191
Proposed(centroid, no RIR)	0.1249	16.7159	0.1421
Proposed(optimal, no RIR)	0.1261	16.5128	0.1458
Proposed(centroid with RIR)	0.1380	19.8219	0.0515

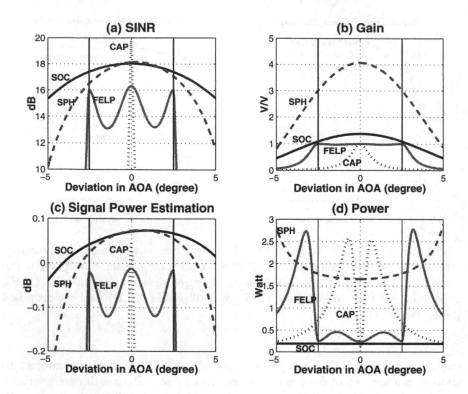

Fig. 7. Performance of a 10-element array against AOA mismatch: Capon MV (CAP), hypersphere (SPH), flat ellipsoid (FELP), and the proposed (SOC) beamformers. The two vertical solid lines in each plot denote the AOA uncertainty range

5 Conclusion

This paper has presented an efficient geometrical approach for designing RMV beamformers utilizing SOC uncertainty bounding. The algorithm exploits the convexity of the optimization constraints and reduces the dimension of the optimization process from a convex hull (covering the uncertainty set) to the circumference of a hyperellipse outside the hull. Extension of this idea to robust interference rejection has been illustrated. Its application has been demonstrated through a generic example of modeling array uncertainties using complex-plane trapezoids. The beamforming task has been transformed into an SOCP problem that can be efficiently solved using either interior point algorithms or the Lagrange multiplier method. Simplification of the proposed scheme using a centroid heuristics and its extension to arbitrary uncertainty geometries have also been discussed. Numerical examples have confirmed that the proposed SOC RMV beamformer exhibits high computational efficiency, better tightness, power requirement, and convergence in signal power estimation over other schemes.

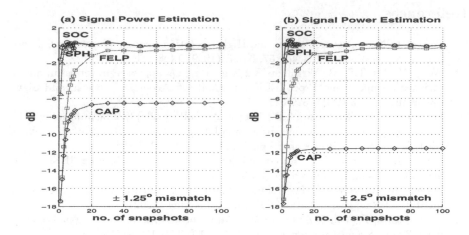

Fig. 8. Signal power estimation for a 10-element array using sample covariance matrix: (a) 1.25° AOA mismatch. (b) 2.5° AOA mismatch. Each point is obtained from the average of 100 Monte Carlo simulations. Featured designs: Capon MV (CAP), hypersphere (SPH), flat ellipsoid (FELP), and the proposed (SOC) beamformers

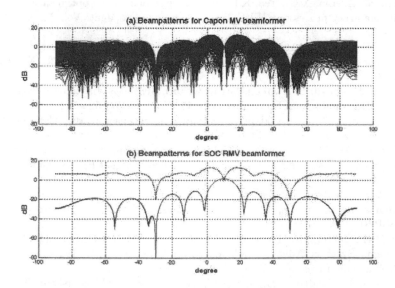

Fig. 9. (a) & (b). 1000 random beampatterns of a 10-element array featuring the Capon MV beamformer (with point interference rejection) and the proposed beamformer (with robust interference rejection), interference signals being at −30°, +50°. Dotted envelope in (b) shows the maximum gain recorded in (a)

References

1. Capon, J.: Nonlinear oscillations and High-resolution frequency-wavenumber spectrum analysis. Proc. IEEE Vol. 57, No. 8 (1969) 1408-1418
2. Wong, N., Ng, T.S., Balakrishnan, V.: A geometrical approach to robust minimum variance beamforming. Proc. IEEE Int. Conf. on Acoustics, Speech, and Signal Processing. Vol. 5 (2003) 329-332.
3. Vorobyov, S.A., Gershman, A.B., Luo, Z.Q.: Robust adaptive beamforming using worst-case performance optimization: a solution to the signal mismatch problem. IEEE Trans. Signal Processing. Vol. 51, No. 2 (2003) 313-324
4. Lorenz, R. G., Boyd, S. P.: Robust minimum variance beamforming. IEEE Trans. Signal Processing, submitted for publication (2001)
5. Lorenz, R. G., Boyd, S. P.: Robust beamforming in GPS arrays. Proc. of the Institute of Navigation, National Technical Meeting (2002)
6. Li, J., Stoica, P., Wang, Z.: On robust Capon beamforming and diagonal loading. IEEE Trans. Signal Processing. Vol. 51, No. 7 (2003) 1702-1715
7. Li, J., Stoica, P., Wang, Z.: Robust Capon beamforming. IEEE Signal Processing Letters. Vol. 10, No. 6 (2003) 172-175
8. Wu, S.Q., Zhang, J.Y.: A new robust beamforming method with antennae calibration errors. IEEE Wireless Comm. and Networking Conf. Vol. 2 (1999) 869-872
9. Wang, F., Balakrishnan, V., Zhou, P., Chen, J., Yang, R., Frank, C.: Optimal array pattern synthesis using semidefinite programming. IEEE Trans. Signal Processing. Vol. 51, No. 5 (2003) 1172-1183
10. Cantoni, A., Lin, X.G., Teo, K.L.: A new approach to the optimization of robust antenna array processors. IEEE Trans. Antennas Propagat. Vol. 41, No. 4 (1993) 403-411
11. Er, M.H., Cantoni, A.: A unified approach to the design of robust narrow-band antenna array processors. IEEE Trans. Antennas Propagat. Vol. 38, No. 1 (1990) 17-23
12. Lebret, H., Boyd, S.: Antenna array pattern synthesis via convex optimization. IEEE Trans. on Signal Processing. Vol. 45, No. 3 (1997) 526-532
13. Gershman, A.B.: Robust adaptive beamforming in sensor arrays. AEU-Int. Journal Electron. Comm. Vol. 53, No. 6 (1999) 305-314
14. Wax, M., Anu, Y.: Performance analysis of the minimum variance beamformer in the presence of steering vector errors. IEEE Trans. Signal Processing. Vol. 44, No. 4 (1996) 938-947
15. Er, M.H., Cantoni, A.: Derivative constraints for broad-band element space antenna array processors. IEEE Trans. Acoust., Speech, Signal Processing. Vol. ASSP-31 (1983) 1378-1393
16. Er, M.H., Cantoni, A.: Techniques in robust broadband beamforming. Control and Dynamic Systems: Advances in Theory and Applications, C. T. Leondes, Ed. Vol. 53, Academic Press, New York (1992) 321-386.
17. Johnson, D., Dudgeon, D.: Array Signal Processing: Concepts and Techniques. Signal Processing Series. P T R Prentice Hall, Englewood Cliffs (1993)
18. Chang, L., Yeh, C.C.: Performance of DMI and eigenspace-based beamformers. IEEE Trans. Antennas Propagat. Vol. 40 (1992) 1336-1347
19. Takao, K., Kikuma, N.: Tamed adaptive antenna array. IEEE Trans. Antennas Propagat. Vol. AP-34 (1986) 388-394
20. Cox, H, Zeskind, R.M., Owen, M.M.: Robust adaptive beamforming. IEEE Trans. Acoust., Speech, Signal Processing. Vol. ASSP-35 (1987) 1365-1376

21. Lee, C.C., Lee, J.H.: Robust adaptive array beamforming under steering vector errors. IEEE Trans. Antennas Propagat. Vol. 45, No. 1 (1997) 168-175

22. Lobo, M., Vandenberghe, L., Boyd, S., Lebret, H.: Applications of second-order cone programming. Linear Algebra and its Applications, Special Issue on Linear Algebra in Control, Signals and Image Processing. **284** (1998) 193-228

23. Alizadeh, F., Goldfarb, D.: Second-Order Cone Programming. RUTCOR RRR Report number 51-2001, Rutgers University (2001)

24. Nesterov, Yu., Nemirovskii, A.: Interior Point Polynomial Algorithms in Convex Programming. Society for Industrial and Applied Mathematics, Philadelphia (1994)

25. Lobo, M., Vandenberghe, L., Boyd, S.: SOCP: Software for second-order cone programming. *http://www.stanford.edu/~boyd/socp/*

26. Sturm, J.F.: Using SeDuMi 1.02, a MATLAB toolbox for optimization over symmetric cones. Optim. Meth. Softw. Vol. 11-12 (1999) 625-653

27. Gahinet, P., Nemirovskii, A., Laub, A., Chilali, M.: The LMI Control Toolbox. The MathWorks, Inc. (1995)

28. Vandenberghe, L., Boyd, S.: Semidefinite programming. SIAM Review. Vol. 38, No. 1 (1996) 49-95

29. Boyd, S., El Ghaoui, L., Feron, E., Balakrishnan, V.: Linear Matrix Inequalities in System and Control Theory. Vol. 15 of Studies in Applied Mathematics, SIAM, Philadelphia, PA (1994)

30. Welzl, E: Smallest enclosing disks (balls and ellipsoids). H. Maurer, editor, New Results and New Trends in Computer Science, Springer-Verlag (1991) 359-370

Joint Optimization of Wireless Communication and Networked Control Systems

Lin Xiao[1], Mikael Johansson[2], Haitham Hindi[3],
Stephen Boyd[4], and Andrea Goldsmith[4]

[1] Dept. of Aeronautics & Astronautics, Stanford University, Stanford CA 94305, USA
[2] Department of Signals, Sensors and Systems, KTH, SE 100 44 Stockholm, Sweden
[3] Systems and Practices Laboratory, PARC, Palo Alto, CA 94304, USA
[4] Electrical Engineering Department, Stanford University, Stanford, CA 94305, USA

Abstract. We consider a linear system, such as an estimator or a controller, in which several signals are transmitted over wireless communication channels. With the coding and medium access schemes of the communication system fixed, the achievable bit rates are determined by the allocation of communications resources such as transmit powers and bandwidths, to different channels. Assuming conventional uniform quantization and a standard white-noise model for quantization errors, we consider two specific problems. In the first, we assume that the linear system is fixed and address the problem of allocating communication resources to optimize system performance. We observe that this problem is often convex (at least, when we ignore the constraint that individual quantizers have an integral number of bits), hence readily solved. We describe a dual decomposition method for solving these problems that exploits the problem structure. We briefly describe how the integer bit constraints can be handled, and give a bound on how suboptimal these heuristics can be. The second problem we consider is that of jointly allocating communication resources and designing the linear system in order to optimize system performance. This problem is in general not convex. We present an iterative heuristic method based on alternating convex optimization over subsets of variables, which appears to work well in practice.

1 Introduction

We consider a linear system in which several signals are transmitted over wireless communication links, as shown in figure 1. All signals are vector-valued: w is a vector of exogenous signals (such as disturbances or noises acting on the system); z is a vector of performance signals (including error signals and actuator signals); and y and y_r are the signals transmitted over the communication network, and received, respectively. This general arrangement can represent a variety of systems, for example a controller or estimator in which sensor, actuator, or command signals are sent over wireless links. It can also represent a distributed controller or estimator, in which some signals (*i.e.*, inter-process communication) are communicated across a network. In this paper, we address the

R. Murray-Smith, R. Shorten (Eds.): Switching and Learning, LNCS 3355, pp. 248–272, 2005.

problem of optimizing the stationary performance of the linear system by jointly allocating resources in the communication network and tuning the parameters of the linear system.

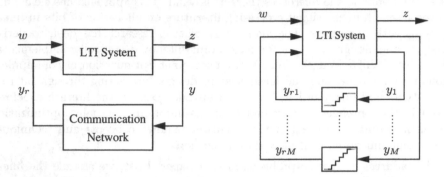

Fig. 1: System set-up (left) and uniform quantization model (right).

Many issues arise in the design of networked controllers and the associated communication systems, including bit rate limitations [WB99, NE00, TSM98], communication delays [NBW98], data packet loss [XHH00], transmission errors [SSK99], and asynchronicity [Özg89]. In this paper we consider only the first issue, *i.e.*, bit rate limitations. In other words, we assume that each communication link has a fixed and known delay (which we model as part of the LTI system), does not drop packets, transfers bits without error, and operates (at least for purposes of analysis) synchronously with the discrete-time linear system.

The problem of control with bit-rate limitations has achieved a lot of attention recently. Much of the research has concentrated on joint design of control and coding to find the minimum bit rate required to stabilize a linear system. For example, [WB99] and [NE98] established various closed-loop stability conditions involving the feedback data rate and eigenvalues of the open-loop system, while [BM97, TSM98] studied control with communication constraints within the classical linear quadratic Gaussian framework. Closely related is also the research on control with quantized feedback information, see [Cur70, Del90, KH94, BL00, EM01].

Our focus is different. We assume that the source coding, channel coding and medium access scheme of the communication system are fixed and concentrate on finding the allocation of communications resources (such as transmit powers and bandwidths) and linear system parameters that yields the optimal closed-loop performance. For a fixed sampling frequency of the linear system, the limit on communication resources translates into a constraint on the number of bits that can be transmitted over each communication channel during one sampling period. We assume that the individual signals y_i are coded using conventional memoryless uniform quantizers, as shown in figure 1. This coding scheme is certainly not optimal (see, *e.g.*, [WB97, NE98]), but it is conventional, easily implemented, and leads to a simple model of how the system performance depends on the bit-rates. In particular, by imposing lower bounds on the number of quantization bits, we ensure that data rates are high enough for stabilization and that the white-noise model for quantization errors introduced by Widrow (see [WKL96] and the references therein) is valid. This approach has clear links to the research in the signal processing literature on allocation of bits in linear systems with quantizers. The main effort of that research has been to derive analysis and design methods for fixed-point filter and controller implementations, (see [Wil85, WK89, SW90]). However, joint optimization of communications resource allocation and linear system design, interacting through bit rate limitations and quantization, has not been addressed in the literature before. Even in the simplified setting under our assumptions, the joint optimization problem is quite nontrivial and its solution requires concepts and techniques from communication, control, and optimization.

We address to specific problems in this paper. First, we assume the linear system is fixed and consider the problem of allocating communication resources to optimize the overall system performance. We observe that this problem is often convex, provided we ignore the constraint that the number of bits for each quantizer is an integer. This means that these communication resource allocation problems can be solved efficiently, using a variety of convex optimization techniques. We describe a general approach for solving these problems based on dual decomposition. The method results in very efficient procedures for solving for many communication resource allocation problems, and reduces to well known water-filling in simple cases. We also show several methods that can be used to handle the integrality constraint. The simplest is to round down the number of bits for each channel to the nearest integer. We show that this results in an allocation of communication resources that is feasible, and at most a factor of two suboptimal in terms of the RMS (root-mean-square) value of critical variable z. We also describe a simple and effective heuristic that often achieves performance close to the bound obtained by solving the convex problem, ignoring the integrality constraints.

The second problem we consider is the problem of jointly allocating communication resources and designing the linear system in order to optimize performance. Here we have two sets of design variables: the communication variables (which indirectly determine the number of bits assigned to each quantizer), and the controller variables (such as estimator or controller gains in the linear system). Clearly the two are strongly coupled, since the effect of quantization errors depends on the linear system, and similarly, the choice of linear system will affect the choice of communication resource allocation. We show that this joint problem is in general not convex. We propose an alternating optimization method that exploits problem structure and appears to work well in practice.

The paper is organized as follows. In §2, we describe the linear system and our model for the effect of uniform quantization error on overall system performance. In §3, we describe a generic convex model for the bit rate limitations imposed by communication systems, and describe several examples. In §4, we formulate the communication resource allocation problem for fixed linear systems, describe the dual decomposition method which exploits the separable structure, and give a heuristic rounding method to deal with the integrality of bit allocations. In §5, we demonstrate the nonconvexity of the joint design problem, and give a iterative heuristic to solve such problems. Two examples, a networked linear estimator and a LQG control system over communication networks, are used to illustrate the optimization algorithms in §4 and §5. We conclude the paper in §6.

2   Linear System and Quantizer Model

2.1   Linear System Model

To simplify the presentation we assume a synchronous, single-rate discrete-time system. The linear time-invariant (LTI) system can be described as

$$z = G_{11}(\varphi)w + G_{12}(\varphi)y_r, \qquad y = G_{21}(\varphi)w + G_{22}(\varphi)y_r, \tag{1}$$

where G_{ij} are LTI operators (i.e., convolution systems described by transfer or impulse matrices). Here, $\varphi \in \mathbf{R}^q$ is the vector of design parameters in the linear system that can be tuned or changed to optimize performance. To give lighter notation, we suppress the dependence of G_{ij} on φ except when necessary. We assume that $y(t)$, $y_r(t) \in \mathbf{R}^M$, i.e., the M scalar signals $y_1, \ldots, y_M$ are transmitted over the network during each sampling period.

We assume that the signals sent (i.e., y) and received (i.e., y_r) over the communication links are related by memoryless scalar quantization, which we describe in detail in the next subsections. This means that all communication delays are assumed constant and known, and included in the LTI system model.

2.2   Quantization Model

Unit Uniform Quantizer   A unit range uniform b_i-bit quantizer partitions the range $[-1, 1]$ into 2_i^b intervals of uniform width 2^{1-b_i}. To each quantization interval a codeword of b bits is assigned. Given a received codeword, the input signal y_i is approximated by (or reconstructed as) y_r, the midpoint of the interval. As long as the quantizer does not overflow (i.e., as long as $|y_i| \leq 1$), the relationship between original and reconstructed values can be expressed as

$$Q_{b_i}(y_i) = \frac{\mathbf{round}(2^{b_i-1}y_i)}{2^{b_i-1}}$$

and the quantization error $y_{ri} - y_i$ lies in the interval $\pm 2^{-b_i}$.

The behavior of the quantizer when y_i overflows (i.e., $|y_i| > 1$) is not specified. One approach is to introduce two more codewords, corresponding to negative and positive overflow, respectively, and to extend Q_{b_i} to saturate for $|y_i| \geq 1$. The details of the overflow behavior will not affect our analysis or design, since we assume by appropriate scaling (described below) that overflow does not occur, or occurs rarely enough to not affect overall system performance.

Scaling To avoid overflow, each signal $y_i(t)$ is scaled by the factor $s_i^{-1} > 0$ prior to encoding with a unit uniform b_i-bit quantizer, and re-scaled by the factor s_i after decoding (figure 2), so that

$$y_{ri}(t) = s_i Q_{b_i}(y_i(t)/s_i).$$

The associated quantization error is given by

$$q_i(t) = y_{ri}(t) - y_i(t) = s_i E_{b_i}(y_i(t)/s_i),$$

which lies in the interval $\pm s_i 2^{-b_i}$, provided $|y_i(t)| < s_i$.

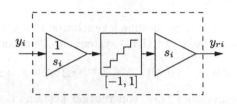

Fig. 2: Scaling before and after the quantizer.

To minimize quantization error while ensuring no overflow (or ensuring that overflow is rare) the scale factors s_i should be chosen as the maximum possible value of $|y_i(t)|$, or as a value that with very high probability is larger than $|y_i(t)|$. For example, we can use the so-called 3σ-*rule*,

$$s_i = 3\,\mathbf{rms}(y_i),$$

where $\mathbf{rms}(y_i)$ denotes the root-mean-square value of y_i,

$$\mathbf{rms}(y_i) = \left(\lim_{t \to \infty} \mathbf{E}\, y_i(t)^2 \right)^{1/2}.$$

If y_i has a Gaussian amplitude distribution, this choice of scaling ensures that overflow occurs only about 0.3% of the time.

White-Noise Quantization Error Model We adopt the standard stochastic quantization noise model introduced by Widrow (see, *e.g.*, [FPW90, Chapter 10]). Assuming that overflow is rare, we model the quantization errors $q_i(t)$ as independent random variables, uniformly distributed on the interval

$$s_i[-2^{-b_i}, 2^{-b_i}].$$

In other words, we model the effect of quantizing $y_i(t)$ as an additive white noise source $q_i(t)$ with zero mean and variance $\mathbf{E}\, q_i(t)^2 = (1/3)s_i^2 2^{-2b_i}$, see figure 3. When allocating bits to quantizers, we will impose a lower bound on each b_i. This value should be high enough for stabilizing the closed-loop system (*cf.* [WB99, NE00]) and make the white noise model a reasonable assumption in a feedback control context (*cf.* [WKL96, FPW90]).

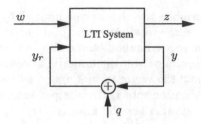

Fig. 3: LTI system with white noise quantization noise model.

2.3   Performance of the Closed-Loop System

We can express z and y in terms of the inputs w and q as

$$z = G_{zw}w + G_{zq}q, \qquad y = G_{yw}w + G_{yq}q,$$

where G_{zw}, G_{zq}, G_{yw} and G_{yq} are the closed-loop transfer matrices from w and q to z and y, respectively. From the expression for z, we see that it consists of two terms: $G_{zw}w$, which is what z would be if the quantization were absent, and $G_{zq}q$, which is the component of z due to the quantization. The variance of z induced by the quantization is given by

$$V_q = \mathbf{E}\,\|G_{zq}q\|^2 = \sum_{i=1}^{M} \|G_{zqi}\|^2 \left(\frac{1}{3}s_i^2 2^{-2b_i}\right) \tag{2}$$

where G_{zqi} is the ith column of the transfer matrix G_{zq}, and $\|\cdot\|$ denotes the $\mathbf{L}^2$ norm (see [BB91, §5.2.3]). We can use V_q as a measure of the effect of quantization on the overall system performance. If w is also modeled as a stationary stochastic process, the overall variance of z is given by

$$V = \mathbf{E}\,\|z\|^2 = V_q + \mathbf{E}\,\|G_{zw}w\|^2. \tag{3}$$

The above expression shows how V_q depends on the allocation of quantizer bits $b_1, \ldots, b_M$, as well as the scalings $s_1, \ldots, s_M$ and LTI system (which affect the a_i's). Note that while the formula (2) was derived assuming that b_i are integers, it makes sense for $b_i \in \mathbf{R}$.

3   Communications Model and Assumptions

3.1   A Generic Model for Bit Rate Constraints

The capacity of communication channels depend on the media access scheme and the selection of certain critical parameters, such as transmission powers and bandwidths or time-slot fractions allocated to individual channels (or groups of channels). We refer to these critical communication parameters collectively as *communication variables*, and denote the vector of communication variables by θ.

The communication variables are themselves limited by various resource constraints, such as limits on the total power or total bandwidth available. We will assume that the medium access methods and coding and modulation schemes are fixed, but that we can optimize over the underlying communication variables θ.

We let $b \in \mathbf{R}^M$ denote the vector of bits allocated to each quantized signal. The associated communication rate r_i (in bits per second) can be expressed as $b_i = \alpha r_i$, where the constant α has the form $\alpha = c_s / f_s$. Here f_s is the sample frequency, and c_s is the channel coding efficiency in source bits per transmission bit. This relationship will allow us to express capacity constraints in terms of bit allocations rather than communication rates.

We will use the following general model to relate the vector of bit allocations b, and the vector of communication variables θ:

$$\begin{aligned}
f_i(b, \theta) &\leq 0, \quad i = 1, \ldots, m_f \\
h_i^T \theta &\leq d_i, \quad i = 1, \ldots, m_h \\
\theta_i &\geq 0, \quad i = 1, \ldots, m_\theta \\
\underline{b}_i &\leq b_i \leq \overline{b}_i, \quad i = 1, \ldots, M
\end{aligned} \tag{4}$$

We make the following assumptions about this generic model.

- The functions f_i are convex functions of (b, θ), monotone increasing in b and monotone decreasing in θ. These inequalities describe capacity constraints on the communication channels. We will show below that many classical capacity formula satisfy these assumptions.
- The second set of constraints describes resource limitations, such as a total available power or bandwidth for a group of channels. We assume the vectors h_i have nonnegative entries. We assume that d_i, which represent resource limits, are positive.
- The third constraint specifies that the communication resource variables (which represent powers, bandwidths, time-slot fractions) are nonnegative.
- The last group of inequalities specify lower and upper bounds for each bit allocation. We assume that $\underline{b}_i$ and $\overline{b}_i$ are (nonnegative) integers. The lower bounds are imposed to ensure that the white noise model for quantization errors is reasonable. The upper bounds can arise from hardware limitations.

This generic model will allow us to formulate the communication resource allocation problem, i.e., the problem of choosing θ to optimize overall system performance, as a convex optimization problem.

There is also one more important constraint on b not included above:

$$b_i \text{ is an integer}, \quad i = 1 \ldots, M. \tag{5}$$

For the moment, we ignore this constraint. We will return to it in §4.2.

3.2   Capacity Constraints

In this section, we describe some simple channel models and show how they fit the generic model (4) given above. More detailed descriptions of these channel models, as well as derivations, can be found in, e.g., [CT91, Gol99].

Gaussian Channel We start by considering a single Gaussian channel. The communication variables are the bandwidth $W > 0$ and transmission power $P > 0$. Let N be the power spectral density of the additive white Gaussian noise at the front-end of the receiver. The channel capacity is given by ([CT91])

$$R = W \log_2 \left(1 + \frac{P}{NW} \right)$$

(in bits per second). The achievable communication rate r is bounded by this channel capacity, *i.e.*, we must have $r \leq R$. Expressed in terms of b, we have

$$b \leq \alpha W \log_2 \left(1 + \frac{P}{NW} \right). \tag{6}$$

We can express this in the form

$$f(b, W, P) = b - \alpha W \log_2 \left(1 + \frac{P}{NW} \right) \leq 0,$$

which fits the generic form (4). To see that the function f is jointly convex in the variables (b, W, P), we note that the function $g(P) = -\alpha \log_2(1 + P/N)$ is a convex function of P and, therefore its *perspective function* (see [BV04])

$$W g(P/W) = -\alpha W \log_2 \left(1 + \frac{P}{NW} \right)$$

is a convex function of (P, W). Adding the linear (hence convex) function b establishes convexity of f. It is easily verified that f is monotone increasing in b, and monotone decreasing in W and P.

Gaussian Broadcast Channel with FdMA In the Gaussian broadcast channel with frequency-domain multiple access (FDMA), a transmitter sends information to n receivers over disjoint frequency bands with bandwidths $W_i > 0$. The communication parameters are the bandwidths W_i and the transmit powers $P_i > 0$ for each individual channel. The communication variables are constrained by a total power limit

$$P_1 + \cdots + P_n \leq P_{\text{tot}}$$

and a total available bandwidth limit

$$W_1 + \cdots + W_n \leq W_{\text{tot}},$$

which have the generic form for communication resource limits.

The receivers are subject to independent white Gaussian noises with power spectral densities N_i. The transmitter assigns power P_i and bandwidth W_i to the ith receiver. The achievable bit rates b are constrained by

$$b_i \leq \alpha W_i \log_2 \left(1 + \frac{P_i}{N_i W_i} \right), \qquad i = 1, \ldots, n. \tag{7}$$

Again, the constraints relating b and $\theta = (P, W)$ have the generic form (4).

Gaussian Multiple Access Channel with FDMA In a Gaussian multiple access channel with FDMA, n transmitters send information to a common receiver, each using a transmit power P_i over a bandwidth W_i. It has the same set of constraints as for the broadcast channel, except that $N_i = N$, $i = 1, \ldots, n$ (since they have a common receiver).

Variations and Extensions The capacity formulas for many other channel models, including the Parallel Gaussian channel, Gaussian broadcast channel with TDMA and the Gaussian broadcast channel with CDMA, are also concave in communications variables and can be included in our framework. It is also possible to combine the channel models above to model more complex communication systems. Finally, channels with time-varying gain variations (fading) as well as rate constraints based on bit error rates (with or without coding) can be formulated in a similar manner; see, $e.g.$, [LG01, CG01].

4 Optimal Resource Allocation for Fixed Linear System

In this section, we assume that the linear system is fixed and consider the problem of choosing the communication variables to optimize the system performance. We take as the objective (to be minimized) the variance of the performance signal z, given by (3). Since this variance consists of a fixed term (related to w) and the variance induced by the quantization, we can just as well minimize the variance of z induced by the quantization error, $i.e.$, the quantity V_q defined in (2). This leads to the optimization problem

$$
\begin{aligned}
\text{minimize} \quad & \sum_{i=1}^{M} a_i 2^{-2b_i} \\
\text{subject to} \quad & f_i(b, \theta) \leq 0, \quad i = 1, \ldots, m_f \\
& h_i^T \theta \leq d_i, \quad i = 1, \ldots, m_h \\
& \theta_i \geq 0, \quad i = 1, \ldots, m_\theta \\
& \underline{b}_i \leq b_i \leq \overline{b}_i, \quad i = 1, \ldots, M
\end{aligned}
\tag{8}
$$

where $a_i = (1/3)\|G_{zqi}\|^2 s_i^2$, and the optimization variables are θ and b. For the moment we ignore the constraint that b_i must be integers.

Since the objective function, and each constraint function in the problem (8) is a convex function, this is a convex optimization problem. This means that it can be solved globally and efficiently using a variety of methods, $e.g.$, interior-point methods (see, $e.g.$, [BV04]). In many cases, we can solve the problem (8) more efficiently than by applying general convex optimization methods by exploiting its special structure. This is explained in the next subsection.

4.1 The Dual Decomposition Method

The objective function in the communication resource allocation problem (8) is separable, $i.e.$, a sum of functions of each b_i. In addition, the constraint functions $f_k(b, \theta)$ usually involve only one b_i, and a few components of θ, since the channel capacity is determined by the bandwidth, power, or time-slot fraction, for example, allocated to that channel. In other words, the resource allocation problem (8) is almost separable; the small groups of variables (that relate to a given link or channel) are coupled mostly through the resource limit constraints $h_i^T \theta \leq d_i$. These are the constraints that limit the total power, total bandwidth, or total time-slot fractions.

This almost separable structure can be efficiently exploited using a technique called dual decomposition (see, e.g., [BV04, Ber99]). We will explain the method for a simple FDMA system to keep the notation simple, but the method applies to any communication resource allocation problem with almost separable structure. We consider an FDMA system with M channels, and variables $P \in \mathbf{R}^M$ and $W \in \mathbf{R}^M$, with a total power and a total bandwidth constraint. We will also impose lower and upper bounds on the bits. This leads to

$$
\begin{aligned}
\text{minimize} \quad & \sum_{i=1}^M a_i 2^{-2b_i} \\
\text{subject to} \quad & b_i \leq \alpha W_i \log_2(1 + P_i/N_i W_i), \quad i = 1, \ldots, M \\
& P_i \geq 0, \quad i = 1, \ldots, M \\
& \sum_{i=1}^M P_i \leq P_{\text{tot}} \\
& W_i \geq 0, \quad i = 1, \ldots, M \\
& \sum_{i=1}^M W_i \leq W_{\text{tot}} \\
& \underline{b}_i \leq b_i \leq \overline{b}_i, \quad i = 1, \ldots, M.
\end{aligned}
\tag{9}
$$

Here N_i is the receiver noise spectral density of the ith channel, and $\underline{b}_i$ and $\overline{b}_i$ are the lower and upper bounds on the number of bits allocated to each channel. Except for the total power and total bandwidth constraint, the constraints are all local, i.e., involve only b_i, P_i, and W_i.

We first form the Lagrange dual problem, by introducing Lagrange multipliers but *only* for the two coupling constraints. The Lagrangian has the form

$$
L(b, P, W, \lambda, \mu) = \sum_{i=1}^M a_i 2^{-2b_i} + \lambda \left(\sum_{i=1}^M P_i - P_{\text{tot}} \right) + \mu \left(\sum_{i=1}^M W_i - W_{\text{tot}} \right).
$$

The dual function is defined as

$$
g(\lambda, \mu) = \inf \left\{ L \mid P_i \geq 0, \ W_i \geq 0, \ \underline{b}_i \leq b_i \leq \overline{b}_i, \ b_i \leq \alpha W_i \log_2(1 + P_i/N_i W_i) \right\}
$$
$$
= \sum_{i=1}^M g_i(\lambda, \mu) - \lambda P_{\text{tot}} - \mu W_{\text{tot}}
$$

where

$$
g_i(\lambda, \mu) = \inf \Big\{ a_i 2^{-2b_i} + \lambda P_i + \mu W_i \ \Big|
$$
$$
P_i \geq 0, \ W_i \geq 0, \ \underline{b}_i \leq b_i \leq \overline{b}_i, \ b_i \leq \alpha W_i \log_2(1 + P_i/N_i W_i) \Big\}.
$$

Finally, the Lagrange dual problem associated with the communication resource allocation problem (9) is given by

$$
\begin{aligned}
\text{maximize} \quad & g(\lambda, \mu) \\
\text{subject to} \quad & \lambda \geq 0, \quad \mu \geq 0.
\end{aligned}
\tag{10}
$$

This problem has only two variables, namely the variables λ and μ associated with the total power and bandwidth limits, respectively. It is a convex optimization problem, since g is a concave function (see [BV04]). Assuming that Slater's condition holds, the optimal value of the dual problem (10) and the primal problem (9) are equal. Moreover, from the optimal solution of the dual problem, we can recover the optimal solution of the primal. Suppose $(\lambda^\star, \mu^\star)$ is the solution to the dual problem (10), then the primal optimal solution is the minimizer $(b^\star, P^\star, W^\star)$ when evaluating the dual function $g(\lambda^\star, \mu^\star)$. In other words, we can solve the original problem (9) by solving the dual problem (10).

The dual problem can be solved using a variety of methods, for example, cutting-plane methods. To use these methods we need to be able to evaluate the dual objective function, and also obtain a subgradient for it (see [BV04]), for any given $\mu \geq 0$ and $\lambda \geq 0$. To evaluate $g(\lambda, \mu)$, we simply solve the M separate problems,

$$\begin{aligned} &\text{minimize} \quad a_i 2^{-2b_i} + \lambda P_i + \mu W_i \\ &\text{subject to} \quad P_i \geq 0, \ W_i \geq 0, \\ &\qquad\qquad \underline{b}_i \leq b_i \leq \bar{b}_i, \\ &\qquad\qquad b_i \leq \alpha W_i \log_2(1 + P_i/N_i W_i), \end{aligned}$$

each with three variables, which can be carried out separately or in parallel. Many methods can be used to very quickly solve these small problems.

A subgradient of the concave function g at (λ, μ) is a vector $h \in \mathbf{R}^2$ such that

$$g(\tilde{\lambda}, \tilde{\mu}) \leq g(\lambda, \mu) + h^T \begin{bmatrix} \tilde{\lambda} - \lambda \\ \tilde{\mu} - \mu \end{bmatrix}$$

for all $\tilde{\lambda}$ and $\tilde{\mu}$. To find such a vector, let the optimal solution to the subproblems be denoted

$$b_i^\star(\lambda, \mu), \quad P_i^\star(\lambda, \mu), \quad W_i^\star(\lambda, \mu).$$

Then, a subgradient of the dual function g is readily given by

$$\begin{bmatrix} \sum_{i=1}^M P_i^\star(\lambda, \mu) - P_{\text{tot}} \\ \sum_{i=1}^M W_i^\star(\lambda, \mu) - W_{\text{tot}} \end{bmatrix}.$$

This can be verified from the definition of the dual function.

Putting it all together, we find that we can solve the dual problem in time linear in M, which is far better than the standard convex optimization methods applied to the primal problem, which require time proportional to M^3.

The same method can be applied whenever there are relatively few coupling constraints, and each link capacity is dependent on only a few communication resource parameters. In fact, when there is only one coupling constraint, the subproblems that we must solve can be solved analytically, and the master problem becomes an explicit convex optimization problem with only one variable. It is easily solved by bisection, or any other one-parameter search method. This is the famous water-filling algorithm (see, *e.g.*, [CT91]).

4.2 Integrality of Bit Allocations

We now come back to the requirement that the bit allocations must be integers. The first thing we observe is that we can always round down the bit allocations found by solving the convex problem to the nearest integers. Let b_i denote the optimal solution of the convex resource allocation problem (8), and define $\tilde{b}_i = \lfloor b_i \rfloor$. Here, $\lfloor b_i \rfloor$ denotes the *floor* of b_i, *i.e.*, the largest integer smaller than or equal to b_i. First we claim that $\tilde{b}$ is feasible. To see this, recall that f_k and h_k are monotone decreasing in b, so since b is feasible and $\tilde{b} \leq b$, we have $\tilde{b}$ feasible.

We can also obtain a crude performance bound for $\tilde{b}$. Clearly the objective value obtained by ignoring the integer constraint, *i.e.*,

$$J_{\text{cvx}} = \sum_{i=1}^{M} a_i 2^{-2b_i},$$

is a lower bound on the optimal objective value J_{opt} of the problem with integer constraints. The objective value of the rounded-down feasible bit allocation $\tilde{b}$ is

$$J_{\text{rnd}} = \sum_{i=1}^{M} a_i 2^{-2\tilde{b}_i} \leq \sum_{i=1}^{M} a_i 2^{-2(b_i-1)} = 4 J_{\text{cvx}} \leq 4 J_{\text{opt}},$$

using the fact that $\tilde{b}_i \geq b_i - 1$. Putting this together we have

$$J_{\text{opt}} \leq J_{\text{rnd}} \leq 4 J_{\text{opt}},$$

i.e., the performance of the suboptimal integer allocation obtained by rounding down is never more than a factor of four worse than the optimal solution. In terms of RMS, the rounded-down allocation is never more than a factor of two suboptimal.

Variable Threshold Rounding Of course, far better heuristics can be used to obtain better integer solutions. Here we give a simple method based on a variable rounding threshold.

Let $0 < t \leq 1$ be a threshold parameter, and round b_i as follows:

$$\tilde{b}_i = \begin{cases} \lfloor b_i \rfloor, & \text{if } b_i - \lfloor b_i \rfloor \leq t, \\ \lceil b_i \rceil, & \text{otherwise.} \end{cases} \tag{11}$$

Here, $\lceil b_i \rceil$ denotes the *ceiling* of b_i, *i.e.*, the smallest integer larger than or equal to b_i. In other words, we round b_i down if its remainder is smaller than or equal to the threshold t, and round up otherwise. When $t = 1/2$, we have standard rounding, with ties broken down. When $t = 1$, all bits are rounded down, as in the scheme described before. This gives a feasible integer solution, which we showed above has a performance within a factor of four of optimal. For $t < 1$ feasibility of the rounded bits $\tilde{b}$ is not guaranteed, since bits can be rounded up.

For a given fixed threshold t, we can round the b_i's as in (11), and then solve a convex feasibility problem over the remaining continuous variables θ:

$$f_i(\tilde{b}, \theta) \leq 0$$
$$h_i^T \theta \leq d_i \qquad (12)$$
$$\theta_i \geq 0$$

The upper and lower bound constraints $\underline{b}_i \leq \tilde{b}_i \leq \bar{b}_i$ are automatically satisfied because $\underline{b}_i$ and $\bar{b}_i$ are integers. If this problem is feasible, then the rounded $\tilde{b}_i$'s and the corresponding θ are suboptimal solutions to the integer constrained bit allocation problem.

Since f_i is monotone increasing in b, hence in t, and monotone decreasing in θ, there exists a t^* such that (12) is feasible if $t \geq t^*$ and infeasible if $t < t^*$. In the variable threshold rounding method, we find t^*, the smallest t which makes (12) feasible. This can be done by bisection over t: first try $t = 1/2$. If the resulting rounded bit allocation is feasible, we try $t = 1/4$; if not, we try $t = 3/4$, etc.

Roughly speaking, the threshold t gives us a way to vary the conservativeness of the rounding procedure. When t is near one, almost all bits are rounded down, and the allocation is likely to be feasible. When t is small, we round many bits up, and the bit allocation is unlikely to be feasible. But if it is, the performance (judged by the objective) will be better than the bit allocation found using more conservative rounding (*i.e.*, with a larger t). A simple bisection procedure can be used to find a rounding threshold close to the aggressive one that yields a feasible allocation.

4.3   Example: Networked Linear Estimator

To illustrate the ideas of this section, we consider the problem of designing a networked linear estimator with the structure shown in figure 4. We want to estimate an unknown point $x \in \mathbf{R}^{20}$ using $M = 200$ linear sensors,

$$y_i = c_i^T x + v_i, \qquad i = 1, \dots, M.$$

Each sensor uses b_i bits to code its measurements and transmits the coded signal to a central estimator over a Gaussian multiple access channel with FDMA. The performance of the estimator is judged by the estimation error variance $J_K = \mathbf{E} \|\hat{x} - x\|^2$. We assume that $\|x\| \leq 1$ and that the sensor noises v_i are

Fig. 4: Networked linear estimator over a multiple access channel

IID with $\mathbf{E} v_i = 0$, $\mathbf{E} v_i^2 = 10^{-6}$. In this example, the sensor coefficients c_i are uniformly distributed on $[0, 5]$. Since $\|x\| \leq 1$, we choose scaling factors $s_i = \|c_i\|$.

The noise power density of the Gaussian multiple access channel is $N = 0.1$, the coding constant is $\alpha = 2$, and the upper and lower bounds for bit allocations are $\underline{b} = 5$ and $\bar{b} = 12$. The total available power is $P = 300$ and the total available bandwidth is $W = 200$.

The estimator is a linear unbiased estimator

$$\hat{x} = Ky_r,$$

where $KC = I$, with $C = [c_1, \ldots, c_M]^T$. In particular, the minimum variance estimator is given by

$$K = \left(C^T(R_v + R_q)^{-1}C\right)^{-1} C^T(R_v + R_q)^{-1} \tag{13}$$

where R_v and R_q are the covariance matrices for the sensor noises and quantization noises, respectively. (Note that the estimator gain depends on the bit allocations.) The associated estimation error variance is

$$J_K(b) = \frac{1}{3}\sum_{i=1}^{M} s_i^2\|k_i\|^2 2^{-2b_i} + \mathbf{Tr}\left(KR_vK^T\right)$$

where k_i is the ith column of the matrix K. Clearly, $J_K(b)$ is on the form (3) and will serve as the objective function for the resource allocation problem (8).

First we allocate power and bandwidth evenly to all sensors, which results in $b_i = 8$ for each sensor. Based on this allocation, we compute the quantization noise variances $\mathbf{E}\,q_i^2 = (1/3)s_i^2 2^{-2b_i}$ and design a least-squares estimator as in (13). The resulting RMS estimation error is 3.676×10^{-3}. Then we fix the estimator gain K, and solve the relaxed optimization problem (8) to find the resource allocation that minimizes the estimation error variance. The resulting RMS value is 3.1438×10^{-3}. Finally, we perform a variable threshold rounding with $t^* = 0.4211$. Figure 5 shows the distribution of rounded bit allocation. The resulting RMS estimation error is 3.2916×10^{-3}. Thus, the allocation obtained from optimization and variable threshold rounding gives a 10% improved performance compared to the unirform resource allocation, which is not very far from the performance bound given by the relaxed convex optimization problem.

We can see that the allocation obtained from optimization and variable threshold rounding give a 10% improved performance compared to the uniform resource allocation, and is not very far from the performance bound given by the relaxed convex optimization problem.

Note that with the new bit allocations, the quantization covariance changes — it is not the one that was used to design K. We will address this issue of the coupling between the choice of the communication variables and the estimator.

5   Joint Design of Communication and Linear Systems

We have seen that when the linear system is fixed, the problem of optimally allocating communication resources is convex (when we ignore integrality of bit

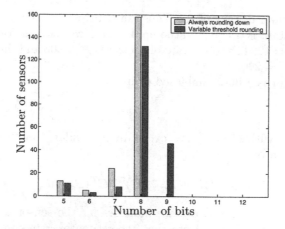

Fig. 5: Bit allocation for networked least-squares estimator.

allocations), and can be efficiently solved. In order to achieve the optimal system performance, however, one should optimize the parameters of the linear system *and* the communication system *jointly*. Unfortunately, this joint design problem is in general not convex. In some cases, however, the joint design problem is bi-convex: for fixed resource allocation the controller design problem is convex, and for fixed controller design and scalings the resource allocation problem is convex. This special structure can be exploited to develop a heuristic method for the joint design problem, that appears to work well in practice.

5.1 Nonconvexity of the Joint Design Problem

To illustrate that the joint design problem is nonconvex, we consider the problem of designing a simple networked least-squares estimator for an example small enough that we can solve the joint problem globally.

An unknown scalar parameter $x \in \mathbf{R}$ is measured using two sensors that are subject to measurement noises:

$$y_1 = x + v_1, \qquad y_2 = x + v_2.$$

We assume that v_1 and v_2 are independent zero-mean Gaussian random variables with variances $\mathbf{E} v_1^2 = \mathbf{E} v_2^2 = 0.001$. The sensor measurements are coded and sent over a communication channel with a constraint on the total bit rate. With a total of b_{tot} bits available we allocate b_1 bits to the first sensor and the $b_2 = b_{\text{tot}} - b_1$ remaining bits to the second sensor. For a given bit allocation, the minimum-variance unbiased estimate can be found by solving a weighted least-squares problem. Figure 6 shows the optimal performance as function of b_1 when $b_{\text{tot}} = 8$ and $b_{\text{tot}} = 12$. The relationship is clearly not convex.

These figures, and the optimal solutions, make perfect sense. When $b_{\text{tot}} = 8$, the quantization noise is the dominant noise source, so one should allocate all 8 bits to one sensor and disregard the other. When $b_{\text{tot}} = 12$, the quantization

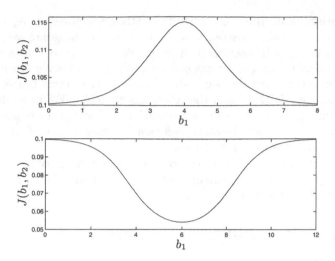

Fig. 6: Estimator performance for $b_1 + b_2 = 8$ (top) and $b_1 + b_2 = 12$ (bottom).

noises are negligible in comparison with the sensor noise. It is then advantageous to use both sensors (*i.e.*, assign each one 6 bits), since it allows us to average out the effect of the measurement noises.

5.2 Alternating Optimization for Joint Design

The fact that the joint problem is convex in certain subsets of the variables while others are fixed can be exploited. For example (and ignoring the integrality constraints) the globally optimal communication variables can be computed very efficiently, sometimes even semi-analytically, when the linear system is fixed. Similarly, when the communication variables are fixed, we can (sometimes) compute the globally optimal variables for the linear system. Finally, when the linear system variables and the communication variables are fixed, it is straightforward to compute the quantizer scalings using the 3σ-rule. This makes it natural to apply an approach where we sequentially fix one set of variables and optimize over the others:

> **given** initial linear system variables $\phi^{(0)}$, communication variables $\theta^{(0)}$, and scaling factors $s^{(0)}$.
>
> $k := 0$
>
> **repeat**
> 1. Fix $\phi^{(k)}$, $s^{(k)}$, and optimize over θ. Let $\theta^{(k+1)}$ be the optimal value.
> 2. Fix $\theta^{(k+1)}$, $s^{(k)}$, and optimize over ϕ. Let $\phi^{(k+1)}$ be the optimal value.
> 3. Fix $\phi^{(k+1)}$, $\theta^{(k+1)}$. Let $s^{(k+1)}$ be appropriate scaling factors.
>   k:=k+1
>
> **until** convergence

Many variations on this basic heuristic method are possible. We can, for example, add trust region constraints to each of the optimization steps, to limit the variables changes in each step. Another variation is to convexify (by, for example, linearizing) the jointly nonconvex problem, and solve in each step using linearized versions for the constraints and objective terms in the remaining variables; see, e.g., [HHB99] and the references therein. . We have already seen how the optimization over θ can be carried out efficiently. In many cases, the optimization over ϕ can also be carried efficiently, using, e.g., LQG or some other controller or estimator design technique.

Since the joint problem is not convex, there is no guarantee that this heuristic converges to the global optimum. On the other hand the heuristic method appears to work well in practice.

5.3   Example: Networked Linear Estimator

To demonstrate the heuristic method for joint optimization described above, we apply it to the networked linear estimator described in §4.3. The design of the linear system and the communication system couple through the weighting matrix Q in (13). The alternating procedure for this problem becomes

> **given** initial estimator gain $K^{(0)}$ and resource allocations $(P^{(0)}, W^{(0)}, b^{(0)})$.
>
> k:=0
>
> **repeat**
> 1. Fix estimator gain $K^{(k)}$ and solve the problem (9) to obtain resource allocation $(P^{(k+1)}, W^{(k+1)}, b^{(k+1)})$.
> 2. Update the covariance matrix $R_q^{(k+1)}$ and compute new estimator gain $K^{(k+1)}$ as in (13) using weight matrix $Q^{(k+1)} = (R_v + R_q^{(k+1)})^{-1}$.
>
>   k:=k+1
>
> **until** bit allocation converges.

Note that the scaling factors are fixed in this example, since neither the bit allocations nor the estimator gain affect the signals that are quantized, hence the scaling factors.

When we apply the alternating optimization procedure to the example given in §4.3, the algorithm converges in six iterations, and we obtain very different resource allocation results from before. Figure 7 shows the distribution of rounded bit allocation. This result is intuitive: try to assign as much resources as possible to the best sensors, and the bad sensors only get minimum number of bits. The RMS estimation error of the joint design is reduced significantly, 80%, as shown in Table 1. In this table, **rms**(e) is the total RMS error, **rms**(e_q) is the RMS error induced by quantization noise, and **rms**(e_v) is the RMS error induced by sensor noise.

We can see that joint optimization reduces the estimation errors due to both quantization and sensor noise. In the case of equal resource allocation, the RMS error due to quantization is much larger than that due to sensor noise. After

RMS values	equal allocation	joint optimization	variable threshold rounding
$\mathbf{rms}(e_q)$	3.5193×10^{-3}	0.3471×10^{-3}	0.3494×10^{-3}
$\mathbf{rms}(e_v)$	1.0617×10^{-3}	0.6319×10^{-3}	0.6319×10^{-3}
$\mathbf{rms}(e)$	3.6760×10^{-3}	0.7210×10^{-3}	0.7221×10^{-3}

Table 1: RMS estimation errors of the networked LS estimator.

the final iteration of the alternating convex optimization, the RMS error due to quantization is at the same level as that due to sensor noise. Also, because the in the relaxed problem, most bits are integers (either $\underline{b} = 5$ or $\bar{b} = 12$; see Figure 7), variable threshold rounding (which gives $t^\star = 0.6797$) does not change the solution, or the performance, much.

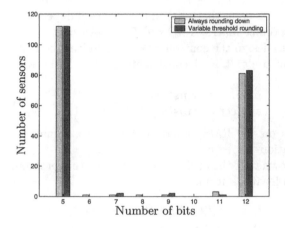

Fig. 7: Joint optimization of bit allocation and least-squares estimator

5.4   Example: LQG Control over Communication Networks

We now give a more complex example than the simple static, open-loop estimator described above. The situation is more complicated when the linear system is dynamic and involves feedback loops closed over the communication links. In this case, the RMS values of both control signals and output signals change when we re-allocate communication resources or adjust the controller. Hence, the alternating optimization procedure needs to include the step that modifies the scalings.

Basic System Setup   First we consider the system setup in figure 8, where no communication links are included. The linear system has a state-space model

$$x(t + 1) = Ax(t) + B\left(u(t) + w(t)\right)$$
$$y(t) = Cx(t) + v(t)$$

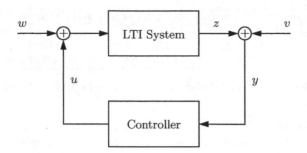

Fig. 8: Closed-loop control system without communication links.

where $u(t) \in \mathbf{R}^{M_u}$ and $y(t) \in \mathbf{R}^{M_y}$. Here $w(t)$ is the process noise and $v(t)$ is the sensor noise. Assume that $w(t)$ and $v(t)$ are independent zero-mean white noises with covariance matrices R_w and R_v respectively.

Our goal is to design the controller that minimizes the RMS value of $z = Cx$, subject to some upper bound constraints on the RMS values of the control signals:

$$\begin{aligned} \text{minimize} \quad & \mathbf{rms}(z) \\ \text{subject to} \quad & \mathbf{rms}(u_i) \leq \beta_i, \quad i = 1, \ldots, M_u \end{aligned} \tag{14}$$

The limitations on the RMS values of the control signals are added to avoid actuator saturation.

It can be shown that the optimal controller for this problem has the standard estimated state feedback form,

$$\begin{aligned} \widehat{x}(t+1|t) &= A\widehat{x}(t|t-1) + Bu(t) + L\left(y(t) - C\widehat{x}(t|t-1)\right) \\ u(t) &= -K\widehat{x}(t|t-1) \end{aligned}$$

where K is the state feedback control gain and L is the estimator gain, found by solving the algebraic Riccati equations associated with an appropriately weighted LQG problem. Finding the appropriate weights, for which the LQG controller solves the problem (14), can be done via the dual problem; see, *e.g.*, [TM89, BB91].

Communications Setup We now describe the communications setup for the example. The sensors send their measurements to a central controller through a Gaussian multiple access channel, and the controller sends control signals to the actuators through a Gaussian broadcast channel, as shown in figure 9.

The linear system can be described as

$$\begin{aligned} x(t+1) &= Ax(t) + B\left(u(t) + w(t) + p(t)\right) \\ y_r(t) &= Cx(t) + v(t) + q(t), \end{aligned}$$

where p and q are quantization noises due to the bit rate limitations of the communication channels. Since these are modeled as white noises, we can include

the quantization noises in the process and measurement noises, by introducing
the equivalent process noise and measurement noise

$$\widetilde{w}(t) = w(t) + p(t), \qquad \widetilde{v}(t) = v(t) + q(t),$$

with covariance matrices

$$R_{\widetilde{w}} = R_w + \mathbf{diag}\left(\frac{s_{a,1}^2}{3} 2^{-2b_{a,1}}, \ldots, \frac{s_{a,M_u}^2}{3} 2^{-2b_{a,M_u}} \right),$$

$$R_{\widetilde{v}} = R_v + \mathbf{diag}\left(\frac{s_{s,1}^2}{3} 2^{-2b_{s,1}}, \ldots, \frac{s_{s,M_y}^2}{3} 2^{-2b_{s,M_y}} \right). \tag{15}$$

Here b_a and b_s are number of bits allocated to the actuators and sensors.

The scaling factors can be found from the 3σ-rule, by computing the variance
of the sensor and actuator signals. Hence, given the signal ranges and numbers
of quantization bits, we can calculate $R_{\widetilde{w}}$ and $R_{\widetilde{v}}$, and then design a controller
by solving (14). Notice that the signal ranges are determined by the RMS values,
which in turn depend on the controller design. This intertwined relationship will
show up in the iterative design procedures.

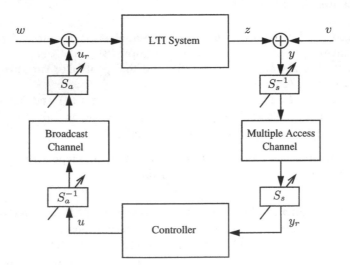

Fig. 9: Closed-loop control system over communication networks.

**Iterative Procedure to Design a Controller with Uniform Bit Alloca-
tion** First we allocate an equal number of bits to each actuator and sensor. This
means that we assign power and bandwidth (in the case of FDMA) uniformly
across all channels. We design a controller for such uniform resource allocation
via the following iterative procedure (iterate on the scaling factors and the controller):

given $\beta_i = \mathbf{rms}(u_i)$ and estimated $\mathbf{rms}(z_j)$.

repeat

1. Let $s_{a,i} = 3\,\mathbf{rms}(u_i)$ and $s_{s,j} = 3\,\mathbf{rms}(z_j)$, and compute $R_{\widetilde{w}}$ and $R_{\widetilde{v}}$ as in (15).
2. Solve problem (14) and compute $\mathbf{rms}(u_i)$ and $\mathbf{rms}(z_j)$ of the closed-loop system.

until stopping criterion is satisfied.

If the procedure converges, the resulting controller variables K and L of this iterative design procedure will satisfy the constraints on the control signals.

The Alternating Optimization Procedure Our goal here is to do joint optimization of bit allocation and controller design. This involves an iteration procedure over controller design, scaling matrices update and bit allocation. The controller and scaling matrices designed for uniform bit allocation by the above iteration procedure can serve as a good starting point. Here is the alternating optimization procedure:

given R_w, R_v, $\beta_i = \mathbf{rms}(u_i)$ and $\mathbf{rms}(z_j)$ from the above iteration design procedure.

repeat

1. Allocate bit rates $b_{a,i}$, $b_{s,j}$ and communication resources by solving a convex optimization problem of the form (8).
2. Compute $R_{\widetilde{w}}$ and $R_{\widetilde{v}}$ as in (15), and find controller variables K and L by solving (14).
3. Compute closed-loop system RMS values $\mathbf{rms}(u_i)$ and $\mathbf{rms}(z_j)$, then determine the signal ranges $s_{a,i}$ and $s_{s,j}$ by the 3σ rule.

until the RMS values $\mathbf{rms}(z_j)$ and bit allocation converges.

The convex optimization problem to be solved in step 1 depends on the communication system setup and resource constraints.

Numerical Example: Control of a Mass-Spring System Now we consider the specific example shown in figure 10. The position sensors on each mass send measurements $y_i = x_i + v_i$, where v_i is the sensor noise, to the controller through a Gaussian multiple access channel using FDMA. The controller receives data $y_{ri} = x_i + v_i + q_i$, where q_i is the quantization error due to bit rate limitation of the multiple access channel. The controller sends control signals u_j to actuators on each mass through a Gaussian broadcast channel using FDMA. The actual force acting on each mass is $u_{rj} = u_j + w_j + p_j$, where w_j is the exogenous disturbance force, and p_j is the quantization disturbance due to bit rate limitation of the broadcast channel. The mechanical system parameters are

$$m_1 = 10, \quad m_2 = 5, \quad m_3 = 20, \quad m_4 = 2, \quad m_5 = 15, \quad k = 1$$

The discrete-time system dynamics is obtained using a sampling frequency which is 5 times faster than the fastest mode of the continuous-time dynamics. The

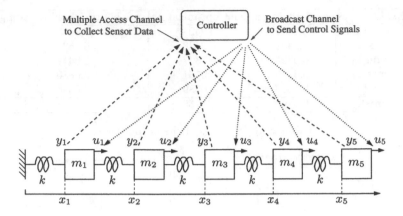

Fig. 10: Series-connected mass-spring system controlled over network.

independent zero mean noises w and v have covariance matrices $R_w = 10^{-6}I$ and $R_v = 10^{-6}I$ respectively. The actuators impose RMS constraints on the control signals:

$$\mathbf{rms}(u_i) \le 1, \quad i = 1, \ldots, 5.$$

For the Gaussian multiple access channel, the noise power density is $N = 0.1$, and the total power available is $P_{\text{mac,tot}} = 7.5$. For the Gaussian broadcast channel, the noise power density at each user is $N_i = 0.1$ for all i's, and the total power available for all users is $P_{\text{bc,tot}} = 7.5$. All users of the multiple access channel and the broadcast channel share a total bandwidth of $W = 10$. The proportionality coefficient α in the capacity formula is set to 2. Finally, we impose a lower bound $\underline{b} = 5$ and an upper bound $\overline{b} = 12$ on the number of bits allocated to each quantizer.[1]

First we allocate power and bandwidth evenly to all sensors and actuators, which results in a uniform allocation of 8 bits for each channel. We then designed a controller using the first iteration procedure based on this uniform resource allocation. This controller yields $\mathbf{rms}(u_i) = 1$ for all i's, and the RMS-values of the output signal z are listed in Table 2.

Finally, we used the second iteration procedure to do joint optimization of bit allocation and controller design. The resulting resource allocation after four iterations is shown in figure 11. It can be seen that more bandwidth, and hence more bits are allocated to the broadcast channel than to the multiple access

---

[1] To motivate our choice of lower bound on the bit allocations, note that our system is critically stable and that the lower bound for stabilization given in [WB99, NE00, TSM98] is zero. In general, if we discretize an open-loop unstable continuous-time linear system using a sampling rate which is at least twice the largest magnitude of the eigenvalues (a traditional rule-of-thumb in the design of digital control systems [FPW90]), then the lower bound given in [WB99, NE00, TSM98] is less than one bit. The analysis in [WKL96] shows that $b_i \ge 3$ or 5 is usually high enough for assuming the white noise model for quantization errors.

RMS values	equal allocation	joint optimization	variable threshold rounding
$\mathbf{rms}(z_1)$	0.1487	0.0424	0.0438
$\mathbf{rms}(z_2)$	0.2602	0.0538	0.0535
$\mathbf{rms}(z_3)$	0.0824	0.0367	0.0447
$\mathbf{rms}(z_4)$	0.4396	0.0761	0.0880
$\mathbf{rms}(z_5)$	0.1089	0.0389	0.0346
$\mathbf{rms}(z)$	0.5493	0.1155	0.1258

Table 2: RMS-values of the output signal.

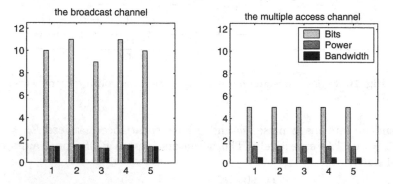

Fig. 11: Joint optimization of bit rates and linear control system.

channel. This means that the closed-loop performance is more sensitive to the equivalent process noises than to the equivalent sensor noises. The joint optimization resulted in $\mathbf{rms}(u_i) = 1$ for all i's, and the RMS-values of the output signal z are listed in Table 2. At each step of the variable threshold rounding, we check the feasibility of the resource allocation problem. The optimal threshold found is $t^\star = 0.6150$. Then we fix the integer bit allocation obtained with this threshold, and used the first iteration procedure to design the controller. We see a 77% reduction in RMS value over the result for uniform bit allocation, and the performance obtained by variable threshold rounding is quite close to that of the relaxed non-integer joint optimization.

6 Conclusions

We have addressed the problem of jointly optimizing the parameters of a linear system and allocating resources in the communication system that is used for transmitting sensor and actuator information. We considered a scenario where the coding and medium access scheme of the communication system is fixed, but the available communications resources, such as transmit powers and bandwidths, can be allocated to different channels in order to influence the achievable communication rates. To model the effect of limited communication rates on the performance of the linear system, we assumed conventional uniform quantization and used a simple white-noise model for quantization errors. We demonstrated that the problem of allocating communications resources to optimize the stationary performance of a fixed linear system (ignoring the integrality constraint) is often convex, hence readily solved. Moreover, for many important channel models, the communication resource allocation problem is separable except for a small number of constraints on the total communication resources. We illustrated how dual decomposition can be used to solve this class of problems efficiently, and suggested a variable threshold rounding method to deal with the integrality of bit allocations. The problem of jointly allocating communication resources and designing the linear system is in general not convex, but is often convex in subsets of variables while others are fixed. We suggested an iterative heuristic for the joint design problem that exploits this special structure, and demonstrated its effectiveness on the two examples: the design of a networked linear estimator, and the design of a multivariable networked LQG controller.

Acknowledgments

The authors are grateful to Wei Yu and Xiangheng Liu for helpful discussions.

References

[BB91]   S. Boyd and C. Barratt. *Linear Controller Design: Limits of Performance*. Prentice-Hall, 1991.

[Ber99]   D. P. Bertsekas. *Nonlinear Programming*. Athena Scientific, second edition, 1999.

[BL00]   R. Brockett and D. Liberzon. Quantized feedback stabilization of linear systems. *IEEE Transactions on Automatic Control*, 45(7):1279–1289, July 2000.

[BM97]   V. Borkar and S. Mitter. LQG control with communication constraints. In A. Paulraj, V. Roychowdhury, and C. D. Schaper, editors, *Communications, Computations, Control and Signal Processing, a Tribute to Thomas Kailath*, pages 365–373. Kluwer, 1997.

[BV04]   S. Boyd and L. Vandenberghe. *Convex Optimization*. Cambridge University Press, 2004. Available at http://www.stanford.edu/~boyd/cvxbook.html.

[CG01]   S. T. Chung and A. J. Goldsmith. Degrees of freedom in adaptive modulation: A unified view. *IEEE Transactions on Communications*, 49(9):1561–1571, 2001.

[CT91]   T. Cover and J. Thomas. *Elements of Information Theory*. John Wiley & Sons, 1991.

[Cur70]   R. E. Curry. *Estimation and Control with Quantized Measurements*. The MIT Press, 1970.

[Del90]   D. F. Delchamps. Stabilizing a linear system with quantized state feedback. *IEEE Transactions on Automatic Control*, 35(8):916–924, August 1990.

[EM01]   N. Elia and S. K. Mitter. Stabilization of linear systems with limited information. *IEEE Transactions on Automatic Control*, 46(9):1384–1400, September 2001.

[FPW90]   G. F. Franklin, J. D. Powell, and M. L. Workman. *Digital Control of Dynamic Systems*. Addison-Wesley, 3rd edition, 1990.

[Gol99]   A. Goldsmith. *Course reader for EE359: Wireless Communications*. Stanford University, 1999.

[HHB99]   A. Hassibi, J. P. How, and S. P. Boyd. A path-following method for solving BMI problems in control. In *Proceedings of American Control Conference*, volume 2, pages 1385–9, June 1999.

[KH94]   P. T. Kabamba and S. Hara. Worst-case analysis and design of sampled-data control systems. *IEEE Transactions on Automatic COntrol*, 38(9):1337–1357, September 1994.

[LG01]   L. Li and A. J. Goldsmith. Capacity and optimal resource allocation for fading broadcast channels: Part I: Ergodic capacity. *IEEE Transactions on Information Theory*, 47(3):1103–1127, March 2001.

[NBW98]   J. Nilsson, B. Bernhardsson, and B. Wittenmark. Stochastic analysis and control of real-time systems with random time delays. *Automatica*, 34(1):57–64, 1998.

[NE98]   G. N. Nair and R. J. Evans. State estimation under bit-rate constraints. In *Proc. IEEE Conference on Decision and Control*, pages 251–256, Tampa, Florida, 1998.

[NE00]   G. N. Nair and R. J. Evans. Stabilization with data-rate-limited feedback: tightest attainable bounds. *Systems & Control Letters*, 41:49–56, 2000.

[Özg89]   Ü. Özgüner. Decentralized and distributed control approaches and algorithms. In *Proceedings of the 28th Conference on Decision and Control*, pages 1289–1294, Tampa, Florida, December 1989.

[SSK99]   K. Shoarinejad, J. L. Speyer, and I. Kanellakopoulos. An asymptotic optimal design for a decentralized system with noisy communication. In *Proc. of the 38th Conference on Decision and Control*, Phoenix, Arizona, December 1999.

[SW90]   R. E. Skelton and D. Williamson. Guaranteed state estimation accuracies with roundoff error. In *Proceedings of the 29th Conference on Decision and Control*, pages 297–298, Honolulu, Hawaii, December 1990.

[TM89]   H. T. Toivonen and P. M. Mäkilä. Computer-aided design procedure for multiobjective LQG control problems. *Int. J. Control*, 49(2):655–666, February 1989.

[TSM98]   S. Tatikonda, A. Sahai, and S. Mitter. Control of LQG systems under communication constraints. In *Proc. IEEE Conference on Decision and Control*, pages 1165–1170, December 1998.

[WB97]   W. S. Wong and R. W. Brockett. Systems with finite communication bandwidth constraints I: state estimation problems. *IEEE Transactions on Automatic Control*, 42:1294–1299, 1997.

[WB99]   W. S. Wong and R. W. Brockett. Systems with finite communication bandwidth constraints – II: Stabilization with limited information feedback. *IEEE Transactions on Automatic Control*, 44:1049–1053, May 1999.

[Wil85]   D. Williamson. Finite wordlength design of digital Kalman filters for state estimation. *IEEE Transactions on Automatic Control*, 30(10):930–939, October 1985.

[WK89]   D. Williamson and K. Kadiman. Optimal finite wordlength linear quadratic regulation. *IEEE Transactions on Automatic Control*, 34(12):1218–1228, December 1989.

[WKL96]   B. Widrow, I. Kóllar, and M.-L. Liu. Statistical threory of quantization. *IEEE Trans. Instrumentation and Measurements*, 45:353–361, April 1996.

[XHH00]   L. Xiao, A. Hassibi, and J. P. How. Control with random communication delays via a discrete-time jump system approach. In *Proc. American Control Conf.*, Chicago, IL, June 2000.

Reconciliation of Inconsistencies in the Theory of Linear Systems

Emanuele Ragnoli and William Leithead

Hamilton Institute,
NUI Maynooth,
Co.Kildare, Ireland
emanuele.ragnoli@may.ie

Abstract. In the last few years some articles have emphasized certain fundamental inconsistencies underlying feedback control theory. The paper of Willems [1] Georgiou and Smith [2], later the works of Makila [3],[4], of Leithead et al. [5] have stressed the inconsistency of standard formalisms of linear time-invariant systems when the signals are double sided and the systems are open loop unstable. We establish a framework for a consistent time domain and frequency domain representation of discrete time linear time-invariant systems and, furthermore, that supports the consistent analysis of discrete time linear time-invariant feedback systems when signals are double sided and the systems are open loop unstable.

1 Introduction

System theory is applied to many branches of engineering. Recently, the boundaries between the traditional disciplines have become blurred with, for example, the application of control ideas to communication systems such as the internet and the use of feedback in signal processing. Consequently, the engineering systems, to which system theory is applied, have become more varied and complex. The extension of the classes of signals and systems to cater for this trend requires a careful choice of mathematical formalism. When inadequate, inconsistencies can arise, see Willems [1], Georgiu and Smith [2], Makila [3], [4] [6],Leithead et al. [5], Jacob [7]. One such inconsistency, that has recently been discussed widely, occurs when when double sided signals are considered. A consistent formalism for discrete time systems that resolves the inconsistency associated with stochastic and double sided signals is presented here.

2 Convoluted Double Trouble

In [4], Makila discusses a discrete time, first order convolution system

$$y(t) = b \sum_{j \geq 0} a^j [u(t - j - 1) + v(t - j - 1)] + d(t) \tag{1}$$

R. Murray-Smith, R. Shorten (Eds.): Switching and Learning, LNCS 3355, pp. 273–289, 2005.
© Springer-Verlag Berlin Heidelberg 2005

where y is the output, u is the input, and v and d are disturbance or noise terms. Let $b \neq 0$ and v and d be double sided square summable real sequences and define q^{-1} to be the backward shift operator, such that

$$q^{-n}y(t) = y(t-n)$$

(1) can be compactly written

$$y(t) = G(q)[u(t) + v(t)] + d(t)$$

where

$$G(q) = bq^{-1}\sum_{j\geq 0} a^j q^{-j}$$

In addition, define $X(z)$ to be the bilateral Z-transform of the square summable sequence $\{x(t)\}$ by

$$X(z) = \sum_{t=-\infty}^{\infty} x(t)z^{-1}$$

where, of course, z is a complex variable.
Makila proceeds by considering a simple proportional controller for (1), with a gain k, such that

$$u(t) = -ky(t)$$

and continues with a standard z-domain analysis.
Hence, $G(z) = bz^{-1}/(1 - az^{-1})$; that is the usual transfer function of a first order system. (1) becomes

$$Y(z) = G(z)[U(z) + V(z)] + D(z)$$

where, respectively $Y(z)$, $U(z)$, $V(z)$ and $D(z)$ are the transfer function of y, u, v and d.
Hence, closing the loop

$$(1 + kG(z))Y(z) = G(z)V(z) + D(z) \tag{2}$$

Using standard transform analysis,

$$Y(z) = \frac{G(z)V(z) + D(z)}{1 + kG(z)} \tag{3}$$

should be stable provided $|kb - a| < 1$ holds, including of course $|kb - a| = 0$. In what follows Makila considers the open loop unstable case, that is when $|a| > 1$. Surprisingly, the stabilizing feedback gain value, defined by the condition $(kb - a) = 0$, does not apply here. In fact, Makila proves that the necessary and sufficient condition for the existence and uniqueness of a solution of the closed loop system is

$$\lim_{t \to -\infty} a^{-t}(-kd(t) + v(t)) = 0$$

Hence, for $|a| > 1$, there are square summable signals d and v such that the closed loop equation has no solutions. Where is the error? Makila's conclusion is that "*We can't invert* $(1 + kG(z))$ *so that* $1/(1 + kG(z))$ *would make (3) well defined for arbitrary square summable v and d[...]*". The inconsistency, typified by the above discussion, is not the only one encountered in feedback systems. Others are discussed in, for example in [2], [3], [6], [7]. Inconsistencies arises also when stochastic processes are considered, see [8].

3 Fundamental Requirements

Any theoretical and systematical analysis must provide a consistent and coherent description of the system to be examined.
The fundamental requirements for a consistent time-domain and frequency-domain description of linear time-invariant systems are:

1) the class of signals must constitute a linear space;

2) the class of systems must constitute an algebra of linear operators mapping the class of signals into themselves;

$$\{class\ of\ linear\ operators\}$$

$$\{linear\ space\} \longrightarrow \{linear\ space\}$$

3) the inverses of the return difference operators must exist and themselves belong to the chosen class of systems.

The class of signals and the class of systems then constitute a suitable context for the consideration of linear time-invariant systems incorporating cascade, parallel path, feedback configurations and double sided signals.

Consider the simple feedback diagram

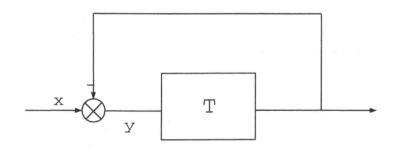

where the linear space of inputs and outputs is $\mathbb{Z}$, the integers, and the linear operator T is I, the identity operator. The linear operator describing the closed loop system would appear to be

$$\frac{1}{2}I$$

Hence, the output is related to the input by

$$output = \frac{input}{2}$$

It is obvious that the output might not belong to $\mathbb{Z}$. The problem is the following. The feedback configuration is equivalent to the operator relationship

$$(I + T)y = x$$

and, furthermore, for the feedback to be well defined, there must exist a solution, a y in the linear space, for all x belonging to the linear space, that is, the inverse operator for $(I + T)$ must exist. This does not happen for the above situation. A natural solution to the problem is easily found, it consists in enlarging the linear space of inputs and outputs, in this case choose the linear space to be $\mathbb{Q}$.

4   Possible Solutions

In both Section 2 and 3 serious difficulties are encountered due to the lack of invertibility of the return difference operator. The solution in Section 3 is to enlarge the class of signals from $\mathbb{Z}$ to $\mathbb{Q}$. The solution of Makila's paradox might be sought in two different ways:

a) only consider stable systems;

b) only consider single sided signals;

(a) is clearly absurd, it would mean dropping most of the problems addressed by control theory. (b) is rather poor, not just from the sake of mathematical completeness in general system theory, but also because the concept of transfer function involves double sided signals. In fact, for any discrete-time linear invariant system with time interval T, when the response exists, $x[n] \rightarrow y[n]$ where

$$x[n] = Me^{\sigma nT}cos(\omega nT + \delta)$$

and

$$y[n] = AMe^{\sigma nT}cos(\omega nT + \delta + \phi)$$

$\forall n \in \mathbb{Z}$. With $z = e^{j(\sigma+j\omega)T}$, the transfer function for the system is defined such that $|G(z)| = A$ and $arg\{G(z)\} = \phi$ for each value of σ and ω.

5 Solution (c), the Fourier Domain

From the above discussion neither (a) nor (b) are satisfactory solutions. A change of the analysis domain is the key for a consistent and coherent approach. For this purpose we must introduce the following notations and conventions:

Mclaurin series and z transform: for $x[k] = 0$, $k < 0$

$$\chi(q) = \sum_{k=0}^{\infty} x[k] q^k$$

for $q \in \mathbb{C}$, with $\chi(q)$ analytic for $|q| \leq R$, provided the summation (the Maclaurin series for $\chi(q)$) exists for $|q| \leq R$. Changing the notation such that

$$X(z) = Z\{x[k]\} = \chi(q)_{q=z^{-1}} = \sum_{k=0}^{\infty} x[k] z^{-k}$$

for $z \leq R^{-1}$ defines the standard (single sided) z transform for the sequence $x[k]$. Its inverse is

$$\{x[k]\} = Z^{-1}\{X(z)\} = \{\frac{1}{2\pi j} \oint_C \frac{\chi(q)}{q^{k+1}} dq\} = \{\frac{1}{2\pi j} \oint_C \frac{X(q^{-1})}{q^{k+1}} dq\}$$

where C is the contour in the complex plane defined by the circle, centered on the origin with radius R, traversed in the anti-clockwise direction.

Laurent series and Fourier series: for the sequence $\{x[k]\}$

$$\chi(q) = \sum_{k=-\infty}^{\infty} x[k] q^k$$

for $q \in \mathbb{C}$, with $\chi(q)$ analytic for $R_1 \leq |q| \leq R_2$, provided the summation (the Laurent series for $X(q)$) exists for $R_1 \leq |q| \leq R_2$. Changing the notation such that

$$X(\omega) = \mathcal{P}\{x[k]\} = \chi(q)_{q=e^{-j\omega T}} = \sum_{k=-\infty}^{\infty} x[k] e^{-j\omega T}$$

with $\chi(\omega)$ a periodic function of period $2\pi/T$. Its inverse is

$$\{x[k]\} = \mathcal{P}^{-1}\{X(w)\} = \{\frac{1}{2\pi j} \oint_C \frac{\chi(q)}{q^{k+1}} dq\} = \{\frac{T}{2\pi} \int_{-\frac{\pi}{T}}^{\frac{\pi}{T}} X(\omega) e^{jkwT} dw\}$$

where C is the contour in the complex plane defined by the circle, centered on the origin with radius R, traversed in the anti-clockwise direction.

In Makila [3], paradox (1) is analyzed in the z-domain using the relationship

$$Y(z) = G(z) X(z)$$

where $G(z)$ is the discrete transfer function for the system and $X(z)$ and $Y(z)$ are the z-transforms of the input, $x[k]$, and of the output, $y[k]$, (to simplify the expression we omit the disturbance terms v and d). We notice the discrepancy between the z-transform defined above (single sided) and the z-transform that the author uses to study (1). The latter is double sided, allowing Makila to deal with double sided signals, but introducing some annoying inconsistencies.

An alternative way of analysis of discrete linear time-invariant systems is proposed by Leithead et al. [5], that is to study them in the periodic frequency domain by the relationship

$$Y(\omega) = K(\omega)X(\omega)$$

where $K(\omega)$, the discrete frequency response function for the system is periodic and $X(\omega)$ and $Y(\omega)$ are periodic functions with Fourier coefficients $x[k]$ and $y[k]$. The new analysis is constructed by analytic extension of the discrete transfer function, and is equivalent to the z-domain analysis for the cases where the signals considered are single sided and causal. However, considering that the Fourier series is double sided, it is natural that signals can now be double sided and, furthermore, the analysis gains the advantage of the full power of Fourier analysis.

6   Causality and Stability

The change of domain suggested and proposed in Leithead et al [5] implies automatically not only a change of mathematical tools, but also a redefinition of goals and context.

We study a discrete linear time invariant system modelled by the linear constant coefficient finite order difference equation

$$y[k] + a_1 y[k-1] + ...a_{n-1} y[k-(n-1)] + a_n y[k-n]$$
$$= b_0 x[k] + b_1 x[k-1] + ... + b_{m-1} x[k-(m-1)] + b_m x[k-m]$$

The corresponding discrete transfer function is

$$G(z) = \frac{b_0 + b_1 z^{-1} + ... + b_{m-1} z^{-(m-1)} + b_m z^{-m}}{1 + a_1 z^{-1} + ...a_{n-1} z^{-(n-1)} + a_n z^{-n}}$$

and the corresponding discrete frequency response function is

$$K(\omega) = \frac{b_0 + b_1 e^{-j\omega T} + ... + b_{m-1} e^{-j(m-1)\omega T} + b_m e^{-jm\omega T}}{1 + a_1 e^{-j\omega T} + ...a_{n-1} e^{-j(n-1)\omega T} + a_n e^{-jn\omega T}}$$

where T is the sampling interval. Formally the discrete frequency response function and the discrete transfer function corresponding to a difference equation are related by

$$K(\omega) = G(z)_{z=e^{j\omega T}}$$

In what follows we consider a particular example and we compare the z-domain analysis with the frequency domain analysis. Let the system be defined by

$$y[k] - ay[k-1] = bx[k]$$

with $a > 1$ and the input be

$$x[k] = \begin{cases} 0 & k < 0 \\ e^{-ckT} & c > 0, \ k \geq 0 \end{cases}$$

In the z-domain the discrete transfer function is

$$G(z) = \frac{b}{1 - az^{-1}}$$

and the z-transforms of input and output

$$X(z) = \frac{1}{1 - e^{-cT}z^{-1}}$$

$$Y(z) = \frac{1}{(1 - az^{-1})(1 - e^{-cT}z^{-1})}$$

Hence, the response of the system is

$$y_z[k] = \begin{cases} 0 & k < 0 \\ \frac{b}{(a - e^{-cT})}(a^{k+1} - e^{-(k+1)cT}) & k \geq 0 \end{cases}$$

In the frequency domain the discrete frequency response function

$$K(\omega) = \frac{b}{1 - ae^{-j\omega T}}$$

and the Fourier series of input and output

$$X(\omega) = \frac{1}{1 - e^{-cT}e^{-j\omega T}}$$

$$Y(\omega) = \frac{b}{(1 - ae^{-j\omega T})(1 - e^{-cT}e^{-j\omega T})}$$

and the response of the system

$$y_P[k] = \begin{cases} -\frac{b}{(a - e^{-cT})}a^{(k+1)} & k < 0 \\ -\frac{b}{(a - e^{-cT})}e^{-(k+1)cT} & k \geq 0 \end{cases}$$

A comparison of the two system responses shows clearly that $\{y_P[k]\} \neq \{y_z[k]\}$. The reason is that a linear constant coefficient finite order difference equation does not define a discrete time linear time-invariant system. Initial conditions must be added. Consequently, the discrete transfer function corresponds to the difference equation with one choice of initial conditions, while the discrete frequency response function corresponds to the difference equation with another choice of initial conditions. Hence, the discrete frequency response function for a particular linear time invariant system is not necessarily obtained from the discrete transfer function of the system by substitution $z = e^{j\omega T}$. In our example the two inverse transfer functions

$$\phi_z[k] = \begin{cases} 0 & k < 0 \\ ba^k & k \geq 0 \end{cases}$$

$$\phi_P[k] = \begin{cases} -ba^k & k < 0 \\ 0 & k \geq 0 \end{cases}$$

are different.

Now consider the general case again. Other than a polynomial in z^{-1}, the discrete transfer function corresponding to the difference equation can be expanded by partial fractions as a sum of terms of the form $b/(z^{-1} + re^{j\theta})$ (the presence of repeated fractions, which are ignored here, does not invalidate what follows). The inverse z-transform of the term is

$$Z^{-1}\left(\frac{b}{(z^{-1} + re^{j\theta})}\right) = \{\phi_z[k]\}$$

where

$$\phi_z[k] = \begin{cases} 0 & k < 0 \\ -b(-r)^{-(k+1)}(cos(k+1)\theta - j sin(k+1)\theta) & k \geq 0 \end{cases}$$

It follows that, for the system corresponding to the discrete transfer function, $\phi_z[k]$ is zero for $k < 0$. It means that the system is causal but not necessarily stable as r may not be strictly greater than one.

The discrete frequency response function corresponding to the difference equation can be expanded by partial fractions as a sum of terms $\Psi_r(\omega)$ where

$$\Psi_r(\omega) = \frac{b}{e^{-j\omega T + re^{j\theta}}}$$

The inverse Fourier series

$$\mathcal{P}^{-1}\{\Psi_r(\omega)\} = \{\phi_r[k]\}$$

where

$$\phi_r[k]_{r<1} = \begin{cases} b(-r)^{-(k+1)}(cos(k+1)\theta - jsin(k+1)\theta) & k < 0 \\ 0 & k \geq 0 \end{cases}$$

$$\phi_r[k]_{r=1} = \begin{cases} \frac{1}{2}b(-r)^{-(k+1)}(cos(k+1)\theta - jsin(k+1)\theta) & k < 0 \\ -\frac{1}{2}b(-r)^{-(k+1)}(cos(k+1)\theta - jsin(k+1)\theta) & k \geq 0 \end{cases}$$

$$\phi_r[k]_{r>1} = \begin{cases} 0 & k < 0 \\ -b(-r)^{-(k+1)}(cos(k+1)\theta - jsin(k+1)\theta) & k \geq 0 \end{cases}$$

It follows that, for the system corresponding to the discrete frequency response function, $|\phi_p[k]|$ is bounded by $(c_1 + c_2 |k|^N)$ for some $N \geq 0$, c_1 and c_2; that means, $\phi_P[k]$ is weakly stable, in the sense that

$$e^{-\gamma|k|}\phi_P[k]$$

tends to zero as k tends to $\pm\infty$ for all $\gamma > 0$, but not necessarily causal as $\phi_P[k]$ may not be zero for $k < 0$. When $r \neq 1$ for all terms in the expansion, $\phi_p[k]$ is stable and tends to zero exponentially as k tends to $\pm\infty$.
It should be noted that

$$Z^{-1}(b/(z^{-1} + re^{j\theta})) = (P)^{-1}\{\Psi_r(\omega)\}$$

only when $r > 1$, that means when the inverse transform is stable and causal. Hence

$$\phi_P[k] = \phi_z[k]$$

if and only if the system is stable and causal.

The change of analysis domain leads us to change the control task, we don't investigate anymore the stability property of $G(z)$, we now investigate the causality property of $K(\omega)$.

7   Reformulation in $\mathcal{U}$

Changing the domain of the analysis we managed to slightly increase (now double sided signals can be treated) the class of signals. But still signals like steps, ramps, unbounded signals and others are excluded. As in Section 3 we proceed by enlarging the linear spaces of inputs and outputs.
A convenient and practical framework is formulated using ultradistributions and periodic ultradistributions. To do that, first we need to introduce the concept of distributions, some related linear spaces relate and of Fourier transform.

Distributions: the value assigned to each $\phi(t) \in D$, the class of good functions of finite support, by the functional $x \in \mathcal{D}$, the class of distributions, is denoted by $x[\phi(t)]$. The symbol for, respectively a regular functional in $\mathcal{D}$ and the ordinary function by which it is defined, e.g. x and $x(t)$, are distinguished by the explicit presence in the latter of the variable.

The following subclasses of $\mathcal{D}$ are required:

$$\mathcal{D}^T = \{x \in \mathcal{D} : x = \sum_{k=-\infty}^{\infty} a_k \delta_{kT}\} \; T > 0$$

$$\mathcal{D}_E^T = \{x \in \mathcal{D}^T : \; with \; \frac{a_k}{(1+|k|T)^N} \; square \; summable \; for \; some \; N \geq 0\} \; T > 0$$

$$\mathcal{D}_{EN}^T = \{x \in \mathcal{D}^T : \; with \; \frac{a_k}{(1+|k|T)^N} \; square \; summable\} \; N \geq 0 \; T > 0$$

$$\mathcal{D}_B^T = \{x \in \mathcal{D}^T : \; with \; |a_k| \leq (1+|k|T)^N \; for \; some \; c > 0 \; and \; N \geq 0\} \; T > 0$$

$$\mathcal{D}_{BN}^T = \{x \in \mathcal{D}^T : \; with \; |a_k| \, (1+|k|T)^N \; for \; some \; c < 0\} \; N \geq 0 \; T > 0$$

where the functional δ_τ in $\mathcal{D}$ is defined by

$$\delta_\tau[\phi(t)] = \phi(\tau)$$

for all $\phi(t) \in D$. In general, a shifted functional is indicated as a subscript; that is, x_τ is defined such that

$$x_\tau[\phi(t)] = x[\phi(t+\tau)]$$

for all $\phi(t) \in D$. When x is a regular functional defined by the function $x(t)$, x_τ is the regular functional defined by the function $x(t-\tau)$. The definitions of $\mathcal{D}^T$, $\mathcal{D}_E^T$, $\mathcal{D}_{EN}^T$, $\mathcal{D}_B^T$ and $\mathcal{D}_{BN}^T$ are specific to some value of the parameter, T, and $\mathcal{D}_E^T$, $\mathcal{D}_{EN}^T$, $\mathcal{D}_B^T$ and $\mathcal{D}_{BN}^T$ are subclasses of $\mathcal{D}_S$, the class of tempered distributions. Clearly $\mathcal{D}_E^T = \mathcal{D}_B^T$.

Fourier transform: for $x(t) \in S$, the class of good functions,

$$X(\omega) = \mathcal{F}\{x(t)\}(\omega) = \int_{-\infty}^{\infty} x(t)e^{-j\omega t}dt$$

with $X(\omega)$ a good function. The inverse is

$$x(t) = \mathcal{F}^{-1}\{X(\omega)\}(t) = \frac{1}{2\pi} \int_{-\infty}^{\infty} X(\omega)e^{j\omega T}d\omega$$

Ultradistributions: ultradistributions are an extension of the definition of Fourier transform. Each functional $x \in \mathcal{D}$ is related by a linear bijection to a functional $X \in \mathcal{U}$ such that

$$x[\phi^*(t)] = 2\pi X[\Phi^*(\omega)]$$

for all $\phi(t) \in D$ with

$$\Phi(\omega) = \mathcal{F}[\phi(t)](\omega)$$

The functional x and X constitute a Fourier transform pair with

$$X = \mathcal{F}\{x\} \; and \; x = \mathcal{F}^{-1}\{X\}$$

$\mathcal{U}^T$, the class of periodic ultradistributions, is the subclass of $\mathcal{U}$ consisting of all periodic functionals in $\mathcal{U}$ of period $2\pi/T$.

Similar to the Fourier transform, the most general extension to the definition of the Fourier series is provided by $\mathcal{U}^T$. For any sequence $\{x[k]\}$, the functional $x \in \mathcal{D}^T$ with $a_k = x[k]$ is related by Fourier transform to the periodic functional $X \in \mathcal{U}^T$ such that

$$X = \mathcal{F}\{x\}$$

$$x = \mathcal{F}^{-1}\{X\} = \sum_{k=-\infty}^{\infty} x[k]\delta_{kT}$$

There thus exists a linear bijection between the class of all sequences and $\mathcal{U}^T$. Furthermore,

$$\mathcal{F}\{\sum_{-\infty}^{\infty} x[k]\delta_{kT}\} = \sum_{-\infty}^{\infty} x[k]e_{kT}$$

where e_{kT} is the regular functional defined by the function $e^{-jk\omega T}$ (the functional e_{0T} is just the identity functional). The functional $\sum_{k=-\infty}^{\infty} x[k]\delta_{kT}$ is the Fourier series and the sequence $\{x[k]\}$ the Fourier coefficients for X with

$$x[k] = \frac{T}{2\pi}X[e^{jk\omega T}h(\omega)]$$

where $h(\omega)$ is any unitary function, with parameter $\frac{2\pi}{T}$, which is the Fourier transform of some function in D.

The subclasses $\mathcal{U}_E^T, \mathcal{U}_{EN}^T, \mathcal{U}_B^T, \mathcal{U}_{BN}^T$ of $\mathcal{U}$ are defined as those for which the members are the Fourier transforms of the members of the corresponding subclass of $\mathcal{D}$. $\mathcal{U}_S$, the Fourier transform of $\mathcal{D}_S$ is itself $\mathcal{D}_S$ and the subclasses $\mathcal{U}_E^T, \mathcal{U}_{EN}^T$, $\mathcal{U}_B^T, \mathcal{U}_{BN}^T$ are subclasses of $\mathcal{U}_S$. Being subclasses of the class of tempered distributions, $\mathcal{D}_E^T, \mathcal{D}_{EN}^T, \mathcal{D}_B^T, \mathcal{D}_{BN}^T$ are also subclasses of $\mathcal{U}_S$.

For a consistent and coherent analysis it is required:

a)the Fourier transform of discrete signals are represented by a class of functionals in $\mathcal{U}$;

b)the discrete frequency response functions are represented by linear operators mapping he linear space of signals into itself.

We notice that $\mathcal{U}$ is a linear space and that the signals themselves are represented by the corresponding class of functionals in $\mathcal{D}$.

A natural choice of representation of the discrete frequency response functions for a discrete system, with sampling time T, would be the class of periodic multipliers on $\mathcal{U}_S$ with period $2\pi/T$. Since the multipliers define linear operators, the requirements (a) and (b) are met. But we still need to establish that the discrete frequency response functions considered before are represented by periodic multipliers on $\mathcal{U}_S$.

The discrete frequency response functions, when non singular, is clearly represented in $\mathcal{U}_S$ by their regular functionals. However, a regular functional in $\mathcal{U}_S$ is not defined by a term $\psi_r(\omega)$, $r = 1$, in the expansion of the discrete frequency response function corresponding to a constant coefficient finite order difference equation system (the presence of repeated fractions continues to be ignored here). Instead, the functional in $\mathcal{U}_S$ corresponding to the function $\psi_r(\omega)$, $r = 1$, is defined by the limit in $\mathcal{U}_S$ as ϵ tends to zero of the regular functional

$$\frac{\psi_{(1+\epsilon)} + \psi_{(1-\epsilon)}}{2}$$

Any term of the form $\psi_r(\omega)$, $r = 1$, in the expansion of the discrete frequency response function corresponding to a constant coefficient difference equation should thus be replaced by a term $(\psi_{(1+\epsilon)} + \psi_{(1-\epsilon)})/2$ with ϵ arbitrary small. With the above modification when singular, any discrete frequency response function considered before is represented in $\mathcal{U}_S$ by the regular functional defined by the function $K(\omega) = G(z)_{z=e^{j\omega T}}$. The following theorem establishes that the functional so defined are multipliers on $\mathcal{U}_S$.

Theorem 1. *With the above modification when singular, let $K(\omega) = G(z)_{z=e^{j\omega T}}$ for a constant coefficient finite order difference equation system with sampling time T then*

(i) $K(\omega)$ is periodic, with period $2\pi/T$, and infinitely differentiable and $K(\omega)$ and all its derivative are bounded;

(ii) $K(\omega)$, the regular functional defined by the function $K(\omega)$, is a periodic multiplier on $\mathcal{U}_S$.

Proof. (i) follows immediately from the definition, modified to be non singular when required, of $K(\omega)$ such that $K(\omega) = G(z)_{z=e^{j\omega T}}$.

(ii) follows immediately from part (i) and the properties of the multipliers on $\mathcal{U}_S$ (a multiplier on $\mathcal{U}_S$ is a regular functional defined by an infinite differentiable function such that the magnitude of the function and all its derivative are bounded by polynomials).

8   Two Definitions of Stability, a Comparison

With the new analysis we study two different ways of approaching the notion of stability. First, we proceed in a natural manner with stability related to square summable signals, second we make the same study when stability is related to bounded signals.

8.1   Reformulation in $\mathcal{U}_E^T$

According to the previous discussion a system corresponding to a discrete frequency response function is required to be stable in some sense such as mapping square summable signals onto square summable signals; that is, they map $\mathcal{D}_{E0}^T$ into itself or more generally $\mathcal{D}_{EN}^T$ into itself for all $N \geq 0$. Hence, an appropriate reformulation is provided by the following definition:

Definition 1. *The Fourier series of signals are represented by the functionals in $\mathcal{U}_E^T$ and the discrete frequency response functions are the functionals in $\mathcal{M}^T$, the class of periodic multipliers on $\mathcal{U}_E^T$ mapping $\mathcal{U}_{EN}^T$ into itself for all $N \geq 0$.*

The inverse Fourier transforms of the discrete frequency response function are convolutes on the class $\mathcal{D}_E^T$, the inverse Fourier transforms of the Fourier series. Like Theorem 1 in the reformulation in $\mathcal{U}$ it only remains to show that the discrete frequency response functions considered before are represented in $\mathcal{M}^T$.

Lemma 1. *Let M be a regular functional defined by the infinitely differentiable function $M(\omega)$ and M^r, the regular functional defined by $M^r(\omega)$, be its rth derivative. Then M is a periodic multiplier on $\mathcal{U}_E^T$ mapping $\mathcal{U}_{EN}^T$ into itself for all $N \geq 0$ provided M is periodic with period $2\pi/T$;*

The proof is a consequence of theorem 15.24 of [9].

Theorem 2. *(i) Let M be a periodic multiplier on $\mathcal{U}_S$, with period $2\pi/T$, then M is a member of $\mathcal{M}^T$.*

(ii) With the usual modification when singular, let $K(\omega) = G(z)_{z=e^{j\omega T}}$ for a constant coefficient finite order difference equation then K, the regular functional defined by the function $K(\omega)$, is a member of $\mathcal{M}^T$.

Proof. (i) follows immediately from Lemma 1. (ii) follows immediately from part (i).

It is well known that addition and multiplication are well-defined for multipliers when considered as linear operators. Then, for M_1 and M_2 it is required that the operator $(M_1 + M_2)$ and $(M_1 M_2)$, defined by the addition and multiplication of operators, themselves correspond to multipliers on $\mathcal{U}_E^T$. Similarly, a feedback configuration is well defined for a multiplier, $M \in \mathcal{M}^T$, when considered as a linear operator, provided the input domain is restricted to the range of $(I + M)$; that is, the input $X \in \mathcal{U}_M^T$ where

$$\mathcal{U}_M^T = \{Y : Y = (I + M)X \ for \ some \ N \geq 0 \ and \ X \in \mathcal{U}_{EN}^T\}$$

When $X \neq 0$ implies $MX \neq -X$ for all $x \in \mathcal{U}_E^T$, the inverse operator

$$(I + M)^{-1} : \mathcal{U}_M^T \to \mathcal{U}_E^T$$

for the operator $(I + M) : \mathcal{U}_E^T \rightarrow \mathcal{U}_M^T$, corresponding to the multiplier $(I + M)$, exists. It is required that the operator $(I+M)^{-1}$ itself corresponds to a multiplier on $\mathcal{U}_M^T$. When such a multiplier exists, it is a left inverse, denoted by $(I + M)^{-1}$, of the multiplier $(I + M)$ on $\mathcal{U}_E^T$. The input domain should be unrestricted, that is $\mathcal{U}_M^T = \mathcal{U}_E^T$, when $(I + M)^{-1}$ is a multiplier in $\mathcal{M}^T$ and an inverse of the multiplier $(I + M)$ on $\mathcal{U}_E^T$. These issues are addresses by the following theorem.

Theorem 3. *(i) $\mathcal{M}^T$ constitutes an algebra;*

(ii) let M be a multiplier on $\mathcal{U}_S$ defined by the function $M(\omega)$ such that $(I + M)$ has no finite zero; then the functional $(I + M)^{-1}$ exists and is a multiplier and inverse on $\mathcal{U}_S$.

Proof. (i) As periodic linear operators, the multiplier define an algebra of periodic operators on $\mathcal{U}_E^T$ mapping $\mathcal{U}_{EN}^T$ into itself for all $N \geq 0$. However, the sum and product of two periodic multipliers are themselves periodic multipliers defined simply by the sum and product, respectively, of the functions defining the original multipliers. Hence, $\mathcal{M}^T$ defines an algebra.

(ii) Since $M(\omega)$ is periodic and everywhere infinitely differentiable and $(I + M(\omega))$ has no finite zeros, $(I + M(\omega))^{-1}$ is periodic and everywhere infinitely differentiable. It follows immediately, by theorem 16.22 of [9] and Theorem 2 part (i), that $(I + M(\omega))^{=1}$ defines a periodic multiplier in $\mathcal{M}^T$. Furthermore, since $(I + M(\omega))^{-1}(I + M(\omega)) = I$, the multiplier is an inverse on $\mathcal{U}_S$ as required.

An implication of Theorem 3 is that compound systems constructed by combining, through cascade, parallel path and feedback configuration, constant coefficient finite order difference equation systems are represented by the systems of Definition 1.

The reformulation of the analysis of discrete time linear invariant systems according to Definition 1 and Section 6 is distinguished from the usual frequency domain analysis for three main reasons:

a) the definition of signals and systems is in terms of functionals;

b) the analysis of systems is centered on causality rather then stability;

c) the periodic frequency response function would usually be expected to correspond to the actual system such that $K(\omega) = \mathcal{F}\{\phi_z\}(\omega)$, which is not necessarily the case here.

8.2   Reformulation in $\mathcal{U}_B$

An alternative definition of stable discrete frequency response functions is to require them to map bounded signals onto bounded signals; it means, they map $\mathcal{D}_{B0}^T$ into itself or more generally $\mathcal{D}_B^T$ into itself for all $N \geq 0$. Hence, an alternative to Definition 1 is provided by the following definition.

Definition 2. *The Fourier series of signals are represented by functionals in* $\mathcal{U}_B^T$ *and the discrete frequency response functions are the functionals in* $\mathcal{M}^{T*}$ *, the class of periodic multipliers on* $\mathcal{U}_B^T$ *mapping* $\mathcal{U}_{BN}^T$ *into itself for all* $N \geq 0$.

The inverse Fourier transforms of the discrete frequency response functions are convolutes on the class $\mathcal{D}_B^T$, the inverse Fourier transforms of Fourier series. The development is similar to that of the previous section, with Theorems 2 and 3 replaced by Theorems 4 and 5 below.

Lemma 2. *Let M be a regular functional defined by the infinitely differentiable function $M(\omega)$ and M^r, the regular functional defined by $M^r(\omega)$, be its rth derivative. Then M is a periodic multiplier on $\mathcal{U}_B^T$ mapping $\mathcal{U}_{BN}^T$ into itself for all $N \geq 0$ provided M is periodic with period $2\pi/T$;*

The proof is a consequence of theorem 15.24 of [9].

Theorem 4. *(i) Let M be a periodic multiplier on $\mathcal{U}_S$, with period $2\pi/T$, then M is a member of $\mathcal{M}^{T*}$.*

(ii) With the usual modification when singular, let $K(\omega) = G(z)_{z=e^{j\omega T}}$ for a constant coefficient finite order difference equation then K, the regular functional defined by the function $K(\omega)$, is a member of $\mathcal{M}^{T*}$.

Proof. (i) follows immediately from Lemma 2. (ii) follows immediately from part (i).

Theorem 5. *(i) $\mathcal{M}^{T*}$ constitutes an algebra*

(ii) Let $M \in \mathcal{M}^{T*}$ be a multiplier on $\mathcal{U}_S$ defined by the function $M(\omega)$ such that $(I + M(\omega))$ has no finite zeros then the functional $(I + M)^{-1}$ exists and is a multiplier and inverse on $\mathcal{U}_S$.

Proof. The proof is similar to the one of theorem 3.

The reformulations of Definition 1 and 2 are closely related. Following Theorem 2(i) and 4(i), $\mathcal{U}_E^T = \mathcal{U}_B^T$ and $\mathcal{M}^T = \mathcal{M}^{T*}$. In other words, the class of systems, stable in the sense of mapping $\mathcal{U}_{E0}^T$ into itself, and the class of systems, stable in the sense of mapping $\mathcal{U}_{B0}^T$ into itself, are identical and have the same domain.

9   Resolution of Makila Paradox

Returning to Makila's paradox we analyze it using a frequency domain analysis according to Definition 1 and 2.
Consider an unstable plant defined by the discrete transfer function

$$G(z) = \frac{bz^{-1}}{1 - az^{-1}}$$

for $a > 1$. By analytic extension the discrete frequency response function is

$$K(\omega) = G(z)_{z=e^{j\omega T}} = \frac{be^{-j\omega T}}{1 - ae^{-j\omega T}}$$

When $(a - kb) \neq 1$, the inverse of $[1 + kK]$ exists and is the multiplier in $\mathcal{M}^T$ defined by the function $[1 + kK(\omega)]^{-1}$. When $(a - kb) = 1$ the inverse of $[1 + kK]$ exists as functional in $\mathcal{U}^T$ but not as a multiplier in $\mathcal{M}^T$, thereby, restricting its domain to some subclass of $\mathcal{U}^T_E$. In this case, the inverse of $[1 + kK]$ is defined by

$$\frac{1}{2} \lim_{\epsilon \to 0} \{[1 + (1 + \epsilon)kK)]^{-1} + [1 + (1 + \epsilon)kK]^{-1}\}$$

that is, the limit of the regular functional defined by the function

$$\frac{1}{2} \lim_{\epsilon \to 0} \{[1 + (1 + \epsilon)kK(\omega))]^{-1} + [1 + (1 + \epsilon)kK(\omega)]^{-1}\}$$

Clearly the inverse of $[1 + kK]$ is causal if and only if $(a - kb) < 1$. Hence, enclosing the plant in a feedback loop with gain k, the closed loop system is stable, in the sense of Definition 1 and 2, if and only if $(a - kb) < 1$. With $(a - kb) < 1$, choose the time domain input to be

$$x[n] = \begin{cases} -1 & n < 0 \\ 1 & n \geq 0 \end{cases}$$

that is, in the periodic frequency domain,

$$X = \lim_{\epsilon \to 0} \{X_{1+\epsilon} + X_{1-\epsilon}\}$$

where $X_{1\pm\epsilon}$ are the regular functionals defined by the function

$$\frac{1}{(1 - \frac{1}{1\pm\epsilon})e^{-j\omega T}}$$

For any sequence, $\{F_n\}$, of functionals in $\mathcal{D}$ such that $\lim_{n \to \infty} F_n = F \in \mathcal{D}$ and any multiplier, M, $MF = M \lim_{n \to \infty} F_n = \lim_{n \to \infty} MF_n$. Hence the corresponding closed loop system output in the periodic frequency domain response is

$$Y = K_C X = K_C \lim_{\epsilon \to 0} \{X_{1+\epsilon} + X_{1-\epsilon}\} = \lim_{\epsilon \to 0} \{K_C(X_{1+\epsilon} + X_{1-\epsilon})\}$$

$$K_C = [1 + kK]^{-1} kK$$

Since, K_C is causal and stable and the time series corresponding to $\mathcal{P}^{-1}\{X_{1\pm\epsilon}\}$ are square summable, the time domain signal equivalent to $K_C(X_{1+\epsilon} + X_{1-\epsilon})$ can be determine in the usual manner, specifically

$$y_\epsilon[n] = \begin{cases} -\frac{kb}{1-(1-\epsilon)(a-kb)}(\frac{1}{1-\epsilon})^{n-1} & n < 1 \\ kb[\frac{1}{1-(1+\epsilon)(a-kb)}(\frac{1}{1+\epsilon})^{n-1} \\ -(\frac{1+\epsilon}{1-(1+\epsilon)(a-kb)} + \frac{1-\epsilon}{1-(1-\epsilon)(a-kb)})(a-kb)^n] & n \geq 1 \end{cases}$$

where

$$Y_\epsilon(\omega) = \frac{kbe^{-j\omega T}}{1 + (kb - a)e^{-j\omega T}} \left(\frac{1}{1 - \frac{1}{1+\epsilon}e^{-j\omega T}} + \frac{1}{1 - \frac{1}{1+\epsilon}e^{-j\omega T}} \right)$$

Hence the closed loop system output in the time domain is

$$y[n] = \lim_{\epsilon \to 0} \begin{cases} -\frac{kb}{1+kb-a} & n < 1 \\ \frac{kb}{1+kb-a} - 2\frac{kb}{1+kb-a}(a - kb)^n & n \geq 1 \end{cases}$$

10   Conclusion

The consistency of time domain and frequency domain description of discrete-time linear time invariant systems is established. The class of signals is enlarged to include double sided signals, random signals, steps, ramps and other unbounded signals of interest in system theory. In addition, the consistency holds when the systems are open loop unstable. A similar framework for continuous time linear time-invariant systems is under development.

Acknowledgements

This work was supported by the European Union funded research training network *Multi-Agent Control*, HPRN-CT-1999-00107.[1]

References

[1] Willems, J.C.: Stability, instability, invertibility and causality. SIAM J. Control **7** (1969) 645–671

[2] Georgiou, T.T., Smith, M.C.: Intrinsic difficulties in using the doubly-infinite time axis for input-output control theory. IEEE Trans. Automat. Control **40** (1995) 516–518

[3] Mäkilä, P.M.: On three puzzles in robust control. IEEE Trans. Automat. Control **45** (2000) 552–556

[4] Mäkilä, P.M.: Convoluted double trouble. IEEE Control System Mag. **22** (2002) 26–31

[5] Leithhead, W.E., J.O'Reilly: A consistent time-domain and frequency-domain representation for discrete-time linear time-invariant feedback systems. In: Proceedings of the American Control Conference, Denver, Colorado (2003) 429–434

[6] Mäkilä, P.M., Partington, J.R.: Input-output stabilization on the doubly-infinite time axis. Internat. J. Control **75** (2002) 981–987

[7] Jacob, B.: An operator theoretical approach towards systems over the signal space $l_2(\mathbb{Z})$. Integral Equations Operator Theory **46** (2003) 189–214

[8] Mäkilä, P.M.: Intrinsic difficulties in stochastic control of unstable convolution operators on $\mathbb{Z}$. IEEE Trans. Automat. Control **48** (2003) 2015–2019

[9] D.C.Champeney: A Handbook of Fourier Theorems. Cambridge University Press, Cambridge (UK) (1987)

---

[1] This work is the sole responsibility of the authors and does not reflect the European Union's opinion.

An Introduction to Nonparametric Hierarchical Bayesian Modelling with a Focus on Multi-agent Learning

Volker Tresp[1] and Kai Yu[2]

[1] Siemens AG, 81730 München, Germany
Volker.Tresp@siemens.com
[2] Siemens AG, 81730 München, Germany
Kai.Yu@siemens.com

Abstract. In this chapter, we address the situation where agents need to learn from one another by exchanging learned knowledge. We employ hierarchical Bayesian modelling, which provides a powerful and principled solution. We point out some shortcomings of parametric hierarchical Bayesian modelling and thus focus on a nonparametric approach. Nonparametric hierarchical Bayesian modelling has its roots in Bayesian statistics and, in the form of Dirichlet process mixture modelling, was recently introduced into the machine learning community. In this chapter, we hope to provide an accessible introduction to this particular branch of statistics. We present the standard sampling-based learning algorithms and introduce a particular EM learning approach that leads to efficient and plausible solutions. We illustrate the effectiveness of our approach in context of a recommendation engine where our approach allows the principled combination of content-based and collaborative filtering.

1 Introduction

There are many occasions where agents should "learn" from one another. As an example, the effectiveness of a treatment for a cardiac disease is a function of the severity of the disease and patient characteristics but might also vary from hospital to hospital (due to hidden factors such as varying patient population, staff training, local expertise, ...). Thus models that predict the outcomes for different hospitals should be quite similar but will also be different to some degree. Despite the differences in the models it would be advantageous if various models could benefit from each other's learned knowledge, in particular in the case that there is only a small data set available for each hospital. A similar situation arises in the design of recommendation engines that predict the interests of users in various items. Essentially each user is an individual and one should learn a personal model for each user. On the other hand if few training data points for the active user are available one would like to benefit from the recommendations of like-minded users, as in collaborative filtering. In machine learning, the scenarios described are known as transfer learning or meta learning.

R. Murray-Smith, R. Shorten (Eds.): Switching and Learning, LNCS 3355, pp. 290–312, 2005.

In the Bayesian literature this framework falls into hierarchical Bayesian (HB) modelling. The basic idea in HB modelling is that information between different models can be exchanged via common hyperparameters. In this chapter, we provide an introduction to HB modelling. We emphasize that, in our view, HB by itself is useful but also severely limited since it is inflexible in the representation of the "learned prior". Additional flexibility is obtained by a process called Dirichlet enhancement in which the prior distribution is specified in terms of a highly flexible multinomial distribution with a Dirichlet prior. Of particular interest is the limit that the number of states in the multinomial becomes infinite in which case we obtain a Dirichlet process and our hierarchical model becomes a Dirichlet process mixture model.[3] Dirichlet process mixture models originated in Bayesian statistics [11] [1] and recently found growing interest in the machine learning community, in particular in the context of infinite mixture models. A particular advantage of Dirichlet process mixture models is that the number of components required for achieving a good overall model is automatically determined by the algorithm. In the problem setting described in this chapter this feature is of minor interest in comparison to the benefits achieved by the transfer of learned knowledge via HB modelling. We describe the standard sampling approach for inference in Dirichlet process mixture models and also introduce a particular expectation maximization (EM) solution that is powerful and efficient in the frameworks addressed in this chapter.

The chapter is organized as follows. In the following section we provide an intuitive motivation for nonparametric HB modelling and present the first algorithmic solution to the problem. In Section 3 we introduce HB modelling more systematically and discuss some of its shortcomings. In Section 4 we introduce the process of a Dirichlet enhancement, which is a first step towards nonparametric HB modelling. The finite-dimensional approach presented in Section 4 is not of great practical interest by itself but provides the basis for the infinite-dimensional nonparametric HB models described in Section 5. We discuss stochastic sampling and EM as approaches towards parameter inference. In Section 6 we illustrate the effectiveness of our approach using the example of a recommendation engine where our approach allows the principled combination of content-based filtering and collaborative filtering. In Section 7 we discuss related work, in particular recent work on infinite models. In Section 8 we provide conclusions.

2 Intuitive Introduction

2.1 Bayesian Modelling

We will develop the ideas based on two-class classification although the same concepts are valid for general probabilistic models, e.g., for regression and density estimation. Readers who want to fresh up on Bayesian statistics may consult

---

[3] Dirichlet process mixture models are also known as mixtures of Dirichlet processes (MDPs).

the excellent tutorial [15]. Let $P(Y = y|x, \theta)$ denote the probability that Y assumes the state $y \in \{0, 1\}$ given features x and given a parameter set $\theta = \{\theta_j\}_j$. In a Bayesian setting one defines an a priori distribution $P(\theta|h_{prior})$ with hyperparameters $h = h_{prior}$. Both prior distribution and hyperparameters specify one's prior belief. The prior belief is typically rather unspecific or non-informative and thus the prior distribution should place nonzero probability on all reasonable model parameters.

As example, in Figure 1A the prior distribution might be specified as a Gaussian distribution with

$$P(\theta|h_{prior}) = \mathcal{N}(\theta|\mu_{prior}, \Sigma_{prior})$$

with $h_{prior} = \{\mu_{prior}, \Sigma_{prior}\}$.

Bayesian learning means updating the parameter distribution based on available training data. Given a data set with N_D data points $D = \{(x_n, y_n)\}_{n=1}^{N_D}$ one can calculate the posterior parameter density using Bayes formula as

$$P(\theta|D, h_{prior}) = \frac{1}{P(D)} P(D|\theta) P(\theta|h_{prior})$$

where, in our classification example, assuming exchangeability,

$$P(D|\theta) = \prod_{n=1}^{N_D} P(y_n|x_n, \theta).$$

Note that in this chapter we do not treat the inputs x probabilistically and focus on the modelling of the condition probability distribution $y|x$.

For classifying a new pattern we obtain the predictive distribution

$$P(Y = y|x, D, h_{prior}) = \int P(Y = y|x, \theta) P(\theta|D, h_{prior}) d\theta.$$

If additional data points become available, the posterior parameter distribution $P(\theta|D, h_{prior})$ now assumes the role of the new "learned prior", i.e., the available knowledge prior to the arrival of the additional data. In the case that the prior distribution is conjugate to the likelihood function, we obtain

$$P(\theta|D, h_{prior}) = P(\theta|h_{post}),$$

i.e., the posterior parameter distribution has the functional form of the prior distribution but with new hyperparameters h_{post}. Returning to our example, we would expect that

$$P(\theta|h_{post}) = \mathcal{N}(\theta|\mu_{post}, \Sigma_{post})$$

with $h_{post} = \{\mu_{post}, \Sigma_{post}\}$ and where $\lim_{N_D \to \infty} \det \Sigma_{post} = 0$, i.e., the posterior distribution become increasingly concentrated (Figure 1B) with an increasing number of data points and asymptotically is locally peaked at the maximum likelihood solution

$$\theta^{ML} := \arg\max_{\theta} P(D|\theta).$$

2.2 Hierarchical Bayesian Modelling

Now assume, that we have obtained M data sets $\{D_j\}_{j=1}^M$ for related but not identical settings and we have trained M *different models* with parameters $\{\theta_j\}_{j=1}^M$ on those data sets. For the sake of argument let's assume that each data set is sufficiently large such that $P(\theta_j|D_j, h_{prior})$ is sharply peaked at the maximum likelihood (ML) estimate θ_j^{ML}. Let $\{\theta_k^{ML}\}_{k=1}^M$ denote the maximum likelihood estimates for the M models. Recall that since the models were trained on different data sets generated from different settings, the maximum likelihood parameter values are not identical. Figure 1C illustrates the set of maximum likelihood parameter estimates. Now, if a new model concerns a related problem, then it makes sense to select new hyperparameters h_{hb} such that $P(\theta|h_{hb})$ approximates the empirical distribution given by the maximum likelihood parameter estimates instead of using the original uninformed prior $P(\theta|h_{prior})$. In this way the new model can inherit knowledge acquired not only from its own data set but also from the other models.

Returning to our example, we would expect that for a new setting with a new model with parameters θ_{M+1}

$$P(\theta_{M+1}|\{D_j\}_{j=1}^M) \approx P(\theta_{M+1}|h_{hb}) \tag{1}$$

where, in the example, $P(\theta_{M+1}|h_{hb}) = \mathcal{N}(\theta_{M+1}|\mu_{hb}, \Sigma_{hb})$, with $h_{hb} = \{\mu_{hb}, \Sigma_{hb}\}$ and where now in the non-degenerate case

$$\lim_{M\to\infty} \det \Sigma_{hb} > 0$$

and the entries of Σ_{hb} converge to fixed typically nonzero values (Figure 1C). What we have just described is the basis for hierarchical Bayesian modelling that we will introduce more formally in Section 3.

2.3 Nonparametric Hierarchical Bayesian Modelling

In more cases than not, the empirical distribution of the maximum likelihood parameters $\{\theta_k^{ML}\}_{k=1}^M$ will not fall into the class of distributions that can be described by $P(\theta|h)$ for any h. If the assumed noninformative prior is too inflexible to truthfully model the learned prior, then this is a severe limitation of the classical HB approach. See for example Figure 1D. Thus we might prefer a nonparametric approximation in the form of the empirical nonparametric distribution of the maximum likelihood parameters

$$P(\theta_{M+1}|\{D_j\}_{j=1}^M) \approx \frac{1}{M} \sum_{k=1}^M \delta_{\theta_k^{ML}},$$

where $\delta_{\theta_k^{ML}}$ is a distribution concentrated at a single point θ_k^{ML}.

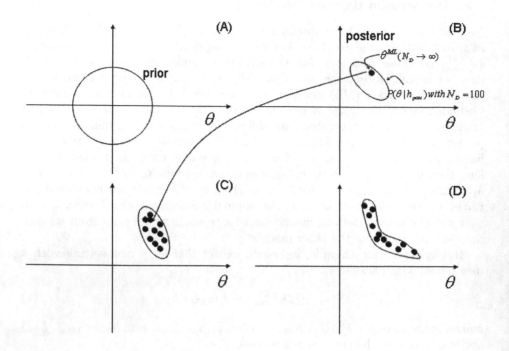

Fig. 1. A: The circle indicates the standard deviation of the prior Gaussian distribution with mean zero representing $P(\theta|h_{prior}) = \mathcal{N}(\theta|\mu_{prior}, \Sigma_{prior})$. B: The posterior parameter distribution $P(\theta|h_{post}) = \mathcal{N}(\theta|\mu_{post}, \Sigma_{post})$ with lets say $N_D = 100$ data points; shape and location of the Gaussian have changed. With $N_D \rightarrow \infty$, $P(\theta|h_{post})$ is concentrated at the maximum likelihood estimate θ^{ML}. C: Set of maximum likelihood estimates $\{\theta_j^{ML}\}_{j=1}^M$ and approximation $P(\theta_{M+1}|\{D_j\}_{j=1}^M) \approx \mathcal{N}(\theta|\mu_{hb}, \Sigma_{hb})$. The implicit assumption in HB modelling is that this distribution can be approximated by a member of the family of distributions assumed for the prior, i.e., in this example a Gaussian distribution. D: Here is an example where this distribution cannot be approximated by a Gaussian distribution. Thus, nonparametric HB with $P(\theta_{M+1}|\{D_j\}_{j=1}^M) \approx \frac{1}{M}\sum_{k=1}^M \delta_{\theta_k^{ML}}$ is more appropriate.

Now if we receive the data set D_{M+1} for the new setting, we predict

$$P(Y_{M+1} = y|x, D_{M+1}) \approx \frac{1}{C} \int P(Y_{M+1} = y|x, \theta) P(D_{M+1}|\theta) \sum_{j=1}^M \delta_{\theta_j^{ML}} \, d\theta$$

$$= \frac{1}{C} \sum_{j=1}^M P(D_{M+1}|\theta_j^{ML}) P(Y = y|x, \theta_j^{ML}) \tag{2}$$

where $C = \sum_{j=1}^{M} P(D_{M+1}|\theta_j^{ML})$ normalizes the distribution. Here and in the following, capital C stands for an appropriate normalization constant. Note that the result (Equation 2) is very intuitive. To make a prediction for setting $M+1$ for input x, each model $1, \ldots, M$ makes a prediction using its maximum likelihood parameter estimate and this prediction is then weighted with the probability that this model explains the data points D_{M+1} of the setting of interest. This means that initially, with only few data points available for setting $M + 1$, the predictions of all previous models are essentially averaged. With more data points available for setting $M + 1$, models that agree well with the data D_{M+1} obtain a higher weight.

3   Hierarchical Bayesian Modelling

In this and the following sections we will introduce HB modelling and non-parametric Bayesian modelling more formally. We start with HB. Recall that in Section 2.2 we essentially learned new hyperparameters h_{hb} to communicate learned knowledge. This is exactly the basis for the knowledge transfer via common hyperparameters in the framework of HB modelling. The joint probabilistic HB model is written as (Figure 2A)

$$P(h) \prod_{j=1}^{M} P(D_j|\theta_j)P(\theta_j|h). \tag{3}$$

The hyperparameters h —now considered to be random variables with prior distribution $P(h)$— are common to all models whereas each model has its own parameters $\{\theta_j\}_{j=1}^{M}$. Given the hyperparameters, the models are exchangeable, which means that the probabilistic model is invariant to a permutation (re-indexing) of the models.[4]

Now, for a model $M + 1$ that did not yet receive any data points, we obtain as a full Bayesian version of Equation 1

$$P(\theta_{M+1}|\{D_j\}_{j=1}^{M}) \propto \int \left[P(\theta_{M+1}|h)P(h) \prod_{j=1}^{M} \int P(\theta_j|h)P(D_j|\theta_j)\, d\theta_j \right] dh. \tag{4}$$

In all but the simplest cases, the inference based on the HB model in Equation 4 does not lead to closed-form solutions and one typically relies on Markov Chain Monte Carlo (MCMC) approximations. We do not want to get deeper into the issues of learning parametric HB models since we already concluded that the conventional HB approach is too limited for many applications. Readers more interested in the basics of HB modelling may consult [12].

---

[4] In contrast to the HB modelling assumption if we would assume that the models are all identical, then all data points are exchangeable and the probabilistic model is $P(h)P(\theta|h) \prod_{j=1}^{M} P(D_j|\theta)$, which would lead to one global model. The other extreme is that all models are independent $\prod_{j=1}^{M} P(h_j)P(D_j|\theta)P(\theta_j|h_j)$, which would result in M independent models.

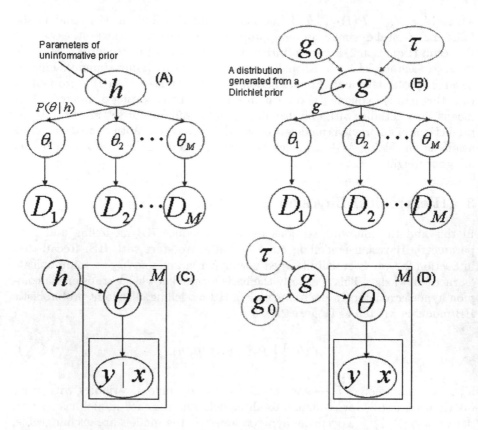

Fig. 2. A: A HB model. B: A Dirichlet enhancement HB model. C: A plate model for HB. The large plate indicates that M samples from $P(\theta|h)$ are generated; the smaller plate indicates that, repeatedly, data points are generated for each θ. D: A plate model for the Dirichlet enhanced HB. In B and D the finite dimensional hyperparameters h are replaced by the distribution g. In the finite-dimensional case, g is finite-dimensional and is generated from a Dirichlet distribution. In the infinite-dimensional case, g is infinite-dimensional and is generated from a Dirichlet process. We also indicate that, in the latter case, the prior distribution for g is defined using a base distribution G_0 with density g_0 and concentration parameter τ (see Section 5)

4   Dirichlet Enhanced Hierarchical Bayesian Modelling

4.1   The Basic Idea

To alleviate the problem of HB we have to specify a parameterization of the prior parameter distribution that on the one hand can represent the assumed noninformative prior knowledge but also is flexible enough to be able to appropriately represent the "learned" prior to be communicated to a new model. The concept we are applying here is sometimes referred to as Dirichlet enhancement [10] and the basic idea is to replace the parametric prior distribution by a finite or infinite multinomial distribution with a Dirichlet prior. The essential features are that, first, the multinomial distribution by itself poses no constraint on the distributions that can be represented and that, second, the noninformative prior knowledge can be encoded in the form of the base distribution of the Dirichlet (which we will introduce further down). In this section we will consider the case that the model parameters can only assume values out of a given finite set of size K. The finite case is mathematically considerably easier and already introduces the main features of Dirichlet enhanced HB modelling. From an application point of view the case that $K \to \infty$ is of greater importance and will be discussed on the the following section.

To represent the model parameters we introduce a random variable Θ_j for each model j that can be in states $\theta_1, \ldots, \theta_K$. We further assume that a particular state is chosen by a multinomial distribution such that, for all j, $P(\Theta_j = \theta_k | g) = g_k$ with $g_k > 0$ and $\sum_{k=1}^{K} g_k = 1$ such that the probabilities $g_k, k = 1, \ldots, K$ play the role of the hyperparameters (previously the h)(Figure 2B). We specify our prior belief in terms of the conjugate prior that in this case is a Dirichlet distribution, i.e.,

$$P(g) = \mathrm{Dir}(g | \tau\alpha_1, \ldots, \tau\alpha_K) = \frac{1}{C} \prod_{k=1}^{K} g_k^{\tau\alpha_k - 1}$$

where $g = \{g_i\}_{i=1}^{K}$, $\alpha = \{\alpha_i\}_{i=1}^{K}$, $\alpha_k \geq 0$, $\sum_{k=1}^{K} \alpha_k = 1$ and with precision parameter $\tau > 0$. A description of the properties of a multinomial model with a Dirichlet prior including most equations used in this section can be found in the already mentioned tutorial [15]. A sample of a Dirichlet distribution is a probability distribution and the precision parameter τ corresponds to an equivalent sample size or weight. We can integrate out g and have $P(\Theta_j = \theta_j) = \alpha_j, j = 1, \ldots, K$.[5] Thus we can specify our non-informative prior belief by defining the $\alpha_j, j = 1, \ldots, K$ and the $\theta_j, j = 1, \ldots, K$ appropriately. The solution used in the following is to randomly select θ_j from $P(\Theta_j)$ and set $\alpha_j = 1/K, j = 1, \ldots, K$ (Figure 3 (top)). This is quite similar to the implementation of the non-informative prior belief in the infinite model of Section 5 where $K \to \infty$.

The joint distribution of the Dirichlet enhanced model is now (compare Equation 3 and Figure 2)

$$P(g) \prod_{j=1}^{M} P(D_j | \Theta_j = \theta_j) P(\Theta_j = \theta_j | g). \tag{5}$$

---

[5] Incidentally, the most likely configuration is also $g = \alpha$.

4.2   Sampling from a Dirichlet Model

First, we consider the simpler model $P(g)P(\Theta|g)$ consisting of a Dirichlet prior for g and a multinomial likelihood. We assume that for a fixed (but potentially unknown) g, N repeated samples of Θ are drawn. These samples form the set D_θ. Let's assume that in D_θ we have N_k instances of θ_k with $N = \sum_{k=1}^{K} N_k$.

Since the Dirichlet distribution is conjugate to the multinomial distribution, we obtain for the posterior distributions for g also a Dirichlet distribution with

$$P(g|D_\theta) = \mathrm{Dir}(g|\tau\alpha_1 + N_1, \ldots, \tau\alpha_K + N_K).$$

A nice property is that one can integrate out g to obtain the posterior predictive density [15]

$$P(\Theta = \theta_k|D_\theta) = \frac{\tau\alpha_k + N_k}{\tau + N}. \tag{6}$$

Equations 6 says that we can conveniently calculate the predictive distribution without the need for the explicit estimation of g. This is of great importance in the next section in the context of Dirichlet processes where g is infinite dimensional and could not explicitly be represented. According to Equation 6, a state becomes more likely if it has previously been observed with high frequency.

Note that Equation 6 also specifies how a new sample can be generated given previously *generated* samples D_θ. This sampling procedure generates data points from a fixed (but potentially unknown) g generated by the Dirichlet prior. Asymptotically, g can be inferred from the samples by noting that $g_k = \lim_{N\to\infty} N_k/N$.

The generation of samples according to Equation 6 is called a Pólya urn sampling process or a Chinese restaurant sampling process (for a recent discussion, see [27]). The essential feature is that if a state is sampled in the past, the probability that the same state is selected in the future is increased. This might be compared to a "Chinese restaurant" where customers select with higher probability a table that is already occupied by customers, or the Pólya urn where, if one draws a ball with a certain color, more than one ball with the same color is replaced and thus the probability of picking the same color in the future is increased.

From Equation 6 it is clear that if the precision parameter τ is large, many samples are generated independently from the base distribution α but if τ is small, the first few samples quickly dominate the sampling procedure and the subsequently generated samples are quite clustered (see Figure 3).

4.3   Gibbs Sampling for Dirichlet Enhanced HB

We now return to the Dirichlet enhanced HB model from Equation 5 where for each setting j we have access to the data sets D_j with likelihood functions $P(D_j|\Theta_j = \theta_k)$. We will discuss two approaches for parameter inference in the HB setting. In this subsection we introduce Gibbs sampling, which is particularly attractive if K is large. Readers, not familiar with Gibbs sampling should

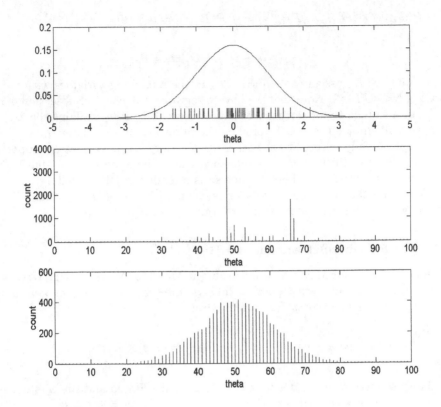

Fig. 3. Top: Subjective non-informative prior (Gaussian) and samples generated from this prior. These samples can be used for Dirichlet enhancement. Center: Samples from a distribution that was generated by a Dirichlet distribution with a Gaussian base distribution with precision $\tau = 10$. Clustering is quite apparent. Although the positions of the samples represent the base distribution, the counts are neither uniform nor follow the base distribution. Counts reflect the Pólya urn process (Section 4.2) or, equivalently, the stick breaking process (Section 5.1). Thus, that the ragged structure is *not* a result of a finite sample size —100000 samples were drawn— but is an inherent property of a distribution generated by a Dirichlet distribution, resp. Dirichlet process. Bottom: Same, but with $\tau = 10000$. With a large precision parameter, samples are drawn predominantly independently from the base distribution. Again, 100000 samples were drawn.

consult [13]. The second approach is an EM solution, which is quite effective for smaller K and will be discussed in the next subsection.

Based on Equation 5 we can derive the conditional distribution of a variable of interest, say Θ_j, given samples from the remaining variables and given the data sets as

$$P(\Theta_j = \theta_k | \{\Theta_l\}_{l \neq j}, \{D_l\}_{l=1}^M) = \frac{1}{C} P(D_j | \Theta_j = \theta_k) \, P(\Theta_j = \theta_k | \{\Theta_l\}_{l \neq j})$$

$$= \frac{1}{C}(\tau \alpha_k + N_k) P(D_j | \Theta_j = \theta_k) \tag{7}$$

where we have N_k assignments of $\Theta_l = \theta_k$ in the remaining variables with $l \neq j$ and $\sum_k N_k = M - 1$. Note that we have integrated out g as in Subsection 4.2.

Thus a sample θ_k for setting j becomes more likely, if θ_k explains the D_j-th data set well and if either it is favored by the prior distribution (large α_k) or if θ_k is a sample already selected by the other models (large N_k). This latter property, that samples for different models influence each other, results in a sharing of information between the different models, as intended in HB modelling.

Note that the representation is upper limited by $\min(M, K)$, thus Gibbs sampling is particularly interesting for large K, i.e., if $K \gg M$.

4.4   EM for Dirichlet Enhanced HB

We now discuss the EM solution to learning in Dirichlet enhanced HB. Here, we treat $\{\Theta_j\}_{j=1}^M$ as unknown variables, that we integrate out, and the goal is to find the MAP estimate of g that is defined as

$$g^{(MAP)} := \arg\max_g P(g) \prod_{j=1}^M \sum_{k=1}^K P(\Theta_j = \theta_k | g) P(D_j | \Theta_j = \theta_k).$$

The EM algorithm iterates for $t = 0, 1, 2, \ldots$ the E-step and the M-step. At iteration t, the E-step estimates [15], for $k = 1, \ldots, M, m = 1, \ldots, K$

$$\hat{P}^{(t)}(\Theta_k = \theta_m | D_k) = \frac{\hat{P}^{(t)}(D_k | \Theta_k = \theta_m) \hat{P}^{(t)}(\Theta_k = \theta_m)}{\sum_{l=1}^K \hat{P}^{(t)}(D_k | \Theta_k = \theta_l) \hat{P}^{(t)}(\Theta_k = \theta_l)} \tag{8}$$

The M-step updates for $k = 1, \ldots, M, m = 1, \ldots, K$

$$\hat{P}^{(t+1)}(\Theta_k = \theta_m) = \hat{g}_m^{t+1}$$

with

$$\hat{g}_m^{t+1} = \frac{1}{\tau + M}\left(\tau \alpha_m + \sum_{j=1}^M \hat{P}^{(t)}(\Theta_j = \theta_m | D_j)\right).$$

After convergence, the prediction of an active model $a \in 1, \ldots, M$ becomes

$$P(Y_a = y | x, \{D_j\}_{j=1}^M) \approx$$

$$\frac{\sum_{k=1}^K \hat{P}(\Theta_a = \theta_k) P(D_a | \Theta_a = \theta_k) P(Y_a = y | x, \Theta_a = \theta_k)}{\sum_{k=1}^K \hat{P}(\Theta_a = \theta_k) P(D_a | \Theta_a = \theta_k)}. \tag{9}$$

Note that this solution is similar to the heuristically motivated solution of Equation 2 in the sense that predictions of the models are weighted by the probability with which those models explain the data set of the active model. The differences are that first, we have an additional weighting constant $\hat{P}(\Theta_a = \theta_k)$ that evaluates the overall relevance of a model and second, we assumed here that the parameters were generated rather unspecifically from the base distribution whereas in the heuristic solution they correspond to maximum likelihood estimates.

5 Hierarchical Bayesian Modelling with Infinite Models

Dirichlet enhancement is of particular importance if we let $K \to \infty$, which is the case we consider in this section.

The transition $K \to \infty$ leads us to nonparametric HB, where, as in the finite-dimensional case, the $\theta_k, k = 1, \ldots$, are sampled randomly from the base distribution. In this context we need to first introduce some properties of the Dirichlet process, which is a generalization of the Dirichlet distribution to infinite dimensions.

5.1 Dirichlet Process

The Dirichlet process (DP) is of central importance in nonparametric Bayesian modelling. A formal definition can be found in the appendix. A DP is written as $DP(G_0, \tau)$ where G_0 is the base distribution with probability density g_0 that corresponds to the α_j in the finite-dimensional case; $\tau \geq 0$ is the concentration parameter. Please, compare this definition to the definition of a Dirichlet distribution in Section 4.1. [6] As in the case of the Dirichlet distribution we can use the Pólya urn representation to sample from a distribution generated by a Dirichlet process. Given previous samples $\{\theta_l\}_{l=1}^{j-1}$ generated from a distribution generated from a DP with base distribution g_0 and precision τ, the $j-th$ sample is generated from the probability density

$$P(\theta_j | \{\theta_l\}_{l=1}^{j-1}) = \frac{\tau g_0(\theta_j) + \sum_{k=1}^{j-1} \delta_{\theta_k}}{\tau + j - 1}. \tag{10}$$

Note that this formula is a direct generalization of the finite-dimensional case, Equation 6. Samples are generated with probability proportional to τ from the base distribution and with increasing probability proportional to $j - 1$ from an already existing sample. Thus, for small τ we observe the same clustering effect as in the finite dimensional case (Figure 3). A mathematical treatment of nonparametric Bayesian modelling and the Dirichlet processes can be found in [14].

Equation 10 specifies how samples are generated from a distribution that is a sample from a DP. It is also possible to generate directly a sample from such a distribution by using the so-called stick breaking process (for a definition consult [26] or [27]) according to which this distribution can be written as an

---

[6] In the literature one often finds the notation α_0 for the concentration parameter.

infinite sum of weighted delta functions placed at samples randomly selected from the base distribution,

$$g = \sum_{k=1}^{\infty} \beta_k \delta_{\theta_k}. \qquad (11)$$

The $\beta_k \geq 0$ and with $\sum_k \beta_k = 1$ only depend on τ and are generated by the stick breaking process, which is based on a sequence of independent beta random variables. Note that even if the base distribution G_0 is smooth, a sample distribution is discrete in nature.

5.2   Nonparametric Bayesian Modelling for Dirichlet Enhanced HB

In the context of an infinite model, i.e. a DP, the HB model of Equation 5 is called a Dirichlet process mixture model. The conditional probability distribution required for Gibbs sampling becomes [10] [32]

$$P(\theta_j | \{\theta_l\}_{l \neq j}, D_j) = \frac{1}{C} \left(\tau g_0(\theta_j) + \sum_{l:l \neq j} \delta_{\theta_l} \right) P(D_j | \theta_j)$$

$$= \frac{1}{C} \left(\tau \tilde{P}(D_j) \tilde{P}(\theta_j | D_j) + \sum_{l:l \neq j} \delta_{\theta_l} P(D_j | \theta_l) \right)$$

where

$$\tilde{P}(D_j) := \int P(D_j | \theta) g_0(\theta) d\theta, \quad \tilde{P}(\theta_j | D_j) := P(D_j | \theta_j) g_0(\theta_j) / \tilde{P}(D_j),$$

and where $\{\theta_l\}_{l \neq j}$ are the values of the remaining models. Note that this is a direct generalization of Equations 7. With probability proportional to $\tau \tilde{P}(D_j)$ a sample is generated from $\tilde{P}(\theta_j | D_j)$ and with probability proportional to $P(D_j | \theta_l)$ we take an existing sample θ_l. Note that our notation hides the fact that several θ_l might be identical, increasing the selection probability accordingly. This parameter clustering is particularly strong if τ is small in which case the number of distinct parameters is typically much smaller than M. Note also that, despite the fact that we are considering infinite models, computational load per round and memory requirements only grow proportional to M. This semi-automated determination of the number of distinct models is an important feature and was the focus of some recent work (see Section 7) but is not of central interest in the HB framework presented here.

The presented Gibbs sampling approach was introduced by Escobar [9]. Since in Gibbs sampling only one parameter is re-sampled at a time, the clustering of the parameters makes it difficult for the sampling procedure to modify parameter values. In the appendix we describe a mixture of models approach introduces by MacEachern [19] that turns out to be equivalent to the presented model. Gibbs sampling based on that model exhibits much better mixing properties.

The blocked Gibbs sampler that is based on a finite stick-breaking prior provides another computational attractive sampling procedure [18]. A comprehensive overview of sampling techniques for Dirichlet process mixture models can be found in [21].

5.3 Variational EM

In a nonparametric setting our EM equations from Subsection 4.4 cannot directly be applied since a distribution generated by a Dirichlet process is infinite-dimensional. In [29] the authors discuss a one-step EM solution. Here, we discuss an EM solution that can be derived from a variational approximation that approximates probability densities of the E-step in Equation 8 by a simpler approximating density [30]. We propose a sum of weighted delta functions defined at the maximum likelihood estimates of the models, i.e.,

$$\hat{P}(\theta_j | D_j) \approx q_j(\theta_j) = \sum_{k=1}^{M} \xi_{j,k} \delta_{\theta_k^{ML}} \quad j = 1, \ldots, M \tag{12}$$

where $\xi_{j,k}$ are the variational parameters with $\xi_{j,k} \geq 0$ and $\sum_{k=1}^{M} \xi_{j,k} = 1$. In each variational E-step, the variational parameters are adapted such that KL-divergence between the variational approximation and $\hat{P}^{(t)}(\theta_j | D_j)$ is minimized.

As a generalization to the finite-dimensional case we propose as update equations for $t = 1, 2, \ldots$:

$$\xi_{j,k}^t = \frac{P(D_j | \theta_k^{ML}) \hat{P}^{(t)}(\theta_k^{ML})}{\sum_{k=1}^{M} P(D_j | \theta_k^{ML}) \hat{P}^{(t)}(\theta_k^{ML})} \quad j = 1, \ldots, M \quad k = 1, \ldots, M. \tag{13}$$

The M-step updates

$$\hat{P}^{(t+1)}(\theta_k^{ML}) = \frac{1}{\tau + M} \left(\tau g_0(\theta_k^{ML}) + \sum_{l=1}^{M} \xi_l \delta_{\theta_l^{ML}} \right) \quad k = 1, \ldots, M$$

with $\xi_l = \sum_{j=1}^{M} \xi_{j,l}$.

Note that the EM iterations are quite simple since many terms, such as $P(D_j | \theta_k^{ML})$, don't change in the iterations. Also note the similarity to the finite-dimensional case in Section 4.4.

Now, the prediction of an active model $a \in 1, \ldots, M$ becomes

$$P(Y_a = y | x, \{D_j\}_{j=1}^{M}) \approx$$

$$\frac{\tau \tilde{P}(D_a) \tilde{P}(Y_a = y | x, D_a) + \sum_{k=1}^{M} \xi_k P(D_a | \theta_k^{ML}) P(Y_a = y | x, \theta_k^{ML})}{\tau \tilde{P}(D_a) + \sum_{k=1}^{M} \xi_k P(D_a | \theta_k^{ML})} \tag{14}$$

where we use

$$\tilde{P}(D_a) := \int g_0(\theta)P(D_a|\theta)\,d\theta$$

$$\tilde{P}(Y_a = y|x, D_a) := \frac{1}{\tilde{P}(D_a)}\int g_0(\theta)P(D_a|\theta)P(Y_a = y|x, \theta)\,d\theta.$$

Note the great similarity of this prediction equation to the prediction equation for the finite dimensional case (Equation 9) and the heuristically defined solution of Equation 2: the second term in the numerator in Equation 14 contains the model predictions using maximum likelihood parameter estimates, weighted by the probability that models agrees with the data set of the active model $P(D_a|\theta_k^{ML})$. Here, additional relevance weights ξ_k are included, which represent the overall relevance of the models. If we look at Equation 13, it becomes clear that the contribution of the j-th setting to the relevance weight ξ_k is essentially determined by the term $P(D_j|\theta_k^{ML})$ which means that a setting j which has received a small number of data points contributes to all ξ_k, whereas a setting j which receives a large number of data points will mostly contribute to ξ_j. In our experiments we found that the weight of a model prediction in Equation 14 is mostly determined by the term $P(D_a|\theta_k^{ML})$ and that the ξ_k are more or less of the same magnitude and thus have only a minor influence. Thus in many applications one might refrain from the fitting of the variational parameters $\xi_{j,k}$ and use Equation 14 with $\xi_k = 1, k = 1, \ldots, M$.

The first term in the numerator of Equation 14 puts additional weight on the prediction of the active model. In particular, it consists of the Bayesian prediction of the active model a based on its own data $\tilde{P}(Y_a = y|x, D_a)$ weighted by τ and the evidence of the data of the active model $\tilde{P}(D_a)$. The latter term evaluates the correctness of the prior modelling assumption.

Equation 14 is equivalent to the Bayesian prediction of the active model if we use a prior proportional to

$$\tau g_0(\theta) + \sum_{k=1}^{M} \xi_k \delta_{\theta_k^{ML}}$$

which illustrates the similarity of this solution to the heuristically defined solution of of Equation 2, in particular with $\tau \to 0$. With $\tau \to \infty$ the prediction of the active model is simply the prediction of the active model trained on its own data, i.e., all models are independent and only rely on their own data. With a finite τ we obtain the HB solution.

If compared to the Gibbs sampling approach, the variational EM solution has several important advantages. Each model can be trained independently from the other models just based on its own data set. Thus the solution is easy to train, modular, and efficient and an additional model can easily be incorporated. From a theoretical perspective, Gibbs sampling might be more appealing but one should note its typically slow convergence and the slow mixing of the sampling process in practice.

An advantage of the sampling approach is that it leads to an automated clustering of the models, a feature that is not achieved in the variational EM solution. On the other hand, if such a model clustering is of prime importance, one can achieve it, for example, by a corresponding postprocessing step.

The variational approximation of Equation 12 uses maximum likelihood parameter estimates. This has the advantage that asymptotically, with an increasingly large number of data points for the active model, the overall prediction converges to the prediction of the active model. The same feature can be achieved if the variational approximation is based on the maximum a posteriori (MAP) parameter estimates of the models. The MAP estimate is more appropriate if only few training data points are available. Alternatively, one could select for the variational approximation sets of samples obtained from the posterior parameter distributions for each model.

6   A Recommendation Engine

In this section we provide a summary of the application of nonparametric hierarchical Bayesian modelling to information filtering. A more detailed description can be found in [30].

Information filtering denotes a family of techniques that try to understand people's information needs, and then help them find the right information items while filtering out undesired ones. In a very wide range of applications, such as spam email filtering, news filtering, recommender systems for products (e.g., books, movies, CDs), and web navigation, information filtering is playing an increasingly important role. Content-based filtering (CBF) and collaborative filtering (CF) represent the two major information filtering technologies.

CBF has its root in the concept of *relevance feedback* in the information retrieval literature (e.g., Rocchio's algorithm [25]). It explores the similarity of contents between information items (e.g., articles, paintings, music), to infer which of the yet unseen items might be of interest for the active user, based on some annotated examples previously given by the user. In contrast, collaborative filtering methods typically accumulate a database of item ratings—explicitly or implicitly—cast by a large set of users. The prediction of ratings for the active user is solely based on the ratings provided by all other users, under the assumption that like-minded users share similar information needs. The method does not rely on a description of the item's content.

It is often difficult in CBF systems to define content features that are sufficiently indicative. There is often a large gap between low-level content features (visual, auditory, or others) and high-level user interests (like or dislike a painting or a CD). In some other circumstances, the features are not available at all.

On the other hand, pure CF only relies on user preferences, without incorporating the actual content of items. CF often suffers from the extreme sparsity of the available data set, in the sense that users typically rate only very few items, thus making it difficult to compare the interests of two users. Furthermore, pure

CF can not handle items for which no user has previously given a rating. Such cases are easily handled in CBF systems, which can make predictions based on the content of the new item.

We combine CF and CBF under the framework of nonparametric hierarchical Bayesian modelling which leads to a model that combines the advantages of both approaches. Essentially a CBF model is formed for every user and the predictions are combined using the nonparametric HB approach using variational EM as described in Section 5.3.

In our application, we focus on a survey of 642 paintings of 30 artists. A web-based online survey is built to gather user ratings. In the survey, each user gave ratings, i.e., "like", "dislike", or "not sure", to a randomly selected set of paintings. Finally we got a total of $N = 190$ users' ratings. On average, each of them had rated 89 paintings.

For each painting, we calculate the *color histogram* (216-dim.), the *correlagram* (256-dim.), the *first and second color moments* (9-dim.) and the *pyramid wavelet texture* (10-dim.) to form a 491-dimensional feature vector.

We will examine the performance of various algorithms in terms of their accuracy in predicting users' interests in paintings. We used as our base user models a probabilistic version of the support vector machine (SVM) [22] with Gaussian kernels. *Hybrid filtering 1* implements the nonparametric HB approach using variational EM as described in Section 5.3; *Hybrid filtering 2* is identical, except that we set $\tau = 0$; for *SVM Content-Based filtering* (CBF) we use $\tau \to \infty$ and obtain independent user models; *Collaborative filtering* (CF) combines a society of advisory users' preferences to predict an active user's preferences. The combination is weighted by the *Pearson correlation* between the active user and the other advisory users' preferences. The algorithm applied here is described in [7].

These algorithms are evaluated using two metrics. One is *Top-L accuracy*, i.e., the proportion of truly liked paintings among L top ranked paintings. Since normal users only care about the quality of the first set of returned items, this quantity reflects the *subjective* quality of an information filter system. Secondly, we evaluated the ROC (receiver operating characteristics) curve, which plots *sensitivity* versus *1-specificity*. Sensitivity is defined as the probability that a good painting is recommended by the system; and specificity is the probability that a disliked painting is rejected by the system. By changing the cut point (e.g., return top 10 or 20 paintings), a curve can be plotted. ROC curve is insensitive to the prior distribution of liked (or disliked) paintings. The area under the curve, called *ROC sensitivity*, measures the *objective* quality of the ranking. A higher ROC sensitivity indicates a better ranking.

In the application it was not required that a user rates all of the 642 paintings in the survey; thus for each user we only partially know the "ground truth" of preferences. As a result, the true top-L accuracy cannot be computed. We thus adopt as accuracy measure the fraction of *known* liked paintings in the top ranked L paintings. The quantity is smaller than true accuracy because *unknown* liked paintings are missing in the measurement. However, in our survey, the

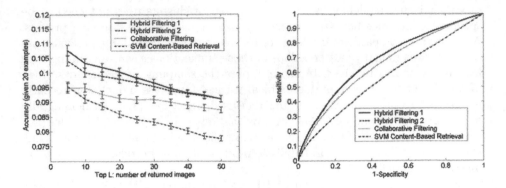

Fig. 4. Left: Top-L accuracy. Right: ROC curves. From [30].

presentation of paintings to users is completely random, thus the distributions of rated/unrated paintings in both unranked and ranked lists are also random. This randomness dose not change the relative values of the compared methods. Thus in the evaluation of the experiment it still makes sense to use the adopted accuracy measurement to compare the three retrieval methods. The ROC curves are insensitive to this problem.

In our experiments, we used a 10-fold cross validation scheme, in which we pick up each fold as a set of active users and treat the rest as users in the data base. We fix the number of given examples for each active user to be 20 (10 positive and 10 negative), and predict the user's interests in the remaining paintings. For each active user, recommendations for 10 different paintings are calculated. Finally, the overall average performances and error bars are computed. 4 shows the results. Both Top-L accuracy and ROC curve clearly indicate that the two hybrid algorithms outperform CF and CBF. We found that the extracted painting features are poor indicators of human interests, which is the reason for the bad performance of CBF. The ROC curves of the two hybrid filtering algorithms are essentially overlapping. However, Top-L accuracy suggests that hybrid filtering 1 is slightly better.

7 Related Work in Machine Learning

Dirichlet process mixture models were introduced into machine learning by Neal [20] [21] who used them to realize infinite mixture models. Dirichlet processes were applied to realize infinite mixtures of Gaussians [23], infinite mixtures of Gaussian experts [24] and infinite hidden Markov models [3]. These models are also based on nonparametric HB modelling but the application focus is different: In these papers, there are no repeated measurements for a given model (i.e., in the plate model of Figure 2D, $N = 1$) and the focus is on model-based soft clustering using an infinite mixture approach and on the realization of an infinite mixture of experts solution. An inherent advantage of Dirichlet process mixture modelling is that the number of clusters does not need to be specified in advance but is automatically determined via the sampling process. A small precision parameter τ leads to few clusters whereas a large precision parameter leads to many clusters. Thus in those applications a sensible tuning —or learning— of τ is required. In those papers the sampling procedure described in the appendix is used. A hierarchical Dirichlet process model was recently introduced to model hierarchical unsupervised structures [27]. Mathematically demanding variational mean-field approximations were applied to Dirichlet processes in [6] and [31]. Some of the work on the development of self-organizing maps for the clustering of probabilistic models can also be related to nonparametric HB modelling [17].

Examples of the application of HB to machine learning are probabilistic clustering [8], the finite-dimensional HB approach by [4] [5] who used HB in the context of a model for latent semantic analysis and information retrieval and the application of neural networks models to HB [16] [2].

8 Conclusions

Nonparametric hierarchical Bayesian modelling is a powerful and flexible approach for multi-agent learning if agents need to share learned knowledge. We introduced the basic background and the common inference approach via Gibbs sampling. We described a variational EM solution that leads to excellent results in a multi-agent learning framework. The main advantages of the EM solution are its modularity, low computational complexity, intuitive plausibility and good performance. Many variants of nonparametric hierarchical Bayesian modelling have been used in the literature with various combinations of model specific parameters, shared parameters and Dirichlet enhanced distributions and with varying levels of hierarchies (see, for example, [28] and [27]). Thus nonparametric hierarchical Bayesian modelling is quite flexible and might find an increasing number of applications in multi-agent learning.

9 Appendix

9.1 Definition of a Dirichlet Process

The theorem asserts the existence of a Dirichlet process and also serves as a definition [14]. Let $(\mathbb{R}, \mathcal{B})$ be the real line with the Borel σ-algebra $\mathcal{B}$ and let $M(\mathbb{R})$ bet the set of probability measures on $\mathbb{R}$, equipped with the σ-algebra $\mathcal{B}_M$.

Theorem 1 *Let α be a finite measure on $(\mathbb{R}, \mathcal{B})$. Then there exists a unique probability measure D_α on $M(\mathbb{R})$ called the Dirichlet process with parameters α satisfying:*
For every partition $B_1, B_2, ..., B_k$ of $\mathbb{R}$ by Borel sets
$(P(B_1), P(B_2), \ldots, P(B_k))$ *is* $Dir(\alpha(B_1), \alpha((B_2), \ldots, \alpha((B_k))$

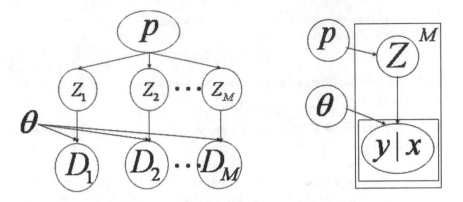

Fig. 5. Left: The mixture model. Right: The plate model. Note that in contrast to the HB model, all parameters are global parameters.

9.2   Equivalence of Dirichlet Enhanced HB to Mixture Models

Finite Mixture Models: In Section 4 we had concluded that the prior distribution in HB must be made flexible and we introduced the process of Dirichlet enhancement. Thus we obtained a highly flexible prior distribution that permitted the sharing of knowledge between models. An alternative formulation is the mixture of models approach presented here (see Figure 5).

The predictive model of the mixture model is

$$P(Y = y|x) = \sum_{k=1}^{K} P(Z = k)P(Y = y|x, k)$$

where Z is a latent variable with states $1, \ldots, K$. It is now uncertain which model generated the data for the active setting such that

$$P(D_a, Y_a = y|x) = \sum_{k=1}^{K} P(Z = k)P(D_a|k)P(Y_a = y|x, k).$$

To classify a new pattern, we thus obtain

$$P(Y_a = y|x, D_a) = \frac{\sum_{k=1}^{K} P(Z = k)P(D_a|k)P(Y_a = y|x, k)}{\sum_{k=1}^{K} P(Z = k)P(D_a|k)}.$$

Please, note the similarity of this equation to Equation 9 that deals with the finite-dimensional Dirichlet enhanced case.

It now turns out that there is an exact equivalence with the finite-dimensional Dirichlet enhanced model in Section 4 if:

– the likelihood models for HB and the mixture approach are identical

$$P(Y_a = y|x, k) = P(Y_a = y|x, \theta_k),$$

- the same parameter vectors $\{\theta_k\}_{k=1}^K$ are selected,
- the prior for Z is a multinomial,

$$P(Z = k) = p_k$$

which is generated from a Dirichlet distribution with

$$p_1, \ldots, p_K \sim \mathrm{Dir}(\tau\alpha_1, \ldots, \tau\alpha_K).$$

Details can be found in [21] and the graphical model and plate model are shown in Figure 5.

Infinite Mixture Models: It turns out that the equivalence also holds if we let $K \to \infty$ in which case we obtain an infinite mixture model, which is equivalent to a Dirichlet process mixture model (Section 5), if we chose a prior parameter distribution as the base distribution

$$P(\theta_k) = g_0(\theta_k) \quad \forall k,$$

and with

$$p_1, \ldots, p_K \sim \mathrm{Dir}(\tau/K, \ldots, \tau/K).$$

Stochastic sampling based on this model can be implemented as follows [21]: One first updates the latent variables $\{Z_j\}_{j=1}^M$. Let consider the update of Z_j, which denotes the latent variable which is associated with the j-th model (Figure 5). As in nonparametric HB, a new sample depends on the states of the latent variables of the remaining variables which might also be clustered. Let N_k be the number of variables in the set $\{Z_l\}_{l=1}^M$, which are in state k, *without counting the state of* Z_j, i.e., $\sum_k N_k = M - 1$.

Then for all states with $N_k > 0$

$$P(Z_j = k | \{Z_l\}_{l \neq j}, D_j, \theta) \propto N_k P(D_j | \theta_k).$$

A new state is generated with probability

$$P(Z_j \neq k \text{ for all } k \neq j | \{Z_l\}_{l \neq j}, D_i, \theta) \propto \frac{1}{C} \tau \tilde{P}(D_j)$$

with $\tilde{P}(D_j) := \int g_0(\theta) P(D_j | \theta) \, d\theta$. In the first case, the j-th model inherits the parameters of the models assigned to state k and in the latter case, a new θ is drawn from $P(\theta | D_j)$.

Typically after one update of all latent variables, the model parameters are all updated. E.g., for all models in state k, a new θ_k is drawn from

$$\frac{1}{C} g_0(\theta_k) \prod_{\{j : Z_j = k\}} P(D_j | \theta_k).$$

The advantage of this sampling scheme is that at each round all parameters are re-sampled and typically assume new values whereas in the sampling schemes described in Section 5.2 it is rather unlikely that clustered parameters will assume new values since only one parameter is re-estimated at a time.

Neal [21] discusses additional advanced sampling techniques.

References

[1] Antoniak, C. E.: Mixtures of Dirichlet Processes with Applications to Bayesian Nonparametric Problems. Annals of Statistics **2** (1974) 1152-1174

[2] Bakker, B., Heskes, T.: Task Clustering and Gating for Bayesian Multitask Learning. Journal of Machine Learning Research, **4** (2003)

[3] Beal, M. J., Ghahramani, Z., Rasmussen, C. E.: The Infinite Hidden Markov Model. Advances in Neural Information Processing Systems **14** (2002)

[4] Blei, D. M., Ng, A. Y., Jordan, M. I.: Latent Dirichlet Allocation. Journal of Machine Learning Research **3** (2003)

[5] Blei, D. M., Jordan, M. I., Ng, A. Y. : Hierarchical Bayesian Modelling for Applications in Information Retrieval. Bayesian Statistics **7**. Oxford University Press (2003)

[6] Blei, D. M., Jordan, M. I.: Variational methods for the Dirichlet process. To appear in Proceedings of the 21st International Conference on Machine Learning (2004)

[7] Breese, J. S, Heckerman, D., Kadie, C.: Empirical Analysis of Predictive Algorithms for Collaborative Filtering. Proceedings of the 14th Conference on Uncertainty in Artificial Intelligence (1998)

[8] Cadez, I., Smyth, P.: Probabilistic Clustering using Hierarchical Models. TR No 99-16, Dept. of Information and Computer Science. University of California, Irvine (1999)

[9] Escobar, M. D.: Estimating the Means of Several Normal Populations by Nonparametric Estimation of the Distribution of the Means. Unpublished PhD dissertation, Yale University (1988)

[10] Escobar, M. D., West, M.: Computing Bayesian Nonparametric Hierarchical Models. Practical Nonparametric and Semiparametric Bayesian Statistics, D. Dey, P. Müller, D. Sinha (eds.), Springer (1998)

[11] Ferguson, T. S.: A Bayesian Analysis of some Nonparametric Problems. Annals of Statistics **1** (1973) 209-230

[12] Gelman, A., Carlin, J. B., Stern, H. S., Rubin, D. B.: Bayesian Data Analysis. CRC press (2003)

[13] Gilks, W. R., Richardson, S., Spiegelhalter, D. J.: Markov Chain Monte Carlo in Practice. CRC press (1995)

[14] Gosh, J. K, Ramamoorthi, R. V.: Bayesian Nonparametrics. Springer Series in Statistics (2002)

[15] Heckerman, D.: A Tutorial on Learning with Bayesian Networks. Technical report MSR-TR-95-06 of Microsoft Research (1995)

[16] Heskes, T.: Empirical Bayes for Learning to Learn. In Proc. 17th International Conf. on Machine Learning, Morgan Kaufmann, San Francisco, CA (2000) 367–374

[17] Holmen, J., Tresp, V., Simula, O.: A Self-Organizing Map for Clustering Probabilistic Models. Proceedings of the Ninth International Conference on Artificial Neural Networks (ICANN'99) **2** (1999)

[18] Ishwaran, H., James, L. F. : Gibbs Sampling Methods for Stick-Breaking Priors. Journal of the American Statistical Association, Vol. 96, No. 453 (2001)

[19] MacEachern, S. M.: Estimating Normal Means with Conjugate Style Dirichlet Process Prior. Technical report No. 487, Department of Statistics, The Ohio State University (1992)

[20] Neal, R, M.: Bayesian Mixture Modeling by Monte Carlo Simulation. Techni-
    cal Teport No. DCR-TR-91-2, Department of Computer Science, University of
    Toronto (1991)
[21] Neal, R, M.: Markov Chain Sampling Methdos for Dirichlet Process Mixture Mod-
    els. Technical report No. 9815, Department of Statistics, University of Toronto
    (1998)
[22] Platt, J. C.: Probabilities for SV machines. In Advances in Large Margin Classi-
    fiers. MIT Press (1999)
[23] Rasmussen, C. E.: The Infinite Gaussian Mixture Model. Advances in Neural
    Information Processing Systems **12** (2000)
[24] Rasmussen, C. E., Ghahramani, Z.: Infinite Mixtures of Gaussian Process Experts.
    Advances in Neural Information Processing Systems **14** (2002)
[25] Rocchio, J. J.: Relevance Feedback in Information Retrieval. The SMART Re-
    trieval System: Experiments in Automatic Document Processing, Prentice Hall
    (1971)
[26] Sethuraman, J.: A Constructive definition of Dirichlet Priors. Statistica Sinica **4**
    (1994)
[27] Teh, Y. W. , Jordan, M. I., Beal, M. J., Blei, D. M.: Hierarchical Dirichlet Proceses.
    Technical Report 653, UC Berkeley Statistics (2004)
[28] Tomlinson, G., Escobar, M.: Analysis of Densities. Talk given at the Joint Statis-
    tical Meeting (2003)
[29] Yu, K., Schwaighofer, A., Tresp, V., Ma, W.-Y., Zhang, H.: Collaborative Ensem-
    ble Learning: Combining Collaborative and Content-Based Information Filtering
    via Hierarchical Bayes. Proceedings of the 19th Conference on Uncertainty in
    Artificial Intelligence (UAI) **19** (2003)
[30] Yu, K., Tresp, V., Yu S.: A Nonparametric Bayesian Framework for Informa-
    tion Filtering. In the proceedings of the 27th Annual International ACM SIGIR
    Conference (2004)
[31] Yu, K., Yu S., Tresp, V. : Dirichlet Enhanced Latent Semantic Analysis. In the pro-
    ceedings of the 10th International Workshop on Artificial Intelligence and Statis-
    tics (2005)
[32] West, M., Müller, P., Escobar, M. D.: Hierarchical Priors and Mixture Models,
    with Application in Regression and Density Estimation. Aspects of Uncertainty:
    A Tribute to D. V. Lindley, A.F.M. Smith and P. Freeman, (eds.), Wiley New
    York (1994) 363–386

Simultaneous Localization and Surveying with Multiple Agents

Sam T. Roweis and Ruslan R. Salakhutdinov

Department of Computer Science, University of Toronto
Toronto, Ontario, M5S3G4, CANADA,
{roweis,rsalakhu}@cs.toronto.edu

Abstract. We apply a constrained Hidden Markov Model architecture to the problem of simultaneous localization and surveying from sensor logs of mobile agents navigating in unknown environments. We show the solution of this problem for the case of one robot and extend our model to the more interesting case of multiple agents, that interact with each other through proximity sensors. Since exact learning in this case becomes exponentially expensive, we develop an approximate method for inference using loopy belief propagation and apply it to the localization and surveying problem with multiple interacting robots. In support of our analysis, we report experimental results showing that with the same amount of data, approximate learning with the interaction signals outperforms exact learning ignoring interactions.

1 Introduction

In the following, we study the problem of analyzing sensor logs created by mobile agents navigating in unknown environments. We assume that the environment is static, so that any variation in the sensors is caused by the movement of the agents in the world. Our goal is to develop algorithms for localization when the environment is known and also for simultaneous localization and identification of the environment, which we dub "surveying". We also consider the situation of multiple agents that have limited but nontrivial interaction as they explore. All of these problems can be cast as statistical estimation computations and approached using techniques of probabilistic inference.

Simultaneous localization and surveying (SLAS) is distinct from the well known simultaneous localization and mapping (SLAM) problem. In SLAS we are not trying to learn the occupancy grid of a world, rather we are trying to learn the values that various sensors (e.g. altitude, temperature, light level, beacon signals) take on as a function of position in the unknown environment. Furthermore, we cannot control the movement of the agent, we can only analyze the sensor logs recorded as it traverses the world. This task is motivated by agents (for example mobile planetary rovers) which generally operate in open spaces, collect temporal histories of multiple sensors, and cannot rely on the odometry of self-locomotion (e.g. because they are navigating extremely rough terrain or not using conventional wheels).

R. Murray-Smith, R. Shorten (Eds.): Switching and Learning, LNCS 3355, pp. 313–332, 2005.
© Springer-Verlag Berlin Heidelberg 2005

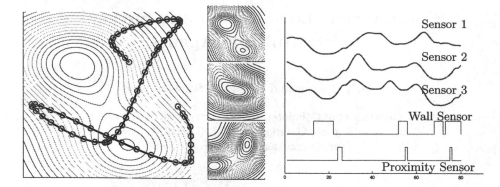

Fig. 1. Typical noisy sensor readings seen by a single agent. The first three sensors are continuous valued sensors; the fourth is a wall-detecting binary sensor. The proximity sensor indicates whether an agent is in the close proximity with other agents.

Figure 1 shows the typical input to a single agent (robot) in the scenario we are studying. Each robot moves through the environment under the control of an external navigation algorithm that we cannot influence. As it proceeds, it logs readings from multiple noisy sensors, some of which may be smoothly varying functions of its position in the world and others of which may be intermittent or discontinuous. No odometry or other information about navigational control signals (either intended or realized) is available to the agent. When there are multiple agents navigating the environment at the same time, and they interact in some way, the problem becomes much more rich. (Without interaction, the pooled data from all agents is almost identical to one longer data stream from a single agent.) In the following, we consider a very simple type of interaction: we assume that agents can detect each other when they are in close proximity. The most general problem we wish to solve is to simultaneously discover the trajectories taken by each robot (localization) and to learn the values of each sensor variable across the environment (surveying).

We approach the localization task as one of state inference in a probabilistic model and the surveying task as one of unsupervised learning of the model parameters. In the inference problem, the location of each agent over time is treated as a hidden variable that is to be inferred, conditioned on the observed sensor readings (and possibly proximity interactions with other agents). Inference on its own assumes we are given the model survey parameters, which specify the distributions of sensor values as a function of position across the space. The learning problem addresses the estimation of these model parameters (sensor map) given the agent locations over time. The two problems are interconnected, since to localize, an agent must know the sensor map, but to contribute to the learning of this map, it must have an estimate of location in the unknown environment.

One very effective way of tackling these problems is to discretize the world into small spatial cells and to identify each such cell with a discrete state in a dynamic Bayesian network such as a Hidden Markov Model (HMM) or a series of coupled HMMs. When the state is low-dimensional (e.g. containing only a two dimensional position and no orientation information), this discretization is expensive but possible. However, in higher dimensions the number of cells required scales prohibitively. In fact, this discretization is simultaneously the source of the algorithm's power and its greatest computational challenge.

Once discretized, we can treat the state as a latent variable and apply standard statistical learning methods for discrete state models. The key insight is that by identifying each state in the hidden Markov model with some small spatial region of the continuous world space, it is possible to naturally define "neighbouring" states as those which correspond to connected regions in the underlying space. The transition matrix of the HMM can then be *constrained* to allow transitions only between neighbors; this means that all valid state sequences correspond to connected paths in the continuous space. The transition matrix does not need to be explicitly stored or learned, it is merely computed by a function that respects the state topology; the remaining parameters of the model scale only linearly with the number of states[12].

For a single agent, or multiple non-interacting agents, the learning and inference algorithms are identical to those for standard HMMs trained on multiple observation sequences, except that the transition matrix of the HMM is fixed by the spatial topology of the problem and is not updated during learning.

Another way to address the problem is to represent the state of each agent using a distribution over the continuous domain of the random variables. Because the distribution on these variables can be highly complex and multimodal, the most natural representation is to store a set of (weighted) samples as an approximation of the distribution. The greatest advantage of this approach is that it does not require discretization of the entire state space and thus does not scale up prohibitively with the number of spatial cells. This general approach is known as "particle filtering", and the central technical challenge is how to update the particles to efficiently represent the state posterior.

In this work, we take the very simple approach of searching only for the mode of the true posterior, and use a single particle at each time to represent the state of the agent. Optimizing this set of particles to find the maximum a posteriori state trajectory is a very difficult search problem. We use the *embedded hidden Markov model* architecture discussed by [9, 10] as our search engine. (Although originally it was proposed as a more sophisticated method for drawing samples from the exact posterior.) The search begins by forming a pool of possible candidate states at each time, candidates, representing the agent's possible locations in the unknown environment. Given the pool, we can restrict the agent's state to only those values represented by candidates at each time, and thus define an "embedded HMM", whose discrete states are the indices within each pool time. By performing efficient Viterbi-style decoding, we can find the most probable trajectory of the agent, constrained to pass only through the existing pool states. If we always include the current best estimate of the state into the pool of candidates at each time, we are guaranteed to either find a new and improved (more probable) state sequence or to retain the same trajectory we currently have. After the dynamic programming, we create a new pool by randomly sampling states at each time in the vicinity of the current best state estimate.

The most difficult and interesting case we explore in this work is that of multiple agents that navigate through the environment simultaneously and interact in some way with each other. In this case, exact inference requires estimating the joint state of all agents, and quickly becomes exponentially expensive because the effective state space is the product of the state spaces of the individual robots. For the case of discretized world, we develop an approximate but efficient and accurate method for solving this multiple-inference problem using belief propagation (BP). We apply our BP algorithm to the localization and surveying problem with multiple interacting robots and show that approximate learning using multiple agents with interaction signals outperforms exact learning using the same amount of data but ignoring the interaction signals that make the problem more difficult.

2   Localization with a Single Robot

In this section we develop techniques to solve the localization problem for a single robot, which navigates in an unknown environment and records some observations (continuous or binary) from its sensors. We denote the observation at (discrete) time t from sensor c by y_{ct}, the entire vector of sensor readings at time t by y_t, and the unknown state (location) of the robot at time t by s_t.

The probabilistic graphical model that relates the empirical observation sequence $Y = \{y_1, ..., y_T\}$ to the hidden state sequence $S = \{s_1, ...s_T\}$ is shown in figure 2. This model specifies a factorization of the joint distribution between the trajectory and the observation sequence as:

$$p(Y, S) = \prod_t p(s_t|s_{t-1})p(y_t|s_t) \tag{1}$$

Our goal in localization is to find the optimal trajectory given a sequence of observations:

$$\arg\max_S \log P(S|Y) = \arg\max_S \sum_t \left[\log P(s_t|s_{t-1}) + \log P(y_y|s_t)\right] \tag{2}$$

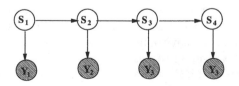

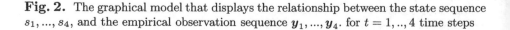

Fig. 2. The graphical model that displays the relationship between the state sequence $s_1, ..., s_4$, and the empirical observation sequence $y_1, ..., y_4$. for $t = 1, .., 4$ time steps

Later, when we are interested in identifying the parameters (survey maps) of the unknown world, our goal will be to maximize the likelihood of the observations given the (survey) parameters, integrating (summing) over all possible paths the robot could have taken:

$$\max_{\theta} \log P(Y|\theta) = \log \sum_{S} \sum_{t} \left[\log P(s_t|s_{t-1}, \theta) + \log P(\boldsymbol{y}_y|s_t, \theta) \right] \qquad (3)$$

As a byproduct of this parameter learning we will end up inferring the marginal posterior of the robot's position at each time, thus also performing a form of average localization. (It is also possible to use our inference about the single most probable (Viterbi) path of each agent to do a form of MAP learning, although in the discussion below we focus on maximum likelihood estimation which sums over all possible paths.)

2.1   Discretizing the World

Our first approach to solving the localization problem is to discretize the world into small spatial cells. By identifying each state in a hidden Markov model with some small spatial region of a continuous space, it is possible to naturally define neighboring states as those which correspond to connected regions in the underlying space, which leads us to the constrained HMM architecture [12]. The transition matrix of the HMM is precomputed to allow transitions only between neighbors; this means that all valid state sequences correspond to connected paths in the continuous space. The transition matrix does not need to be explicitly stored or learned, it is merely computed by a function that respects the state topology; the remaining parameters of the model scale only linearly with the number of states. Given these constraints, localization reduces to inference in this sparsely connected HMM, and can be solved using the well known Viterbi decoder.

To represent the world map of each continuous sensor c, we assume a conditional Gaussian model given the index of the discrete state: $P(y_{ct}|s_t = i) = \mathcal{N}(y_{ct}; m_i^c, \sigma_i^c)$. For binary sensors d we assume a simple Bernoulli model: $P(y_{dt} = 1|s_t = i) = m_i^d$. We assume that the noise in the sensor observations is uncorrelated from sensor to sensor and also over time (white).

An example of using this approach for localization (given knowledge of the sensor maps) is presented in figure 3 (left panel). Note that inference using this approach will always be only approximate due to the discretization error.

The discretization of the continuous state space can be very expensive and presents the model's greatest computational challenge. In higher dimensions or for very large areas requiring fine spatial resolution to reduce the discretization error, the number of cells required scales prohibitively. To alleviate this problem, we can represent the state of the robot using continuous random variables, as discussed in the following subsection. However, the optimization (search) problem of finding the best path given a history of sensor readings become extremely difficult.

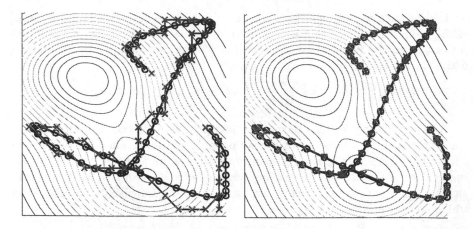

Fig. 3. Display of the discretized approach (left panel) and embedded HMM approach (right panel) to the robot localization problem, given the true map of sensor readings. The "o"s indicate the time slices when the true observation was taken, and "x"s indicate the inferred path taken by the robot.

2.2 Continuous Representation

To avoid the exponential cost of discretization as well as the unavoidable estimation error it imposes on localization, we can attempt to represent the trajectories of each agent using truly continuous random variables. However, two major difficulties arise when taking this approach: first, how should we represent a joint distribution over states at all times and second, how should we optimize this distribution given the sensor readings and the maps? These issues are central to the study of inference in all nonlinear dynamical systems and various approximation solutions have been proposed. Here, we follow the common programme of representing the distribution using sample trajectories, this approach is often called "particle filtering" or "Monte Carlo filtering".

We consider the robot's continuous trajectory to be a sequence of unknown positions $L = \{\ell_1, ..., \ell_T\}$ and our goal is to infer something about L given the empirical observation sequence $Y = \{y_1, ..., y_T\}$ and the known observation functions $p(y_c|\ell)$ for each sensor. In the most ambitious setting, we could attempt to infer the full joint posterior over L given Y, but as a first step we consider only finding the mode of L, i.e. the most likely trajectory given the observations.

For this search, we apply the embedded hidden Markov model[9, 10] as a simple optimization technique to solve the localization problem without the need to discretize the world. The optimization starts with some initial guess $\hat{L} = \{\hat{\ell}_1, ..., \hat{\ell}_T\}$ of the state trajectory, perhaps from a very coarse discretization or other approximation. At each step of the search, we create a pool of candidate states at each time. The pool contains K members, representing possible locations of the agent at each time step t. The candidate states within each pool are generated according to a proposal distribution $Q_t(\hat{\ell}_t)$. In our experiments we

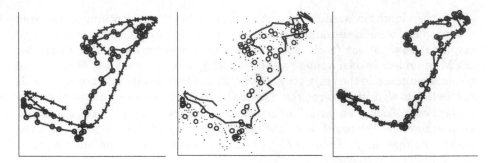

Fig. 4. Display of one iteration of the embedded HMM optimizer. Left panel shows part of the robot's true and currently inferred trajectory, where the "x"s indicate the time slices when the true observation was taken, and "o"s indicate the currently inferred path taken by the robot. The middle panel displays the candidate states (dots) within each pool along with the more probable path after performing Viterbi decoding. The right panel shows the true along with the newly inferred trajectory.

have used simple Gaussian proposals with the mean set to the current guess $\hat{\ell}_t$ and a fixed (isotropic) covariance equal at all time steps: $Q_t = \mathcal{N}(\hat{\ell}_t, \sigma I)$. This encourages the points placed into the pool to represent plausible alternatives to the current guess $\hat{\ell}_t$ about the state at time t. Crucially, we also include the current guess in the pool.

If we now constrain our search to only consider locations represented by pool states, the collection of pools across time define an "embedded HMM", whose $K+1$ states are the indices within each pool. By performing Viterbi decoding on this resulting embedded HMM, we can efficiently search through an exponential number of trajectories to find the best one. As well, we are guaranteed to always find a trajectory that increases $P(\hat{L}|Y)$ or leaves it the same, since the current guess at each time is always included in the pool. The optimization is repeated until several steps have passed with no change to our best trajectory estimate. The complete inference algorithm using embedded HMM is given below:

Localization Algorithm using Embedded HMM:

- Initialize $\hat{L} = \{\hat{\ell}_1, ..., \hat{\ell}_T\}$
- Repeat until a better trajectory cannot be found:
  - Form pools of candidate states for each time step t by:
    - Including the current state $\hat{\ell}_t$ at time t into the pool for time t
    - Sampling K other candidate states for each pool from $Q_t(\hat{\ell}_t)$
  - Define an embedded HMM, whose $K+1$ states are the indices within each pool
  - Perform Viterbi decoding to select the new trajectory $\hat{L}_{new}$ through the pool states that increases $P(L|Y)$ or leaves it the same.
  - Set $\hat{L} = \hat{L}_{new}$

Of course, the initial trajectory estimate greatly affects the quality of the final path returned by the search. If we were to randomly initializing the initial guess, the algorithm would often get stuck in a poor local optimum. To avoid this problem, we first subsample time by forming blocks of contiguous sequential empirical observations (each block containing N observations). The embedded HMM optimizer is then applied as above, using only one pool per block to find a coarse estimate of the trajectory, effectively subsampled by a factor of N. This will help our algorithm to roughly locate the parts of the sensor maps that best explain each block of empirical observations. Once convergence of the search has been achieved at the coarse level, the resulting estimate is then used as an initial trajectory input into a finer level of grouping, and so on until the full resolution of the problem is reached.

An example of using embedded HMM approach for localization (given the sensor maps of the world) is presented in figure 3 (right panel). First, the embedded HMM optimizer is applied for the blocks of $N=20$ sensor observations, then for blocks of length $N=10$ (using the final trajectory from $N=20$ as the initialization); then $N=5$ and finally at the full resolution of the problem.

From the results in figure 3, it can be seen that the embedded HMM is capable of achieving arbitrarily good accuracy in reconstructing the agent trajectories, while the discrete state constrained HMM approach is ultimately limited by the discretization error of the grid. However, the discrete state representation allows us to easily represent and compute a distribution over state trajectories, for example by computing the marginal uncertainty of state occupations at each time. Such a distribution will be useful for learning the parameters of an unknown world, as discussed below.

3   Simultaneous Localization and Surveying

We now turn our attention to the more ambitious problem of analyzing sensor logs from agents operating in unknown environments. Here, our goal is to simultaneously learn (identify) the sensor maps describing the world and localize the agents by estimating their trajectories. We have dubbed this problem Simultaneous Localization and Surveying (SLAS), in contrast to the related problem of Simultaneous Localization and Mapping (SLAM).

Of course, the SLAS problem is only solvable up to certain identifiability limitations. The absolute rotation and reflection of the true world map can never be recovered since rotating or flipping the world and simultaneously rotating and flipping our trajectory estimates will result in identical likelihood of the observed sensor logs. Similarly, the scale of the world cannot be recovered unless we have prior knowledge about the agents velocities. However, up to these degeneracies, the problem is still worth investigating, as we show below.

3.1   Single Agent

For the case of discretized world, and a single agent, we can use the well known EM algorithm for learning the parameters of the effective Hidden Markov Model being used to model the world. In the case of HMMs, the EM algorithm is known as Baum-Welch, and the associated equations are very well known. In our case, the HMM is highly constrained, which makes learning much easier since we do not need to estimate the state transition matrix. This matrix is fixed by the spatial topology of states, allowing transitions only between neighbors, and is not updated during learning. In fact, the sparsity of the spatial topology results in very efficient inference and learning, since very few entries of the transition matrix are nonzero and sums need be performed only over these. The learning equations for the output distributions of each state, which represent the sensor maps, are the same as for a regular HMM. Inference is performed using the standard forward-backwards recursions, which are a special case of belief propagation applied to the graphical model of the HMM.

An example of simultaneously learning the survey parameters and estimating the agent's trajectory is shown in the second row of figure 7.

When the sensor maps are known a priori, localization can be performed optimally, and given only a small amount of data it is usually possible to discover the trajectory of the agent quite accurately. However, when simultaneously learning the maps and performing localization the problem is much harder. With small amounts of data (short trajectories), parts of the environment map that have similar patterns of sensor readings are difficult to distinguish from one another. Therefore a single robot may have difficulty accurately localizing itself.

The problem can be alleviated by having multiple robots which explore the environment simultaneously and interact with each other, for example through proximity sensors. (If the robots were not interacting, the problem would be exactly equivalent to that of a single robot who explored the environment on several independent excursions.)

In the interacting agents case, exact learning requires inferring the joint state of all robots, and quickly becomes exponentially expensive because the effective state space of the HMM is the product of the state spaces of the individual robots. In the next section we develop an approximate but efficient method for inference in this case using belief propagation and apply it to the SLAS problem with multiple interacting robots.

3.2 Multiple Interacting Agents

Consider a scenario in which multiple robots explore the environment simultaneously and interact with each other by communicating signals between them. In what follows, we only consider a very simple form of interaction: the robots are equipped with proximity sensors which notify them when they are near another robot. The proximity signal includes the identity of the other agent encountered, but not a relative heading.

The new graphical model relating empirical observation sequences, interaction signals, and hidden state sequences is shown in figure 5 for two robots and four time steps. This model is very similar to the Factorial HMM[4], except that there are both private outputs from each chain (in our case the sensor readings of each robot) as well as shared outputs (in this case the interaction signals).

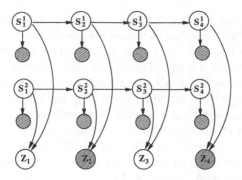

Fig. 5. The graphical model, that displays the relationship between state chains of two agents $S^1 = \{s_1, ..., s_4\}$, $S^2 = \{s_1, ..., s_4\}$, their empirical observations, and the interaction signals $z_1, ..., z_4$ for 4 time steps. The interaction signal z_t is shaded if it is on, which forces the state chains to become coupled.

Note that, even though the state chains are a priori independent, once we condition on the interaction evidence, the chains become coupled. This makes inference much more difficult (and similarly learning, which requires inference), since we can no longer run the simple forward-backward recursions independently on each robot's chain. Of course, the factorial representation of the coupled HMM can always be transformed into a regular HMM, whose effective state space is the Cartesian product of the state spaces of the individual robots. However, inference in this "flattened" model requires working in the joint state space of all robots and quickly becomes exponentially expensive.

Several approaches can be taken to tackle this challenging inference computation. Stochastic sampling algorithms, usually based on importance sampling or Markov Chain Monte Carlo[5] can provide randomized (but often unbiased) estimates of state occupation statistics. One can also employ structured variational approximations to the posterior over hidden states, similar to the ones discussed in [4], and proceed to optimize a lower bound on the likelihood.

Another alternative is to apply a class of approximate (i.e. biased) inference algorithms that are based on belief propagation [11]. Of course, for the uncoupled HMM, the standard HMM inference algorithms are exactly equivalent to belief propagation on a particular junction tree constructed from the original graphical model. For the coupled HMM, we derive below an approximate method for inference which uses loopy belief propagation (LBP) on the equivalent junction tree. Loopy BP passes messages exactly as in regular belief propagation, ignoring the cycles in the graph. The messages are passed according to a predetermined schedule and beliefs are updated in the standard way. Although approximate, LBP has proved to be very successful in practice in many other domains[3, 2]

In theory, LBP runs the risk of "overcounting" information, and thus may not converge or may converge to the wrong answer. However, for our particular problem we find the algorithm very suitable. Indeed, it has been observed in

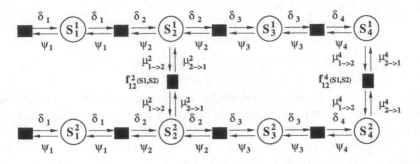

Fig. 6. Factor graph representation of the original Bayes net along with message propagation.

practice that if the original graph does not contain dense loops, LBP usually converges and produces very good approximation. In terms of our setup, if the interaction signals are infrequent (robots do not meet each other very often), then our graphical model will have exactly this characteristic of large, loosely coupled cycles; which is well matched to loopy propagation. In the experimental section we confirm this intuition: loopy BP on our graphs almost always converges to sensible beliefs; and when it is used as the inference step in learning the resulting maps accurately survey the true world sensor maps.

4   Localization with Multiple Agents: Loopy Propagation

The multiple agent localization problem focuses on inferring the most probable trajectories (state sequences) of each agent conditioned on the observed sensor readings of all agents, and the proximity interactions between agents.

To solve this problem, we first pursue only the mode of the joint distribution over agent trajectories. That is, we try to find the single set of trajectories (one for each agent) that simultaneously explain the interaction signals and the sensor logs of each agent as well as possible.

Our approach is to develop below the variant of LBP known as max-product loopy belief propagation [13]. This is exactly the equivalent of Viterbi decoding for coupled HMMs. To derive the necessary equations, we first convert the original Bayes net (fig. 5) to its factor graph (fig. 6) representation [7]. The factor graph contains both variable nodes and factor nodes and is bipartite: edges exist only between variables and factors. Messages flow only from factor nodes to variables and back, but never between factors or between variables. Once a variable node has received messages from all other neighboring factor nodes it takes the product of these messages and delivers it to the destination factor node. The message that a factor node f sends to a variable node x is the maximum over all quantities not present in x of the product of all the incoming messages to f from other neighboring variable nodes y. The incoming messages to f are also multiplied by the potential function defined at f before the max is taken:

$$\mu_{x \to f}(x) = \prod_{h \in neigh(x) \backslash \{f\}} \mu_{h \to x}(x) \tag{4}$$

$$\mu_{f \to x}(x) = \max_{X \backslash x} f(X) \prod_{y \in neigh(f) \backslash x} \mu_{y \to f}(x) \tag{5}$$

(Note that if we ignore the proximity interactions, the max-product algorithm reduces to performing Viterbi decoding separately on each robot's state chain given its sensor observations, i.e. treating it as a regular HMM.) All that remains is to specify a message passing schedule. In principle, one could apply any schedule including running the above updates in parallel. We choose a message passing schedule in which we cycle through chains, passing messages across time for one chain, then from that chain to all others (using the proximity potentials), and then across time in the next chain. In effect, we are performing Viterbi on each chain, taking into account both its observations and the effect of the messages it receives from other chains (which appear as pseudo-observations). This schedule is sometimes known as "chainwise Viterbi".

To formalize the algorithm, we must define the proximity (interaction) potential function $f_{qr}(s^q, s^r)$ that couples robots r and q. In general, one would like to account for noisy interaction signals which are functions of the true separation of the robots. In many applications, the proximity sensors have an extremely low false positive rate, and a moderately low false negative rate. To make inference efficient, and keep the loops in our graphical model as large as possible, we approximate the proximity potential using the assumption that the false positive rate is zero. Thus, at time slice t if a proximity signal between q and r is detected, we set

$$f_{qr}^t(s_t^q, s_t^r) = N(s_t^q, s_t^r)$$

$$\text{where } N(s_t^q, s_t^r) = \begin{cases} 1 \text{ if } s_t^q \text{ and } s_t^r \text{ are neighbouring states} \\ 0 \text{ otherwise} \end{cases}$$

Otherwise, if a proximity signal between q and r sensor is not observed ($p_{qr}^t = 0$), we define $f_{qr}(s^q, s^r)$ to be a constant.

Note that this definition implies that when a proximity signal is observed, both robots *must* be in neighbouring (or identical) states, which according to our problem corresponds to neighbouring or identical discretized spatial cells. (Ideally, the definition of the potential function could be more sophisticated, but this would increase the complexity of inference.) The message at time slice t that a variable node s_t^q node sends to the factor node f_{qr}^t takes the form:

$$\mu_{q \to f_{qr}}^t = \prod_{h \in neigh(s_t^q) \backslash \{f_{qr}^t\}} \mu_{h \to s_t^q} \tag{6}$$

Then the message that a factor node f_{qr}^t sends to the variable node s_t^r is:

$$\mu_{f_{qr} \to r}^t = \max_{s_t^q} f_{qr}^t(s_t^q, s_t^r) \prod_{y \in neigh(f_{qr}) \backslash s_t^r} \mu_{y \to f_{qr}}^t \tag{7}$$

Initially, all messages are set to one.

Max-Product Loopy Belief Propagation executed by each robot r:

- Define: $\delta_t^r(i) = \max_{s_1^r, \ldots, s_{t-1}^r} p(s_1^r, \ldots, s_t^r = i, y_1^r, \ldots, y_t^r | \Theta)$;
  $\psi_t^r(i) = \max_{s_T^r, \ldots, s_{t+1}^r} p(s_T^r, \ldots, s_t^r = i, y_{t+1}^r, \ldots, y_T^r | \Theta)$

  - Initialize $\delta_1^r(i) = \pi_i^r p(y_1^r | s_1^r = i)$, $\psi_T^r(i) = 1$ $1 \le i \le N$,
    with N being the number of states.

  - Induction for $1 \le j \le N$ and $1 \le i \le N$:

$$\delta_{t+1}^r(j) = \left[\max_i \delta_t^r(i) T_{ij} \prod_{q:\text{proximity}} \mu_{q \to r}^t(i) \right] p(y_{t+1}^r | s_t^r = j); \quad 1 \le t \le T-1$$

$$\psi_t^r(i) = \max_j T_{ij} p(y_{t+1}^r | s_t^r = j) \psi_{t+1}^r(j) \prod_{q:\text{proximity}} \mu_{q \to r}^{t+1}(j) \quad t = T-1, \ldots, 1$$

  - Termination: Compute local beliefs and a new set of messages.

$$\gamma_t^r(i) = \frac{\delta_t^r(i) \psi_t^r(i) \prod_q \mu_{q \to r}^t(i)}{\sum_j \delta_t^r(j) \psi_t^r(j) \prod_q \mu_{q \to r}^t(j)}$$

  For proximity signals, where Z is the normalization constant

$$\mu_{r \to q}^t(i) = \frac{1}{Z} \max_{s_t^r} \left[f_{rq}^t(s_t^r, s_t^q = i) \delta_t^r(i) \psi_t^r(i) \prod_{p \ne q} \mu_{p \to r}^t(i) \right] \quad \forall q$$

The final max-product algorithm run by each robot is given above. Our message passing schedule is quite simple: we iterate through robots $r = 1, 2, \ldots, R$ sequentially, running the above max-product algorithm (which includes the effect of all incoming messages). Afterwords, the algorithm computes local beliefs $\gamma_t^r(i)$ and all outgoing messages $\mu_{r \to q}^t$ are sent to all other agents. We monitor the convergence of this LBP by the absolute difference between successive local beliefs and continue passing messages until these stabilize which typically takes 4-5 iterations We have also experimented with running the the above message updates in parallel, and obtained exactly the same results, although with slightly slower convergence.

5   Multi-SLAS: Learning with Multiple Agents

The final problem we discuss is the most difficult: simultaneous localization and surveying using sensor logs from multiple interacting robots. Our approach to this problem is to employ a loopy belief propagation (LBP) method very similar to the one from the previous section as the inference engine, and to alternate between approximate inference using this new version of LBP and parameter (sensor map) estimation based on the results of this inference.

To develop the new LBP equations, which we will use for learning, we focus not on the mode of the distribution but on its marginals. In other words, we use sum-product instead of max-product in an attempt to integrate over all possible paths that the multiple agents could have taken. This is analogous to the forward-backward (alpha-beta) procedure for a single HMM but now in the case of our coupled HMM chains. Once again, we convert the original Bayes net to its factor graph representation. As before, once a variable node has received messages from all other neighboring factor nodes it takes the product of these messages and delivers it to the destination factor node. However, in contrast to the max-product algorithm, the message that a factor node sends to a variable node is the *marginalized* product of all the incoming messages from its other neighboring variable nodes, multiplied by its current potential function:

$$\mu_{x \to f}(x) = \prod_{h \in neigh(x) \backslash \{f\}} \mu_{h \to x}(x) \tag{8}$$

$$\mu_{f \to x}(x) = \sum_{X \backslash x} f(X) \prod_{y \in neigh(f) \backslash x} \mu_{y \to f}(x) \tag{9}$$

Such algorithms are thus often termed "sum-product" algorithms. We employ a message passing schedule identical to the one used for max-product loopy propagation. Using the same potential functions as before, the message at time slice t that a variable node s_t^q node sends to the factor node f_{qr}^t takes the form:

$$\mu_{q \to f_{qr}}^t = \prod_{h \in neigh(s_t^q) \backslash \{f_{qr}^t\}} \mu_{h \to s_t^q} \tag{10}$$

Then the message that a factor node f_{qr}^t sends to the variable node s_t^r is:

$$\mu_{f_{qr} \to r}^t = \sum_{s_t^q} f_{qr}^t(s_t^q, s_t^r) \prod_{y \in neigh(f_{qr}) \backslash s_t^r} \mu_{y \to f_{qr}}^t \tag{11}$$

Sum-Product Loopy Belief Propagation executed by each robot r:

- Define: $\alpha_t^r(i) = p(\boldsymbol{y}_1^r, ..., \boldsymbol{y}_t^r, s_t^r = i | \Theta);$ $\beta_t^r(i) = p(\boldsymbol{y}_{t+1}^r, ..., \boldsymbol{y}_T^r, s_t^r = i | \Theta)$
  $\gamma_t^r(i) = p(s_t^r = i | Y, \Theta)$

  - Initialize $\alpha_1^r(i) = \pi_i^r p(\boldsymbol{y}_1^r | s_1^r = i),$ $\beta_T^r(i) = 1$ $1 \le i \le N,$
    with N being the number of states.

  - Induction for $1 \le j \le N$ and $1 \le i \le N$:

    $$\alpha_{t+1}^r(j) = \left[\sum_i \alpha_t^r(i) T_{ij} \prod_{q:proximity} \mu_{q \to r}^t(i) \right] p(\boldsymbol{y}_{t+1}^r | s_t^r = j); \quad 1 \le t \le T - 1$$

    $$\beta_t^r(i) = \sum_j T_{ij} p(\boldsymbol{y}_{t+1}^r | s_t^r = j) \beta_{t+1}^r(j) \prod_{q:proximity} \mu_{q \to r}^{t+1}(j) \quad t = T - 1, ..., 1$$

  - Termination: Compute marginal beliefs and a new set of messages

    $$\gamma_t^r(i) = \frac{\alpha_t^r(i) \beta_t^r(i) \prod_q \mu_{q \to r}^t(i)}{\sum_j \alpha_t^r(j) \beta_t^r(j) \prod_q \mu_{q \to r}^t(j)}$$

    For proximity signals, where Z is the normalization constant

    $$\mu_{r \to q}^t(i) = \frac{1}{Z} \sum_{s_t^r} \left[f_{rq}^t(s_t^r, s_t^q = i) \alpha_t^r(i) \beta_t^r(i) \prod_{p \ne q} \mu_{p \to r}^t(i) \right] \quad \forall q$$

Initially, all messages are set to one. Note that in our problem setting, the messages that are being passed from one robot to another can be interpreted as a local marginal belief about the state distribution that different robots have. LBP essentially tries to insure the consistency of these different local beliefs at the times when proximity signals are observed.

We iterate through robots $r = 1, 2, ..., R$ sequentially, running the above inference algorithm. When completed, the algorithm has computed $\gamma_t^r(i)$ and all outgoing messages $\mu_{r \to q}^t$ are sent. We monitor the convergence of this LBP by the absolute difference between successive marginal beliefs and continue passing messages until these stabilize (which typically takes 15-20 iterations) or until a maximum number of iterations, which we set to 25, has been reached. After inference has converged, we perform an M-step to update the model parameters Θ and then repeat the iterative inference procedure before updating the parameters again.

The final learning and inference algorithm is given below.

Learning Algorithm for Multiple Interacting Robots :

- Repeat until parameters converged or maximum number of learning iterations

  - E-step: Perform Inference Step (see above box)
    - While inference not yet converged and below maximum inference iterations
      * Run modified FB recursion for each robot $r = 1, ..., R$, which
        · takes into account the incoming messages from other robots
        · computes marginal beliefs about robot's state occupation
        · sends appropriates messages to other robots.
    - If inference does not converge, Run standard FB on disconnected state chains.
    - Compute marginal beliefs
  - Perform an M-step to update parameters

In rare cases, when LBP fails to converge, we may make some further approximations. (However, in all of the examples presented here this never occurred.) It is possible to employ "damped" versions of LBP [8] which often converge empirically, or resort to more tedious and slow double-loop algorithms that are always guarantee to converge [6]. However, it is generally believed that the accuracy of the answers returned by these damped or double-loop algorithms in cases where regular LBP has trouble converging may be quite poor. In more complex examples when LBP failed to converge, we simply ignore all robot interactions for one iteration, and just run the standard forward-backward (FB) recursions on the disconnected state chains in parallel. We then perform an M-step as usual and return to loopy propagation at the next inference step.

6   Experimental Results

We experimented with single and multiple robots in a 15x15 grid world using simulated logs from 3 continuous valued and 1 binary valued sensor. The functions defining the 3 continuous sensors were generated at random using mixtures of small numbers of Gaussians. The binary sensor measures contact with the wall (world edge), using the output model $P(y_{wall} = 1|s_t = i) = 1 - \epsilon$ if i is a wall state and zero if i is not a wall state. In our simulations, $\epsilon = 0.1$ Smooth, continuous trajectories of nonconstant velocity were generated and sampled at regular time intervals. The values of the 3 sensors at these continuous positions and discrete times was corrupted with Gaussian noise of standard deviation 0.1, with the scale of sensor readings being from 0 to 1. (In our experiments we assumed that the output model for the binary sensor only was known to the robot: in effect this lets the robot guess when it has reached the limits of the region it is exploring, although it does not know which of the four wall it might be contacting.)

Figure 7 (top panel) shows the true sensor maps for the continuous sensors, along with a subsequence of one continuous trajectory. It also shows the state discretization of the world as dotted lines. In total, we generated 4 sequences of 2500 noisy observations where each observation consisted of 3 continuous valued and 1 binary valued sensor (fig 1).

Figure 7 (second from top panel) shows the results of applying our SLAS algorithm assuming that these 4 sequences were generated by 4 separate excursions by a single robot (or 4 excursions by non-interacting robots). The reconstructed maps have been flipped vertically for the display, since of course the algorithm cannot recover absolute orientation. (Of course the sensor maps we estimate are piecewise constant at the scale of the grid resolution; but graphically some smoothing has also taken place when drawing the contour lines.)

The average RMS localization error between our reconstructed trajectories (computed using Viterbi decoding in this case) and the true trajectories is only 1.02 times the grid size (averaged over all reconstructed trajectories). This implies that, on average, we can estimate the agents' locations to within our discretization error limit.

Figure 7 (middle panel) shows the results of applying our multi-SLAS algorithm, assuming that the 4 trajectories were executed in two excursions by two interacting robots. Notice that the same total amount of data is used, except for the inclusion of the proximity sensors (in fact exactly the same data traces are used, we just pretend they came from two robots instead of four). The proximity signals were generated with probability $1 - \delta$ if the true (continuous) positions of the robots were within a distance of 1.0 grid units and with probability zero otherwise. (Notice that this is a slightly different process than the one assumed by the robots during inference.) In our experiments, $\delta = 0.1$. This resulted in proximity signals being observed at 3-5% of timesteps for each agent on average. Keep in mind that the proximity signals for each robot are noisy and this noise is independent; this means that at time t robot p may detect robot q but not vice versa.

We can see that by trying to enforce consistency between the robots using LBP, our approximate algorithm improves the survey map as well as the trajectory reconstructions as compared to exact inference without interaction signals. (The RMS position error in this case went down to 0.81.)

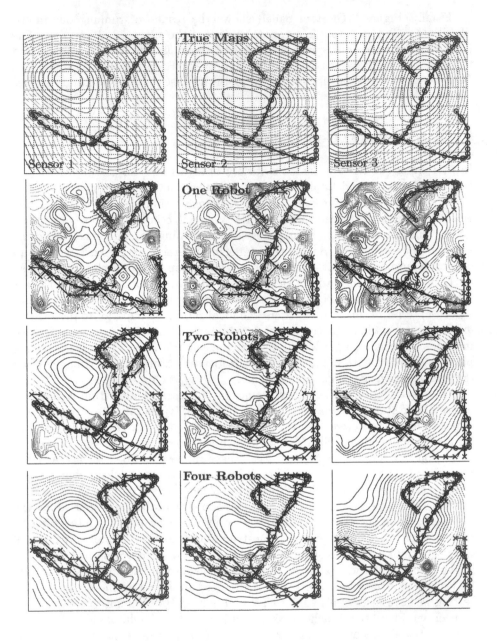

Fig. 7. Display of the true three (left to right) maps of sensor readings along with the part of the true trajectory taken by the robot (top panel), reconstructed maps with inferred trajectories for one (second), two (third) and four (bottom) robots. The "o"s indicate the time slices when the true observation was taken, and "x"s indicate the inferred path taken by the robot.

Finally, Figure 7 (bottom panel) shows the results of running our multi-SLAS algorithm assuming four robots navigated the space simultaneously. The pairwise proximity signals were generated as above. In this case, the survey and trajectory reconstructions have further improved. (RMS position error is 0.76.) Because there were more agents, the graph in this case contains proximity signals at 10-12% of the timesteps.

7   Discussion & Conclusions

In this paper, we have presented a variety of algorithms for solving the simultaneous surveying and localization problem when an unknown environment is explored by multiple interacting agents. Although a simple discretization of the world leads to a tractable constrained HMM architecture in the case of a single agent, multiple interacting agents cause exact learning and inference to become exponentially expensive. Rather than ignoring interactions, we have derived an efficient approximate multi-agent inference algorithm for this architecture based on Loopy Belief Propagation. Although our algorithm does not perform exact inference, we have shown on simple grid world experiments that its performance – both in terms of survey parameters and localization – is superior to performing exact inference while ignoring agent interactions. Other approximate inference methods, especially those based on applying particle filters [1], have been applied to mobile robotics, but this work has focused on mapping occupancy grids (SLAM), and on the single agent setting.

One particularly intriguing byproduct of our learning algorithm is that the intermediate state marginals $\Gamma(i) = \sum_{r=1}^{R} \sum_{t=1}^{T} \gamma_t^r(i)$ contain the estimated occupancy numbers for each grid state. These estimates can potentially be used for traditional mapping (SLAM): states with very low occupation numbers likely correspond to inaccessible regions. Also, these values could be returned as feedback signals to the control algorithm driving the robots to indicate which areas of the world need to be explored further, in the consensus opinion of all agents. Figure 8 displays this statistic for single-robot, two-robot and four-robot experiments using the same data as in the other experiments. It is interesting to note that the map and trajectory reconstruction are inaccurate precisely in areas where we think we are most uncertain about the world according to Γ.

For localization with a single agent, we have also investigated a continuous trajectory representation which avoids the need to discretize the world. In this setting we have successfully employed the embedded HMM architecture as an optimizer and found that it achieves excellent localization results avoiding both the computational cost and the discretization performance limit of our constrained HMMs.

We are currently developing SLAS algorithms based on the discretization-free continuous representation and using the embedded HMM optimizer. We are particularly interested in possible extensions to the multiple robots case both for localization and for multi-SLAS. Ultimately, we hope to apply these algorithms to real data from teams of mobile agents, for example planetary rovers.

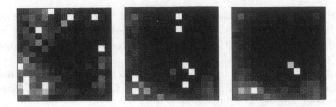

Fig. 8. Display of uncertainty about the current state distribution for 1 robot (left), 2 robots (middle), and 4 robots(right). The uncertainty is measured by Γ. For visualization purposes we display $1/\Gamma$, so white cells correspond to the states which robots are most uncertain about.

Acknowledgments

We thank Tim Barfoot, Martin Wainwright and Max Welling for useful discussions. STR & RRS are supported in part by the Learn Project of IRIS Canada and by NSERC.

References

[1] D. Fox, S. Thrun, W. Burgard, and F. Dellaert. Particle filters for mobile robot localization. In *Sequential Monte Carlo Mehods in Practice*. Springer, 2001.

[2] W. Freeman, E. Pasztor, and O. Carmichael. Learning low-level vision. In *Int. J. Computer Vision*, 2000.

[3] Brendan J. Frey and David J. C. MacKay. A revolution: Belief propagation in graphs with cycles. In Michael I. Jordan, Michael J. Kearns, and Sara A. Solla, editors, *Advances in Neural Information Processing Systems*, volume 10. The MIT Press, 1998.

[4] Zoubin Ghahramani and Michael I. Jordan. Factorial Hidden Markov Models. *Machine Learning*, 29:245–273, January 1996.

[5] W. R. Gilks, S. Richardson, and D. J. Spiegelhalter. *Markov Chain Monte Carlo in Practice*. Chapman_Hall, 1996.

[6] Tom Heskes. Stable fixed points of loopy belief propagation are minima of the bethe free energy. In *Advances in Neural Information Processing Systems*, volume 15, 2003.

[7] Frank R. Kschischang, Brendan J. Frey, and Hans-Andrea Loeliger. Factor graphs and the sum-product algorithm. *IEEE Transactions on Information Theory*, 47(2):498–519, 2001.

[8] Kevin Patrick Murphy, Yair Weiss, and Michael I. Jordan. Loopy Belief Propagation for Approximate Inference: An Empirical Study. In *Proceedings of UAI*, pages 467–475. Morgan Kaufmann Publishers, July 1999.

[9] Radford Neal. Markov chain sampling for non-linear state space models using embedded hidden Markov models. Technical Report 0304, Dept. of Statistics, University of Toronto, April 2003.

[10] Radford Neal, Matthew Beal, and Sam Roweis. Inferring state sequences for non-linear systems with embedded hidden Markov models. In *Advances in Neural Information Processing Systems*, volume 16, 2004.

[11] J. Pearl. *Probabilistic Reasoning in Intelligent Systems.* Morgan-Kaufman, 1988.

[12] S. T. Roweis. Constrained hidden Markov models. In *Advances in Neural Information Processing Systems*, volume 12, pages 782–788, Cambridge, MA, 1999. MIT Press.

[13] Weiss and Freeman. On the optimality of solutions of the max-product belief-propagation algorithm in arbitrary graphs. *IEEETIT: IEEE Transactions on Information Theory*, 47, 2001.

Hex: Dynamics and Probabilistic Text Entry

John Williamson[1] and Roderick Murray-Smith[1,2]

[1] Department of Computing Science, University of Glasgow,
Glasgow G12 8QQ, Scotland, UK
`jhw,rod@dcs.gla.ac.uk`
[2] Hamilton Institute,
National Univ. of Ireland, Maynooth, Co. Kildare, Ireland

Abstract. We present a gestural interface for entering text on a mobile device via continuous movements, with control based on feedback from a probabilistic language model. Text is represented by continuous trajectories over a hexagonal tessellation, and entry becomes a manual control task. The language model is used to infer user intentions and provide predictions about future actions, and the local dynamics adapt to reduce effort in entering probable text. This leads to an interface with a stable layout, aiding user learning, but which appropriately supports the user via the probability model. Experimental results demonstrate that the application of this technique reduces variance in gesture trajectories, and is competitive in terms of throughput for mobile devices. This paper provides a practical example of a user interface making uncertainty explicit to the user, and probabilistic feedback from hypothesised goals has general application in many gestural interfaces, and is well-suited to support multimodal interaction.

1 Introduction

Text entry is an important part of all human computer interfaces, and is particularly important for communication between humans via computer. However, entry on mobile devices can be problematic compared to established keyboard-based desktop systems. The restricted size and reduced processing power of mobile devices, and the changing contexts in which the devices are used are all obstacles to efficient text entry. Current approaches include virtual keyboards [1], handwriting recognition [2], and gesture-based interfaces; the latter is of interest here.

Various interfaces for gestural text entry have been devised, including fixed layout approaches such as [3] and [4], and dynamic layout approaches, as in Dasher [5] and in [6]. Of these, only the latter systems support a probabilistic model for increasing accuracy and throughput. Support for a probabilistic model is vital for optimal performance; in particular the best use of the limited bandwidth available can be made only if the uncertainty in language is adequately represented.

Systems that dynamically optimize letter layouts can impair learning, as the constantly changing interface lacks the stability needed to learn to perform automatic movements. Although initial learning times may be very short, the transition from a novice to an expert user is often slow or impossible. At the novice level, the user is totally dependent on feedback, while at an expert level, control is more open-loop with rapid, learned

R. Murray-Smith, R. Shorten (Eds.): Switching and Learning, LNCS 3355, pp. 333–342, 2005.

responses producing desired control actions. If the configuration changes significantly with varying contexts, learning and producing such automatic high-speed responses is more difficult.

In this paper, we describe a system which uses continuous gestures to produce text. The handling qualities of the system are dynamically altered according to a time-varying probability model. The gestures for letter sequences remain stable, supporting user learning and high-speed, open-loop gesturing. However, the changing control properties reduce the effort required to choose highly probable sequences, and so there is a direct relation between the information content of a sequence and the effort that must be expended by the user.

2 Design

2.1 Layout

Each symbol is coded as a pair of primitive gestures, these "gestures" being movements in one of six directions, allowing thirty-six symbols. Such a division leads to a hexagonal layout, with two stages: selecting a letter group and selecting a letter. As hexagons form a regular tiling, gestures for letter sequences are represented as paths through such a plane (see Figure 1). Recognition involves selecting points in space such that the Voronoi tessellation (using the L2 norm) is hexagonal. Crossing cell boundaries in this tessellation triggers transitions in a finite-state model, which outputs symbols in response.

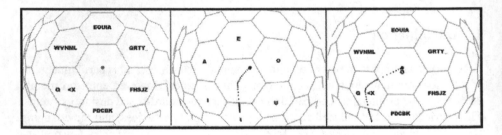

Fig. 1. The hexagonal layout. Each letter is assigned to a group, within which it is associated with a particular edge. In this example, producing "o" requires an upwards then up-right movement.

2.2 Control

The interface can be controlled with a number of input devices; mice and accelerometers are used in the prototypes presented here. The tessellation must be effectively unbounded to permit all combinations of symbols to be entered, and so input deflection is mapped to velocity allowing apparently infinite range of movement. A nonlinear

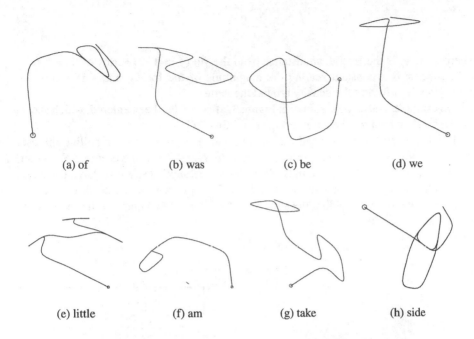

(a) of   (b) was   (c) be   (d) we

(e) little   (f) am   (g) take   (h) side

Fig. 2. Eight common English words, and paths through the hexagonal space that will produce them. These paths are generated via cubic splines (see Section 2.5)

transfer function, with a dead-zone around zero (see [7]) is used to help stabilize the control.

The handling qualities of the system are manipulated by simulating a nonlinear landscape on the selection plane. The local system dynamics are altered by a vector force-field which is computed from the current probabilities. This field is conditioned on the current context, where the context may include position, velocity, acceleration, and the probability of a letter given the current prefix. Given the state x of the system, we have

$$\dot{x} = A(c)x + B(c)u \tag{1}$$

where u is the control action, and $A(c)$ and $B(c)$ are context-varying state and control matrices, conditioned on the context c.

The force vectors require greater control effort on the part of the user to move into low probability areas; this can be thought of as a system of hills and valleys guiding the user away from improbable regions of the state space.

Given six discrete probabilities for each possible transition, $p_1, .., p_6$, the force at any point is given by a function with squared-exponential decay from the vertices. Each force is applied from the two vertices which form the boundary across which the transition can occur. The magnitude of the force applied at each point is given by:

$$f = \frac{k}{2} \sum_{i,j} e^{\frac{d(v_{i(j)})^2}{s}} p_i \qquad (2)$$

where $d(v_{i(j)})$ is the Euclidean distance from the jth ($j = 1..2$) vertex of the ith ($i = 1..6$) edge, k is a constant scaling the magnitude of the forces, and s is a constant specifying the width of the density around the vertex.

An example landscape is shown in Figure 3, after "q" has been entered, which shows the deep valley towards the letter group containing "u".

In addition to this field, fixed forces are applied at the vertices, repelling the user from these points. This limits ambiguous transitions, and forces the user to make a conscious choice at these decision points. These forces are applied as above, having squared exponential decay, but with constant magnitude. The forces act along the direction from the user to the vertex, avoiding the slingshot effect that would occur if the forces were normal to the density.

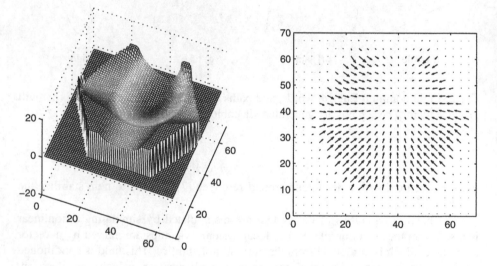

Fig. 3. The vector field (right) produced after the letter "q" is entered, and its magnitude, shown as the surface plot on the left. The group boundary leading to "u" is at the top of these diagrams.

2.3   Probability Model

A simple language model (based upon partial predictive matching, see [8, 9]) is used to produce $p(\text{letter}|\text{prefix})$ (referred to as $p(l|pr)$) on a per-word basis. A tree with probability information is generated from a corpus (in this case texts from Project Gutenberg [10]). For simplicity, no grammar or word-level model is used, although this would be likely to improve performance significantly [11]. More complex language models can easily be incorporated in this framework.

The probability model is extended to include the dynamics of the cursor. The velocity and acceleration of the cursor are numerically estimated. The probability of heading into a hexagon is then given by

$$\frac{\cos\theta + 1}{2}, \tag{3}$$

where θ is the angle between the movement vector and the center of the hexagon being tested. The probability of each hexagon is given by

$$p(h) = p(h|pr)p(h|v)p(h|a). \tag{4}$$

If a transition into hexagon h represents a single letter l_h then

$$p(h|pr) = p(l_h|pr), \tag{5}$$

otherwise

$$p(h|pr) = \sum_{i=1}^{6} p(l_{hi}|pr), \tag{6}$$

where l_{hi} is the ith letter in the letter group selected by a transition into h.

2.4   Autocompletion and Prediction

Potential autocompletions can be predicted using Monte-Carlo sampling. Starting from the current prefix a potential symbol is selected randomly, weighted according to $p(l)$. This letter is concatenated to the prefix, and the process repeated until the end-of-word symbol is produced. The probability of the sequence is evaluated as a by-product of this process.

Each of these word/probability pairs is stored in a list ranked by probability. The sampling is repeated k times, with $k \approx 300$ in the current implementations. The top autocompletion is then presented, and the autocomplete action can be initiated either by a specific button press (in the case of a mouse) or a simple shake gesture (for orientation sensors).

The display can show the path which would generate the current autocomplete possibilities. Fitting a cubic spline through the medians (the centers of edges) of the hexagons gives a smooth path which will generate a given letter sequence. Displaying these splines for the top autocompletes shows the paths of possible completions, giving a background awareness of the "word density" at any point in state space. It also facilitates the learning of smooth trajectories for words. We are currently extending this feedback to an audio display, based on ideas we presented in [12], where we describe a system for audio display of time-varying probabilities. The use of audio is important for mobile devices, where screen space is at a premium, and users visual load is often already high.

2.5   Layout Optimization

The layout used in preceding examples was chosen to aid learning , by grouping letters in a logical manner (such as grouped vowels). Given a source corpus, it is possible to optimize the layout to minimize some cost function, given a model of the user's movement. The cost of a particular layout is

$$c_t = \sum_{i=1}^{n} p(w_i) c(w_i) \tag{7}$$

where n is the number of words in the dictionary, and c is the cost for each word. For the sake of computational efficiency implementations prune the cost evaluation, letting n be the top ranked few hundred words from the corpus.

The cost function used should minimize some aspect of effort on the part of the user; here we penalize the sum of squared j-th derivatives of the trajectory representing the word, i.e we have:

$$c(w_i) = \int_0^t \left(\alpha_j \left(\frac{d^j x}{dt^j} \right)^2 + \beta_j \left(\frac{d^j y}{dt^j} \right)^2 \right) dt. \tag{8}$$

In the implementations the third derivative is penalized. This is based on a minimum-jerk model [13], in contrast to the linear-segment model proposed in [14]. Finally, a model of the user's movement is required; we approximate it with a cubic spline path. This simple approximation is justified experimentally in Section 3.1. We then numerically optimize the layout to minimize c_t.

2.6   Implemenations

The system has been implemented running on a desktop PC with a mouse and with an InterTrax accelerometer, and on the PocketPC platform with an accelerometer (see Figure 4).

3   Results

In throughput testing, one of the authors achieved around 10–12 words per minute with earlier versions of the system. This is the rate for perfect transcription of a hundred words of written text (rather than groups of five characters per minute), including error-correction time. In this case, the user had around 30 hours of use with the layout used for the test. Speeds of around 17wpm are achievable with current versions, for free-form text entry. It should be borne in mind that the layout used for these tests was not optimized (see Section 2.5).

3.1   Spatial Effects

Figure 5 shows twenty trajectories for the word "hello", as performed by one of the authors as force model is adjusted. The four experimental conditions are: forces applied as

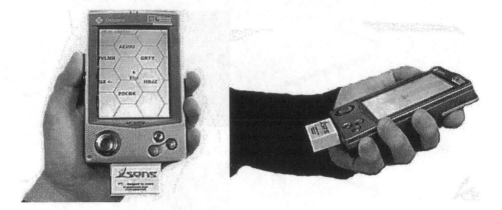

Fig. 4. The system running on Cassiopeia E115 with a miniature accelerometer

previously described; forces not applied; forces applied with double magnitude; forces applied as normal but with probabilities inverted.

Also shown is a cubic spline fit through the medians of the hexagons. It is apparent that the spline fit is a reasonable approximation; the cubic spline is within the distribution of points on the trajectory for most of the path. Exceptions occur at significant decision points where the user follows a less constrained path.

The intention of the force model is to increase accuracy and speed in performance. If the hypothesis that accuracy would increase is to be verified, then the distribution of the trajectories should be narrower for the cases where forces are present than when they are not. This can be seen when comparing Figure 5(a) (no forces) with Figure 5(b) (with forces). Increasing the forces should amplify these effects; this is apparent in Figure 5(c).

To illustrate the effect of the choice of language model on the performance of the system, Figure 5(d) shows the result of inverting the probabilities in the language model (p becomes $1 - p$). This results in a significant increase in the deviation from the ideal path, particularly towards the end when the model is confident of its predictions, and so is opposing most strongly. The vertex and friction forces are as in the other tests, and so all changes of performance can be attributed to the change in the language model.

3.2   Temporal Effects

The right-hand panel in Figure 5 shows the effect of the forces on the timing of the gesture. Without forces applied (Figure 5(b)) the path is smooth, without any significant pauses or accelerations (except at the start). The two runs with forces applied normally and at double strength show a strongly periodic movement. This periodicity is significantly diminished in the example with inverted forces, even though the forces are of the same magnitude as Figure 5(a). This enforced periodicity may be due either to a change in the control strategy pursued by the human, or may simply be a by-product of the changing system dynamics; more testing will be required to separate these issues.

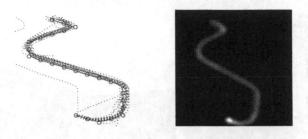

(a) No forces applied

(b) Forces applied as described previously

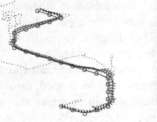
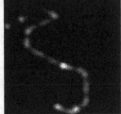

(c) Forces with double magnitude

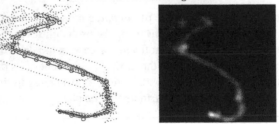

(d) Forces with probabilities inverted

Fig. 5. Trajectories from twenty repetitions of the gesture for "hello" with varying forces. On the left panel, dashed lines show measured trajectories, circles indicate the centers and medians of the hexagons, and the solid line indicates a cubic spline fit through the medians. The right panel shows a density plot produced by summing each of the data points, after convolving with a smoothing window, onto a mesh (higher density areas are lighter). When velocity is lower the local density will increase, assuming equal path density.

Whichever is the case, it is a potentially powerful feature. A periodic interface allows for task interleaving; this is important for mobile devices where interaction may be occurring while occasional attention is required elsewhere. The periodicity of motion may be a useful metric for estimating performance in an adaptive system – it seems possible that confident users will produce more regularly timed movements than users who are relying more heavily on feedback control. Rhythmic movement can also be of use in feedback presentation, particularly in the audio modality, allowing for structured output which requires less constant attention.

4   Conclusions

We have created a text entry system based on continuous gestures performed on a regular tessellation, and demonstrated how dynamically altering the handling qualities of the system given a probabilistic model of context can improve performance. Testing shows that the variance of trajectories for probable sequences can be reduced using this method. Further systematic user trials will be required to establish the effects at the various stages of learning.

This control-based approach supports users without constraining them, resisting low probability actions but not preventing them. This creates a correspondence between the information content of a sequence and the expenditure of energy on the part of the user. It also facilitates a smooth transition from unskilled, feedback-dependent users, to skilled users performing automatic, open-loop movements.

Our system uses the probability of the hypothesised goals compatible with the current context, to provide feedback directly to the user or by adapting the local dynamics of interaction. This is a general technique of interest to the whole area of gesture-interface design, improving throughput and supporting exploration and learning in new users.

Acknowlegements   Both authors are grateful for support from EPSRC grant Modern statistical approaches to off-equilibrium modelling for nonlinear system control GR/M76379/01, and Audioclouds: three-dimensional auditory and gestural interfaces for mobile and wearable computers GR/R98105/01. The Multi-Agent Control Research Training Network EC TMR grant HPRN-CT-1999-00107. We thank Xsens for the use of the P^3C accelerometer.

References

[1] Kolsch, M., Turk, M.: Keyboards without keyboards: A survey of virtual keyboard implemenations. In: Proceedings of Sensing and Input for Media-centric Systems. (2002)
[2] Plamondon, R., Srihari, S.N.: On-line and off-line handwriting recognition: A comprehensive survey. IEEE Transactions on Pattern Analysis and Machine Intelligence **22** (2000) 63–84
[3] Mankoff, J., Abowd, G.D.: Cirrin: A word-level unistroke keyboard for pen input. In: ACM Symposium on User Interface Software and Technology. (1998) 213–214
[4] Perlin, K.: Quikwriting: Continuous stylus-based text entry. In: ACM Symposium on User Interface Software and Technology. (1998) 215–216

[5] Ward, D.J., Blackwell, A.F., MacKay, D.J.C.: Dasher - a data entry interface using continuous gestures and language models. In: UIST'00. (2000) 129–137

[6] Bellman, T., MacKenzie, I.S.: A probabilistic character layout strategy for mobile text entry. In: Proceedings of Graphics Interface '98. (1998) 168–176

[7] Jagacinski, R., Flach, J.: Control theory for humans : quantitative approaches to modeling performance. L. Erlbaum Associates, Mahwah, N.J. (2003)

[8] Cleary, J., Witten, I.: Data compression using adaptive coding and partial string matching. IEEE Transactions on Communications 32 (1984) 396–402

[9] Cleary, J., Teahan, W., Witten, I.H.: Unbounded length contexts for ppm. In: Proceedings DCC'95. (1995) 52–61

[10] Hart, M.: Project gutenberg (2003) Available at http://promo.net/pg/.

[11] Lesher, G., Rinkus, G.: Leveraging word prediction to improve character prediction in a scanning configuration. In: Proceedings of the RESNA 2002 Annual Conference. (2002)

[12] Williamson, J., Murray-Smith, R.: Audio feedback for gesture recognition. Technical Report TR-2002-127, Dept. Computing Science, University of Glasgow (2002)

[13] Flash, T., Hogan, H.: The coordination of arm movements: an experimentally confirmed mathematical model. Journal of Neuroscience 5 (1985) 1688–1703

[14] Isokoski, P.: Model for unistroke writing time. In: CHI. (2001) 357–364

Author Index

Lecture Notes in Computer Science

For information about Vols. 1–3272

please contact your bookseller or Springer